Computer System
and
Programming in C

for BTech and MCA First Semester Students

Computer System
and
Programming in C

for BTech and MCA First Semester Students

Manish Varshney
Professor and Head
Computer Science and Engineering Department
Shri Siddhi Vinayak Institute of Technology
Bareilly

Neha Singh
Assistant Professor
Computer Science and Engineering Department
Shri Sidchi Vinayak Institute of Technology
Bareilly

CBS

CBS Publishers & Distributors Pvt Ltd

New Delhi • Bengaluru • Chennai • Kochi • Mumbai • Pune
Hyderabad • Kolkata • Nagpur • Patna • Vijayawada

Computer System
and
Programming in C

ISBN: 978-81-239-24007

First Edition   2014

Published by Satish Kumar Jain for
CBS Publishers & Distributors Pvt Ltd
4819/XI Prahlad Street, 24 Ansari Road, Daryaganj, New Delhi 110 002, India.
Ph: 23289259, 23266861, 23266867   Fax: 011-23243014   Website: www.cbspd.com
e-mail: delhi@cbspd.com; cbspubs@airtelmail.in.
Corporate Office: 204 FIE, Industrial Area, Patparganj, Delhi 110 092
Ph: 4934 4934   Fax: 4934 4935   e-mail: publishing@cbspd.com; publicity@cbspd.com

Branches

- **Bengaluru:** Seema House 2975, 17th Cross, K.R. Road,
  Banasankari 2nd Stage, Bengaluru 560 070, Karnataka
  Ph: +91-80-26771678/79   Fax: +91-80-26771680   e-mail: bangalore@cbspd.com
- **Chennai:** 20, West Park Road, Shenoy Nagar, Chennai 600 030, Tamil Nadu
  Ph: +91-44-26260666, 26208620   Fax: +91-44-42032115   e-mail: chennai@cbspd.com
- **Kochi:** 36/14 Kalluvilakam, Lissie Hospital Road, Kochi 682 018, Kerala
  Ph: +91-484-4059061-65   Fax: +91-484-4059065   e-mail: kochi@cbspd.com
- **Mumbai:** 83-C, Dr E Moses Road, Worli, Mumbai-400018, Maharashtra
  Ph: +91-22-24902340/41   Fax: +91-22-24902342   e-mail: mumbai@cbspd.com
- **Pune:** Bhuruk Prestige, Sr. No. 52/12/2+1+3/2 Narhe, Haveli
  (Near Katraj-Dehu Road Bypass), Pune 411 041, Maharashtra
  Ph: +91-20-64704058, 64704059, 32392277 Fax: +91-20-24300160   e-mail: pune@cbspd.com

Representatives

- **Hyderabad** 0-9885175004
- **Nagpur** 0-9021734563
- **Kolkata** 0-9831437309, 0-9051152362
- **Patna** 0-9334159340
- **Vijayawada** 0-9000660880

Printed at

Preface

Technological advancements have shown a substantial growth concerned with each and every field. This has aroused the need to learn computer concepts with the use of programming methodologies. Personal computers, once predicted to be owned by only a few individuals, are now everywhere. Practical advances in computer technology revolve around the increased ability of computers to manage specific tasks. This is achieved through the increased processing power of computer. Concepts of computer programming help in understanding the problem and finding the suitable solution.

Efforts are made in order to introduce the students with computer programming concepts, taking C language as the medium. The book lays emphasis on foundations and basic principles and on concepts of computer programming. Beginning with very elementary concepts related to computer fundamentals and operating system, students will be presented with detailed discussions on a variety of programming constructs and features of the C language.

The various concepts are arranged in different chapters providing detailed description of the topic. This book constitutes 17 chapters, each revealing the basic and fundamental concept. No prior knowledge of computers and the programming language is needed.

Chapter 1 describes the various internal functional components like memory, CPU and other peripheral devices of computer along with the evolution and generation of computers.

Chapter 2 defines the term operating system and its functionality describing the various categories of it. It also gives a brief overview of DOS, Windows and UNIX operating system along with its commands.

Chapter 3 gives a brief overview about the fundamentals of programming methodologies like algorithm, flowchart, their complexities and various programming constructs along with programming languages like COBOL, Pascal, Java,etc.

Chapter 4 reveals the logic behind the various number system conversions, paying emphasis on binary numeration system which in turn describes the representation of numbers in computer memory.

Chapter 5 provides a brief introduction of the programming language C elaborating its history, features, various constructs, program structure and its uses.

Chapter 6 elaborates the concepts of variables, keywords and basic data types used in C language. Data types define the initial memory requirements of C variable. The chapter also explains the storage classes in C which determines the existence of a variable in memory.

Chapter 7 explains the syntax involved in constructing expressions using operators in C. This chapter discusses different types of operators required for performing various arithmetic and logical operations.

Chapter 8 defines the functions used in various input/output operations. The chapter focuses on some predefined functions involved in these operations.

Chapter 9 describes the way by which iterations/looping and decision control statements are used in C. It explains the basic control statements that are required by any procedural language.

Chapter 10 defines the term modular programming and explains the approach through which they are implemented. The concept of Functions lies at the center of this programming.

Chapter 11 gives an overview about an arrays and its classification. Arrays are entities that allocate memory non-dynamically.

Chapter 12 deals with the concept of string handling in C. The chapter focuses on their declarations and the various operations that can be done using some predefined string functions.

Chapter 13 describes the logic to declare user-defined data types such as structures and union. The chapter also elaborates the way they are declared and used in programming with help of various examples.

Chapter 14 focuses on the concept of pointers and how they can be used efficiently. The various peculiarities of the pointers are discussed in this chapter.

Chapter 15 deals with the allocation of memory dynamically. The chapter discusses various data structures and algorithms that are used in dynamic memory allocation.

Chapter 16 discusses some basic peculiarities involved with programming. The preprocessor directives are responsible for making the code manageable and maintainable so that they can be easily altered.

Chapter 17 elaborates the concepts of file handling using C programming.

Manish Varshney
Neha Singh

Acknowledgements

We are extending our warm thanks to Shri Anupam Kapoor, Chairman, Shri Siddhi Vinayak Group of Institutions, Bareilly, and Shri Dev Murti, Chairman, SRMS Group of Institutions, for giving us an opportunity to write the book. Authors are also thankful to Shri Saurabh Mehrotra, Secretary, Shri Siddhi Vinayak Group of Institutions, and Shri Aditya Murti, Secretary, SRMS Group of Institutions, Bareilly, for the motivation.

Authors are also thankful to Dr Anil Kumar, Principal, SSVIT, and Shri Sudhakar Jain, Dean–Academics, for the valuable suggestions in the quality improvement of this book. We also acknowledge the consistent support and cooperation received from our faculty members of SSVIT. We are greatly thankful to our families for their patience, support, motivation and inspiration during the compilation of the book.

Finally thanks are due to Mr YN Arjuna of CBSP&D, New Delhi, for not only cooperating and supporting this venture, but also showing great sense of commitment.

Manish Varshney
Neha Singh

Contents

Syllabus

NCS-101/NCS-201 Computer System and Programming in C—LTP 310

Unit 1 (10 Lectures)

Basics of computer: Introduction to digital computer, basic operations of computer, functional components of computer, classification of computers.

Introduction to operating system: [DOS, Windows, Linux and Android] purpose, function, services and types.

Number system: Binary, octal and hexadecimal number systems, their mutual conversions, Binary arithmetic.

Basics of programming: Approaches to problem solving, concept of algorithm and flow charts

Types of computer languages: Machine Language, Assembly Language and High Level Language, Concept of Assembler, Compiler, Loader and Linker.

Unit 2 (8 Lectures)

Standard I/O in C: Fundamental data types—character type, integer, short, long, unsigned, single and double floating point, storage classes—automatic, register, static and external, operators and expression using numeric and relational operators, mixed operands, type conversion, logical operators, bit operations, assignment operator, operator precedence and associativity.

Fundamentals of C programming: Structure of C program, writing and executing the first C program, components of C language. Standard I/O in C.

Unit 3 (10 Lectures)

Conditional program execution: Applying if and switch statements, nesting if and else, use of break and default with switch, program loops and iterations: use of while, do while and for loops, multiple loop variables use of break and continue statements. Functions: Introduction, types of functions, functions with array, passing values to functions, recursive functions.

Unit 4 (6 Lectures)

Arrays: Array notation and representation, manipulating array elements, using multi-dimensional arrays. Structure, union, enumerated data types.

Unit 5 (8 Lectures)

Pointers: Introduction, declaration, applications

File handling, standard C preprocessors, defining and calling macros, conditional compilation, passing values to the compiler.

Fundamentals of Computer

With the fast growing pace of development in the computer hardware field, it is important to learn the basics and keep up-to-date with the latest information. This chapter will enable the students to understand the basic concepts of computer systems and their components. These days computers are being put to use for all sorts of applications in every field of science and technology. The word *computer* refers to a system composed of many components. A computer system has both *hardware* and *software* components.

DEFINITION OF COMPUTER

Technically, a computer is a programmable machine. This means it can execute a programmed list of instructions and respond to new instructions that it is given. Today, however, the term is most often used to refer to the desktop and laptop computers that most people use. When referring to a desktop model, the term "computer" technically only refers to the computer itself not the monitor, keyboard, and mouse. Still, it is acceptable to refer to everything together as the computer. If you want to be really technical, the box that holds the computer is called the "system unit."

Some of the major parts of a personal computer (or PC) include the motherboard, CPU, memory (or RAM), hard drive, and video card. While personal computers are by far the most common type of computers today, there are several other types of computers too. For example, a "minicomputer" is a powerful computer that can support many users at once. A "mainframe" is a large, high-powered computer that can perform billions of calculations from multiple sources at one time. Finally, a "supercomputer" is a machine that can process billions of instructions a second and is used to make extremely complex calculations.

Graphically a computer system may be represented as shown in Fig. 1.1.

OVERVIEW OF A COMPUTER SYSTEM

All physical parts of the computer (or everything that we can touch) are known as **hardware**. All of the components of a computer system can be summarized with the simple equations below:

Computer System # Hardware + Software

Software gives "intelligence" to the computer.

Hardware # Internal devices +
Peripheral devices

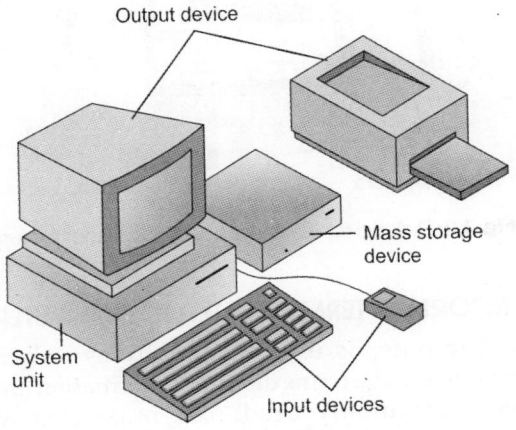

Fig. 1.1: A computer system

1

Basically all computers, regardless of their size, have the same general design which consist of the following units: the central processing unit (CPU), memory, and input/output circuitry which are situated on the **printed circuit board**, also called the **system board** or **motherboard.**

Figure 1.2 describes the relationships between the components of the computer system. Pictorially the computer system may be represented as shown in Fig. 1.3. The block diagram of the computer system is represented as shown in Fig. 1.4.

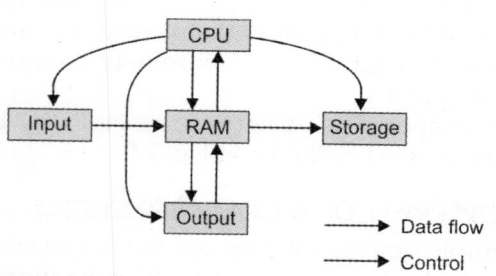

Fig. 1.2: Relationship between the components of computer system

Fig. 1.3: Pictorial representation of the computer system

IMPORTANT TERMS RELATED TO COMPUTER

A computer is used essentially as a data processor. The terms data and information are very commonly used. You must clearly understand the difference between the two.

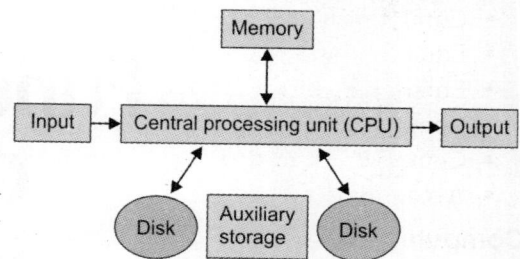

Fig. 1.4: Block diagram of the computer system

Data

Data in computer terminology means raw facts and figures. In computing, data is information that has been translated into a form that is more convenient to move or process. Relative to today's computers and transmission media, data is information converted into binary digital form.

For example, Manish, 1978, 'B', –163.89, +75.2 are data because here the format is not given.

Information

After processing data we get meaningful information, therefore information is the result of processing, manipulating and organizing data in a way that adds to the knowledge of the receiver. In other words, it is the context in which data is interpreted.

BASIC APPLICATIONS OF A COMPUTER SYSTEM

In the 1950s, computers were giant, special-purpose machines that were used by huge institutions like government and universities and they were used for performing complex numerical tasks such as calculating the precise orbit of Mars or planning the trajectories of missiles, etc. In the 1960s, business world started using computers for commercial purposes. In the 1970s, the invention of personal computer revolutionized the world and today computers are being used in nearly every field of life.

The following are the basic areas of computer's applications:

- Business and corporate computing
- Weather forecasting
- Medicines and health care

- Communications
- Education and reference
- Entertainment
- Computer-aided manufacturing
- Computer-aided design
- Accounting

Computers in Business and Corporate Computing

Corporate computing has revolutionized the business environment with the assistance of the computer that perform many time-consuming, complex tasks such as accounting, inventory control, customer databases, shipping control, and financial analysis. Today's desktop computers are versatile and relatively inexpensive. Networking of all the company's desktop computers allows access to and storage of the full range of company resources and information, reducing paper-work and increasing the productive flow of information.

Weather Forecasting

Before a forecast can be made of what the weather is likely to be in the future, knowledge of the present situation is essential. Therefore, regular, reliable and accurate measurements are required. These have to be rapidly sent around the world using a telecommunications system dedicated to weather information. The observations are fed into the computer and used to analyze the weather patterns at a particular time. Once the analysis has been carried out, the computer produces a forecast of the weather for specified times in the future. The forecaster uses the output from the computer to produce weather forecasts that are tailored to a wide range of customers.

Medicine and Health Care

The role of computers in medicine, use of computers in health care and importance of computers in medicine fields. Computers are being used today to diagnose diseases.

Communications

Computers can be effectively used for communicating to different parts of the world. Today computers are capable of exchanging information over the Internet. You can send a message from your computer to the computer of a neighbour or that of a friend on the other side of the planet. Students can use computers to communicate with their classmates to perform homework assignments, group projects, or other school-related activities. They may also use them to submit homework assignments and presentations to their teachers.

Education and Reference

Computers are effectively used in classrooms as an interactive learning pool. In libraries, people can directly read magazines and journals directly from a computer terminal. They do not need to search the shelves for paper originals. Nowadays, online learning and training is possible through networks. A student sitting at home can speak to his/her teacher and get response for his/her queries. Virtual classrooms are a reality these days.

Entertainment

In today's electronic era, computers have a hand in almost everything. Entertainment is no exception in fact, with the coming of digital information entertainment has made one of its greatest leaps. Movies, games, music, even books have been impacted greatly by computers. With the coming of DVDs (Digital video disks), computers have moved into a big portion of the video industry. But with it has come the ability to take a movie off of the disk and distribute it over user sharing programs. Many companies have tried to solve this problem by adding programs and such to the disks so that the information cannot be stolen.

Computer-aided Manufacturing

Computer-aided manufacturing (CAM) is the use of computer-based software tools that assist engineers and machinists in manufacturing or prototyping product components. CAM is a programming tool that allows you to make 3D models using computer-aided design (CAD). CAM was first used in 1971 for car body design and tooling.

Computer-aided Design

Computer-aided design (CAD) is the use of computer technology to aid in the design of a product. Current packages range from 2D vector base drafting systems to 3D solid and surface modelers.

Accounting

Computers play an important part in the recording of financial information. There are many accounting packages available and so businesses are able to use computerized accounting systems. A computerized accounting system provides the same functions as a manual accounting system. Most businesses therefore use computer systems instead of manual systems to record financial information, because a lot faster files can be shared more easily and changes made more conveniently.

CHARACTERISTICS OF COMPUTER

Let us now identify the major characteristics of the computer. These can be discussed under the headings of speed, accuracy, diligence, versatility and memory.

Speed

As you know computer can work very fast. It takes only a few seconds for calculations that we take hours to complete. Suppose you are asked to calculate the average monthly income of one thousand persons. For this you have to add income from all sources for all persons on a day to day basis and find out the monthly average for each one of them. How long will it take for you to do this? One day, two days or one week? But your small computer can finish this work in a few seconds? The weather forecasting that you see everyday on the TV is the result of compilation and analysis of huge amount of data on temperature, humidity, pressure, etc. of various places. It takes only a few minutes for the computer to process this huge amount of data and give the result.

You will be surprised to know that the computer can perform millions (1,000,000) of instructions and even more per second.

Therefore, we determine the speed of computer in terms of microsecond (10^{-6} part of a second) or nano-second (10^{-9} part of a second). From this you can imagine how fast the computer performs work.

Accuracy

Suppose someone calculates faster but commits a lot of errors in computing. Such result is useless. There is another aspect. Suppose you want to divide 15 by 7. You may work out up to 2 decimal places and say the dividend is 2.14. Another person may calculate up to 4 decimal places and say that the result is 2.1428. Someone else may go up to 9 decimal places and say the result is 2.142857143. Hence, in addition to speed, the computer should have accuracy or correctness in computing.

The degree of accuracy of computer is very high and every calculation is performed with the same accuracy. The accuracy level is determined on the basis of design of computer. The errors in computer occur only due to human mistake and inaccurate data.

Diligence

A computer is free from tiredness, lack of concentration, fatigue, etc. It can work for hours without committing any error. If millions of calculations are to be performed, a computer will perform every calculation with the same accuracy. Due to this capability it overpowers human beings in routine type of work.

Versatility

It means the capacity to perform completely different types of work. You may use your computer to prepare payroll slips. Next moment it may be used for inventory management or to prepare electricity bills.

Power of Remembering

Computer has the power of storing any amount of information or data. Any information can be stored and recalled as long as you require it, for any numbers of years. It depends entirely upon you how much data you want to store in a computer and when to lose or retrieve these data.

No IQ

Computer is a dumb machine and it cannot do any work without instructions from the user. It performs the instructions at tremendous speed and with accuracy. It is for you to decide what you want to do and in what sequence. So a computer cannot take its own decisions as you can.

Feeling

Computer does not have feelings or emotions, taste, knowledge and experience. Thus, it does not get tired even after long hours of work. It does not distinguish between users.

Storage

The computer has an in-built memory where it can store a large amount of data. You can also store data in secondary storage devices such as disks, which can be kept outside your computer and can be carried to other computers.

FUNCTIONAL COMPONENTS OF A COMPUTER SYSTEM

Computer systems ranging from a controller in a microwave oven to a large supercomputer contain components providing five functions. A typical personal computer has hard, and CD-ROM disks for storage, memory and CPU chips inside the system unit, a keyboard and mouse for input, and a display, printer and speakers for output. The arrows in Fig. 1.5 represent the direction of information flows between the functional units.

Graphically it is represented as follows:

The components of a computer system can be divided into three different units:

- Input unit
- Processing unit
- Output unit

The input unit of a computer system is responsible for input phase, CPU for process phase and output unit for output phase.

Now let us discuss the functioning of each unit one by one.

The Input Unit

The input unit deals with how to take the input from the user. Any machine that feeds

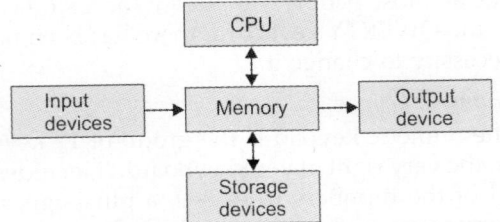

Fig. 1.5: Functional components of computer system

data into a computer, for example, a keyboard is an input device, whereas a display monitor is an output device. Input devices other than the keyboard are sometimes called *alternate input devices*. Mouse, trackballs, and light pens are all alternate input devices. Input devices are the devices that connect you to your computer. Input devices let you input data and other information into your computer and they also let you give your computer special instructions so that it will know what to do. For example, you can type in data by using a keyboard, or you can input data in picture form by using a scanner.

Keyboard

Keyboard is one of the most important input devices for a computer. They serve as your primary means of communication with your computer. Keyboards are a type of keypad similar to a typewriter with a set of keys that you press to input data into your computer.

Most keyboards have an ASCII (American Standard Code for Information Interchange) layout. This layout includes a numeric keypad, special function keys, special control keys, as well as typing keys.

Typing keys

The typing keys are just the keys containing the twenty-six letters of the alphabet. These keys have a QWERTY setup. QWERTY stands for the first six letters in the top row of the standard letter keys, and was first designed by American Christopher Sholes in 1890. It was designed so that the most commonly used letters would not be near each other. This is to prevent people to type too fast, because fast typing often results in keyboard jamming. Of course, jamming is no longer a problem today, but as most people are already accustomed to the QWERTY keyboard, there has been no necessity to change it.

Numeric keypad

The numeric keypad is the group of 17 keys on the very right of your keyboard. It includes all of the numbers from 0–9, a plus sign, a minus sign, and a few other keys. If you look closely, you will realize that the arrangement of the keys on a numeric keypad on your keyboard is quite similar to the arrangement of the keys on a calculator! Because many computers are used for business purposes and to input numbers into a computer, the numeric keypad was designed. This made it easier and faster to input numbers.

Function keys

The function keys can be found in a row at the very top of your keyboard. They are labeled "F1," "F2," "F3," all the way up to "F12." These function keys are mainly used to simplify the use of applications. They perform special commands, depending on which application is open. For example, pressing F7 in a word document opens the spell check. In almost all applications, pressing F1 opens a help menu. Also, the function keys can be combined with other keys, like the control keys to perform different tasks!

Control keys

The control keys are special keys like del (delete), ctrl (control), alt (alternate), shift, and enter. The delete (or backspace) key is used to erase characters in a word document. The esc (escape) key is generally used to exit out of a program or application, while when you strike the enter key, a function or command is carried out. The tab key is primarily used to indent in a word document. The shift key is often used in conjunction with other keys. For example, pressing shift and then a letter will make that letter capital, while pressing shift and a number will display the character shown right above that number. Finally, there are the ctrl and alt keys. Like the shift key, these are also used in conjunction with other keys, and they usually perform functions similar to those of the function keys. There

are also the cursor movement keys. These are the arrows, shaped like an inverted T located between the typing keys and the numeric keypad. These keys are used to move the cursor around on your screen. For example, if you were in a word document, the blinking cursor will move up and down to different lines if you press the up and down cursor movement keys.

Working of a keyboard

The following steps explain how the keyboard when we press any key on the keyboard.

1. You press the "a" key on the keyboard.

2. The keyboard sends an electrical signal to the central processing unit (CPU) telling it that a key was pressed.

3. The CPU is almost ALWAYS busy. Thus the letter "a" is temporarily stored in a special memory cell until the CPU is ready to work with it.

4. The keyboard controller tells the central processing unit that there is information waiting to be processed.

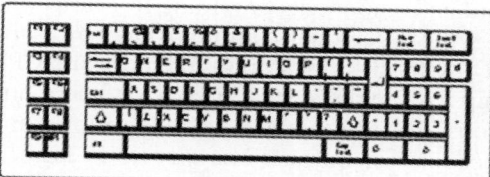

XT keyboard

AT (standard)

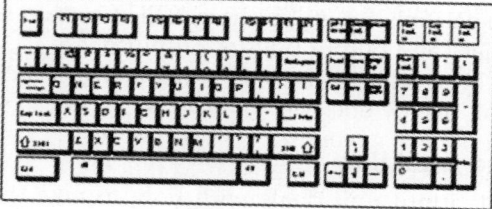

AT (enhanced)

5. The CPU figures out which application you were in when you pressed the key. For example, you might have an Internet browser and a word document open at the same time. The processor will determine whether you pressed the "a" in the word document or in the Internet browser.

6. Assuming that you pressed the key in a word document browser, the letter will be added to the computers randomaccess memory (it takes up one byte of space).

7. Then, the letter "a" will be displayed on the computer screen.

Mouse

A device that controls the movement of the cursor or pointer on a display screen is known as mouse. A mouse is a small object that you can roll along a hard, flat surface. Its name is derived from its shape, which looks a bit like a mouse, its connecting wire that one can imagine to be the mouse's tail, and the fact that one must make it scurry along a surface. As you move the mouse, the pointer on the display screen moves in the same direction. Mouse contains at least one button and sometimes as many as three, which have different functions depending on what program is running. Some newer mice also include a *scroll wheel* for scrolling through long documents.

Invented by Douglas Engelbart of Stanford Research Center in 1963, and pioneered by Xerox in the 1970s, the mouse is one of the great breakthroughs in computer ergonomics because it frees the user to a large extent from using the keyboard. In particular, the mouse is important for graphical user interfaces because you can simply point to options and objects and click a mouse button. Such applications are often called *point-and-click* programs. The mouse is also useful for graphics programs that allow you to draw pictures by using the mouse like a pen, pencil, or paintbrush.

Working of a mouse

The working of a mouse is as follows:

1. There is a ball on the underside of the mouse that spins as you move the mouse around.

2. Inside the mouse there are two rollers. In most mice, these rollers look like black bars. If you open up the disk on the bottom of the mouse and take the ball out, you will be able to see these rollers. One of the rollers detects the horizontal movement of the mouse, while the other detects the mouse's vertical movement. Working together, the two rollers can determine which direction the mouse is moving!

3. When the ball rolls, it causes the rollers to roll. When the rollers roll, they cause a special disk to spin. The disk is circular with rectangular holes spaced out along its outer edge.

4. On one side of the disk there is an infrared LED, while on the other side of the disk, opposite the LED, there is an infrared sensor. When the disk spins, the light from the LED can go through the rectangular holes periodically.

5. The sensor is able to sense these pulses of light.

6. A processor directly built into the mouse converts these pulses into binary digits that the computer can understand, and the information is transferred to the central processing unit through the mouse's cord.

The working of a mouse is graphically represented in Fig. 1.6.

There are three basic types of mouse:

1. **Mechanical:** It has a rubber or metal ball on its underside that can roll in all directions. Mechanical sensors within the mouse detect the direction the ball is rolling and move the screen pointer accordingly.

2. **Optomechanical:** It is same as a mechanical mouse, but uses optical sensors to detect the motion of the ball.

3. **Optical:** It uses a laser to detect the mouse's movement. You must move the mouse along a special mat with a grid so that the optical mechanism has a frame of reference. Optical mice have no

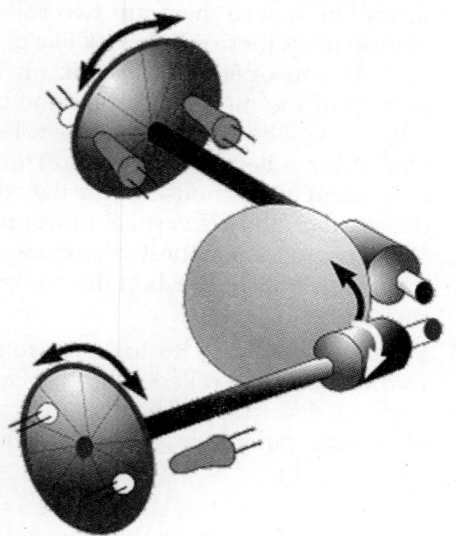

Fig. 1.6: Graphical representation of working of a mouse

mechanical moving parts. They respond more quickly and precisely than mechanical and optomechanical mice, but they are also more expensive.

Mouse connects to PCs in one of several ways:

1. **Serial mice** connect directly to an RS-232C serial port or a PS/2 port. This is the simplest type of connection.

2. **PS/2 mice** connect to a PS/2 port.

3. USB mice.

Cordless mice are not physically connected at all. Instead they rely on infrared or radio waves to communicate with the computer. Cordless mice are more expensive than both serial and bus mice, but they do eliminate the cord, which can sometimes get in the way.

The picture of a mouse is depicted in Fig. 1.7.

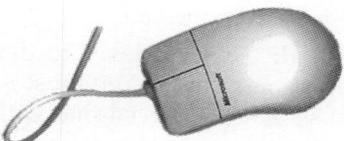

Fig. 1.7: Picture of a mouse

Scanner

Scanners have slowly made their way into the everyday lives of the general public. They are ways in which graphical images can be inputted into the computer for touch-ups or to be distributed via the World Wide Web.

Types of Scanner

There are four types of scanners:

Flatbed scanners

This scanner provides a flat, glass surface to hold pages of paper, books and other objects for scanning. The scan head is moved under the glass across the page. Sheet feeders are usually optionally available that allow multiple sheets to be fed automatically. These are the most commonly used and most versatile scanners. The picture of a flatbed scanner is shown in Fig. 1.8

Sheet-fed scanners

A scanner that allows only paper to be scanned rather than books or other thick objects. It moves the paper across a stationary scan head. Sheet-fed scanners are similar to flatbed scanners. The only difference is that in flatbed scanners the image you wish to scan does not move, in sheet-fed scanners the document moves, like how the piece of paper moves in a printer. The picture of a Sheet-fed scanner is shown in Fig. 1.9.

Handheld scanners

A scanner that is moved across the image to be scanned by hand. Handheld scanners are small and less expensive than their desktop counterparts, but rely on the dexterity of the

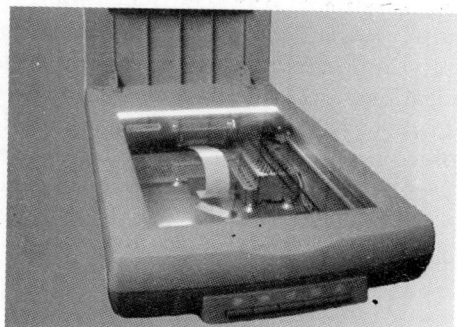

Fig. 1.8: Flatbed scanner

Fig. 1.9: Sheet-fed scanner

speeds exceeding 1000 rpm during the scanning operation. A light source that focuses on one pixel is beamed onto the drum and moves down the drum a line at a time. For transparencies, light is directed from the center of the cylinder. For opaque items, a reflective light source is used. Mirrors filter out the RGB values and send them to the drum scanner's photomultiplier tube (PMT), which is more sensitive than the CCDs (charged-couple devices) used in flatbed and sheet-fed scanners and can produce resolutions exceeding 10,000 dpi. If one PMT is used, three passes across the image are required. When three PMTs are used, a faster single-pass scan is performed. Drum scanners can be used to produce very detailed graphics. The picture of a drum scanner is shown in Fig. 1.11.

user to move the unit across the paper. Trays are available that keep the scanner moving in a straight line. These scanners operate like flatbed scanners except that user has to move the scan head (the part of a scanner responsible for scanning the image) in the place of some motorized machine. The images produced by these handheld scanners are not of very good quality. The picture of a handheld scanner is shown in Fig. 1.10.

Drum scanner
A type of scanner used to capture the highest resolution from an image. Photographs and transparencies are taped, clamped or fitted into a clear cylinder (drum) that is spun at

Optical Character Recognition
Optical character recognition (OCR) is the recognition of printed or written text characters by a computer. This involves photo scanning of the text character-by-character, analysis of the scanned-in image, and then translation of the character image into character codes, such as ASCII, commonly used in data processing.

In OCR processing, the scanned-in image or bitmap is analyzed for light and dark areas in order to identify each alphabetic letter or numeric digit. When a character is recognized, it is converted into an ASCII code. Special circuit boards and computer chips designed expressly for OCR are used to speed up the

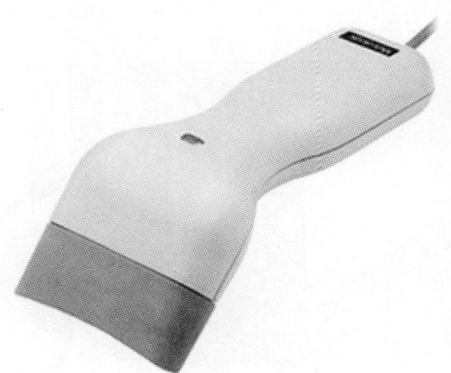

Fig. 1.10: Handheld scanner

Fig. 1.11: Drum scanner

recognition process. OCR is being used by libraries to digitize and preserve their holdings. OCR is also used to process cheques and credit card slips and sort the mail. Billions of magazines and letters are sorted every day by OCR machines, considerably speeding up mail delivery. The picture of a OCR is shown in Fig. 1.12.

Magnetic Ink Character Recognition

Magnetic ink character recognition (MICR) is a character recognition system that uses special ink and characters. When a document that contains this ink needs to be read, it passes through a machine, which magnetizes the ink and then translates the magnetic information into characters.

MICR technology is used by banks. Numbers and characters found at the bottom of cheques (usually containing the cheque number, sort number, and account number) are printed using magnetic ink. To print magnetic ink, you need a laser printer that accepts MICR toner. MICR provides a secure, high-speed method of scanning and processing information.

⑆1234567890⑆ ⑈1234567890⑈ ⑇1234567890⑇ ⑉1234567890⑉

An example of the E-13B MICR font. Shown are the 14 characters of the **E-13B** font. The control characters bracketing each numeral block are (from left to right) *transit*, *on-us*, *amount*, and *dash*.

Light Pen

A light pen is an input device that utilizes a light-sensitive detector to select objects on a display screen. A light pen is similar to a mouse, except that with a light pen you can move the pointer and select objects on the display screen by directly pointing to the objects with the pen.

A light-sensitive stylus is wired to a video terminal used to draw pictures or select menu options. The user brings the pen to the desired point on screen and presses the pen button to make contact. Contrary to what it looks like, the pen does not shine light onto the screen; rather, the screen beams into the pen. Screen pixels are constantly being refreshed. When the user presses the button, the pen senses light, and the pixel being illuminated at that instant identifies the screen location.

The picture of a light pen is shown in Fig. 1.13.

Barcode Reader

A barcode reader also called a price scanner or point-of-sale (POS) scanner is a handheld or stationary input device used to capture and read information contained in a bar code. A barcode reader consists of a scanner, a decoder (either built-in or external), and a cable is used to connect the reader with a computer. Because a barcode reader merely captures and translates the barcode into numbers and/or

Fig. 1.12: OCR reader

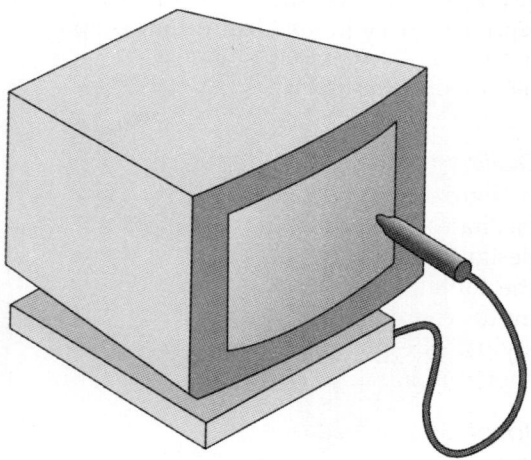

Fig. 1.13: Light pen

letters, the data must be sent to a computer so that a software application can make sense of the data. Barcode scanners can be connected to a computer through a serial port, keyboard port, or an interface device called a wedge. A barcode reader works by directing a beam of light across the bar code and measuring the amount of light that is reflected back. (The dark bars on a barcode reflect less light than the white spaces between them.) The scanner converts the light energy into electrical energy, which is then converted into data by the decoder and forwarded to a computer. The picture of a barcode reader is shown in Fig. 1.14.

Fig. 1.15: Joystick

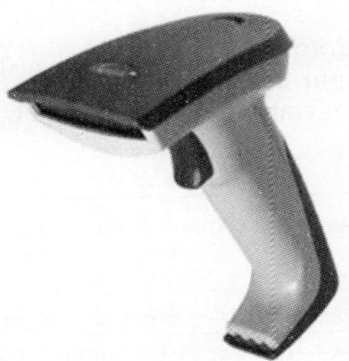

Fig. 1.14: Barcode reader

Joystick

The joystick is a rotary lever. Similar to an aircraft's control stick, it enables you to move within the screen's environment, and is widely used in the computer games industry. The picture of a joystick is shown in Fig. 1.15.

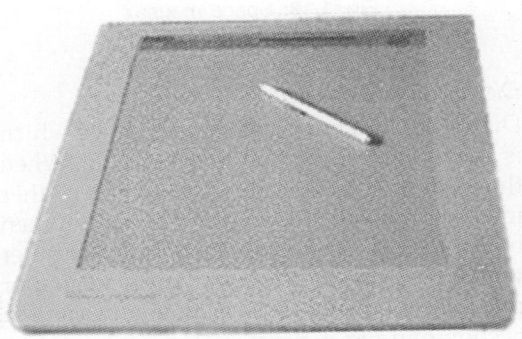

Fig. 1.16: Digitizing tablet

Digitizing Tablet

A digitizing tablet is a pointing device that facilitates the accurate input of drawings and designs. A drawing can be placed directly on the tablet, and the user traces outlines or inputs coordinate positions with a hand-held stylus. The picture of a digitizing tablet is shown in Fig. 1.16.

Touch Sensitive Screen

A touch sensitive screen is a pointing device that enables the user to interact with the computer by touching the screen. There are three forms of touch screen: pressure-sensitive, capacitive surface and light beam. The picture of a touch sensitive screen is shown in Fig. 1.17.

Space Mouse

The space mouse is different from a normal mouse as it has an X axis, a Y axis and a Z axis. It can be used for developing and moving around 3-D environments. The picture of a space mouse is shown in Fig. 1.18.

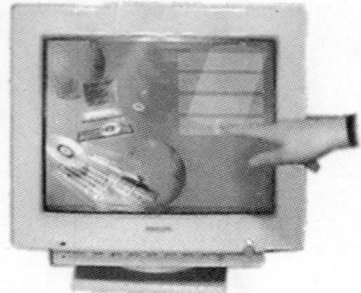

Fig. 1.17: A touch sensitive monitor

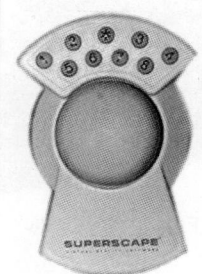

Fig. 1.18: Space mouse

Digital Stills Camera

Digital stills cameras capture an image which is stored in memory within the camera. When the memory is full it can be erased and further images captured. The digital images can then be downloaded from the camera to a computer where they can be displayed, manipulated or printed. The picture of a digital stills camera is shown in Fig. 1.19.

Optical Mark Reader

The optical mark reader (OMR) can read information in the form of numbers or letters and put it into the computer. The marks have to be precisely located as in multiple choice question papers (Fig. 1.20)

Magnetic Reader

This input device reads a magnetic strip on a card. Handy for security reasons, it provides quick identification of the card's owner. This method is used to run automatic teller machine (ATM) of banks or to provide quick identification of people entering buildings. A credit card showing the magnetic stick is illustrated in Fig. 1.21a.

Smart Cards

This input device stores data in a microprocessor embedded in the card (Fig. 1.21b).

Block VII

1	A	B	C	D	E	F	G
2	A	B	C	D	E	F	G
3	A	B	C	D	E	F	G
4	A	B	C	D	E	F	G
5	A	B	C	D	E	F	G
6	A	B	C	D	E	F	G

Fig. 1.20: A sample multiple choice answer paper

Fig. 1.19: A digital stills camera

Fig. 1.21a: A credit card showing the magnetic strip

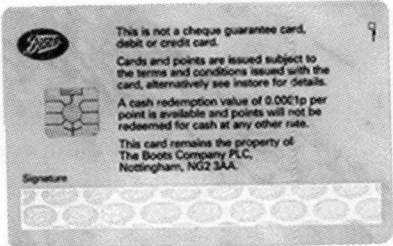

Fig. 1.21b: A store card showing the square micro-processor chip

Fig. 1.23: A sound card

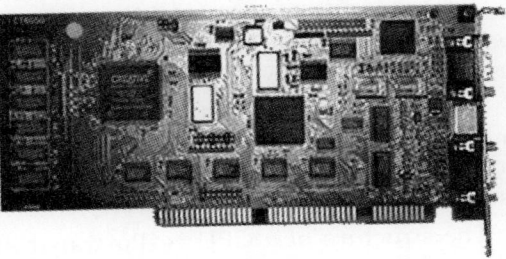

Fig. 1.24: A video capture card

This allows information, which can be updated, to be stored on the card. This method is used in store cards which accumulate points for the purchaser, and to store phone numbers for cellular phones.

Voice Data Entry

This system accepts the spoken word as input data or commands. Human speech is very complex, involving emphasis and facial expressions, so complete voice recognition will not get developed for some more time. However, simple commands from one user can be used to control machines. In this way a paralysed person can operate a wheelchair or control heating and lighting. A microphone used for data input is shown in Fig. 1.22.

Sound Capture

With the addition of a sound card in one of the expansion slots of the computer you can "record" voice or music. The sound card (Fig. 1.23) digitizes the information into a form that the computer can understand.

Video Capture

With a video capture board (Fig. 1.24) in one of the computer's expansion slots, you can capture video (photographic) images through a video camera. The video capture board digitizes the image.

Processing Unit

Processing unit of a computer system holds the central processing unit which is like the mind of a computer system.

Central Processing Unit (CPU)

The CPU is the brain of the computer. Sometimes referred to simply as the *processor* or *central processor*, the CPU is where most calculations take place. In terms of computing power, the CPU is the most important element of a computer system.

It carries out all the processing in the computer. It itself consists of three main subsystems. The first one is control unit, the second is registers, and the third is arithmetic and logic unit (ALU).

A CPU works in a fetch execute cycle. On power on, the CPU fetches the first instruction from a location specified by the program

Fig. 1.22: A microphone used for data input

counter. This instruction is brought into instruction register which is decoded by the control unit. Based on the instruction, the control unit would either fetch the operand and or carry out arithmetic or logical operations on it, or store the result of such an operation into a specified memory location. After one instruction is executed the next instruction is fetched by the processor and executed. This process goes on till the processor does not come to a halt instruction. A real life processor would have a large number of registers, a sophisticated micro program control unit and a sophisticated arithmetic and logic unit. Most powerful processors currently popular are from Intel Pentium IV and core to dual.

On large machines, CPUs require one or more printed circuit boards. On personal computers and small workstations, the CPU is housed in a single chip called *microprocessor*.

The structure of a CPU is illustrated in Fig. 1.25.

Functioning of a CPU

The CPU performs two basic functions:

1. It controls and coordinates the activities of the other parts of the computer and of its own component parts. This control function is performed by circuits which send out sequences of control signals that cause the various components of the computer to act at the correct times.

2. It processes information that it contains or that is sent to it by the other units. This processing function is carried out by circuits capable of performing arithmetic and logical operations on data contained within the CPU.

Components of CPU

Two typical components of a CPU are as follows:
- The arithmetic logic unit (ALU), which performs arithmetic and logical operations.
- The control unit (CU), which extracts instructions from memory and decodes and executes them, calling on the ALU when necessary.

Arithmetic Logic Unit (ALU)

The arithmetic logic unit carries out as the name suggests arithmetic and logical operations on the data made available to it. Basic arithmetic functions which an ALU can carry out are addition and subtraction. More powerful CPU can support additional mathematical operations like multiplication and division. The logical operations which it can carry out are greater than, equal to, less than comparison between two numbers. Besides these operations some processors also support operations which check if particular bits are on or off.

The structure of ALU is shown in Fig. 1.26.

Control Unit

The control unit can be thought of as the brain of the CPU itself. It controls based on the instructions it decodes, how the other parts of the CPU and in turn, the rest of the computer system should work in order that the instruction gets executed in a correct manner. There are two types of control units, the first type is called hardwired control unit. Hardwired control units are constructed

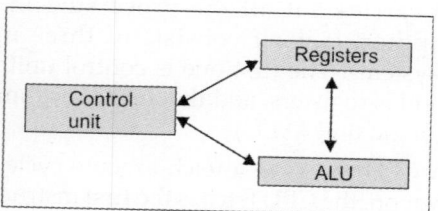

Fig: 1.25: Structure of a CPU

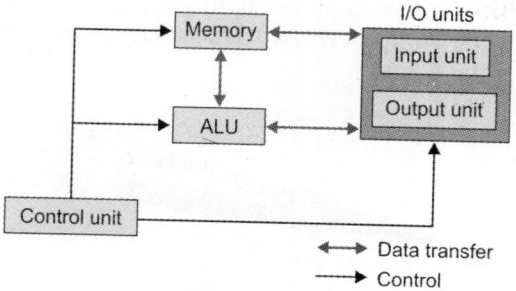

Fig: 1.26: Structure of ALU

using digital circuits and once formed can not be changed. The other type of control unit is the microprogrammed control unit. A micro programmed control unit itself decodes and execute instructions by means of executing microprograms.

The **control unit** coordinates the work of the whole computer system. It has three main tasks:

1. It follows a set of instructions called the computer 'program'.
2. It controls the hardware attached to the system. The control unit monitors the hardware to make sure that the commands given to it by the current program are activated.
3. It controls the input and output of data, so all the signals go to the right place at the right time.

The control units reads a single instruction from the computer program and executes it. Once that instruction is carried out, it moves on to the next one and so on. This often means that it has to fetch some data from memory (RAM) and then do some arithmetic on it with the help of the ALU. The operation of control unit is represented in Fig. 1.27.

The Output Unit

The output unit is responsible for presenting the output in user readable form. Various output devices like monitor (also called VDU, i.e. Visual Display Unit), printer, plotter, etc. make the output unit of a computer.

The function of an output device is to present the processed data to the user. The computer sends the output to the monitor. The output can also be sent to the printer whenever the output is needed in hardcopy (printed form). The sound output is produced with the help of speakers. Graphic output is produced with the help of plotters.

Output Devices

These devices display information that has been held or generated within a computer. We will discuss the following output devices:

- VDU or Monitor
- Printer
- Impact printer
- Dot matrix printer
- Daisywheel printer
- Non-impact printer
- Thermal printer
- Laser printer
- Ink jet printer
- Plotter
- Flatbed plotter
- Drum plotter
- Electrostatic plotter

Visual Display Units

Visual display units (VDUs) or monitors are used to visually interface with the computer and are similar in appearance to television. Visual display units display images and text which are made up of small blocks of colored light called pixels. The resolution of the screen improves as the number of pixels is increased. Most monitors have a 4:3 width to height ratio. Figure 1.28 depicts a cathode ray monitor and a plasma monitor.

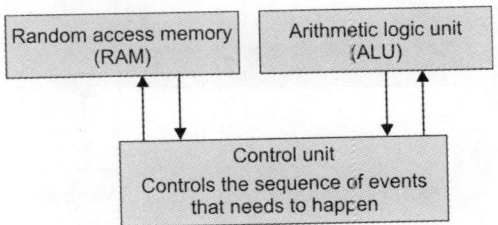

Fig. 1.27: Operation of control unit

(a) A cathode ray monitor   (b) A plasma monitor

Fig. 1.28: Visual display units

Screen Resolution of a VDU

The standard user interface on the PC was originally a text-only mode. With the increased resolution of VDUs applications are now written in graphics mode using individual pixels. A recent standard has been 640 × 480 pixels on the screen (this is called VGA) and the present standard is 800 × 600 (called SVGA). This has enabled the use of an increasingly sophisticated visual interface, utilizing graphical user interfaces (GUIs) such as Microsoft Windows and MAC OS as well as more highly developed user friendly software. Figure 1.29 depicts the different screen resolution.

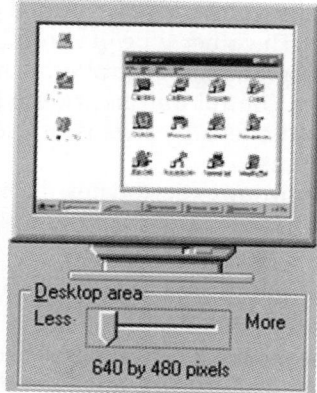

Fig. 1.29: Diagram of different screen resolutions

Printer

You can print out information that is in the computer onto paper. By printing you create what is known as a 'hard copy'. There are different kinds of printers which vary in their speed and print quality. A most convenient and useful method by which the computer can deliver information is by means of printed characters.

Types of Printer

The two main types of printer are:
1. Impact printer
2. Non-impact printer

Impact printer

Impact printers (Fig. 1.30) use a print head containing a number of metal pins which strike an inked ribbon placed between the print head and the paper. Some print heads have only 9 pins to make the dots to build up a character; some have 24 pins which produce a better resolution.

Fig. 1.30: An impact printer

Graphically it is represented as follows:

In this category we will discuss the dot matrix printer and daisywheel printer.

Dot matrix printer: In dot matrix printer characters are formed from a matrix of dot (Fig. 1.31). The speed is usually 30–550 characters per second (cps). This is the cheapest and noisiest of the printer family. The standard of print obtained is poor. These printers are cheap to run and relatively fast. They are useful for low quality carbon copy printing.

Fig. 1.31: A dot matrix printer

Daisywheel printer

Molded metal characters like those in a typewriter (Fig. 1.32) are mounted on extensions attached to a rotating wheel and are printed onto the paper by means of a hammer and print ribbon. This results in a great deal of movement and noise during the printing of documents, so printing is slow (less than 90 cps). The standard of print is similar to that produced by an electric typewriter. As the characters on the wheel are fixed, the size and font can only be changed by using a different wheel. However, this is very rarely done.

Non-impact printer

Non-impact printers are much quieter than impact printers as their printing heads do not strike the paper. Most non-impact printers produce dot-matrix patterns. Several different technologies have been used to provide a variety of printers. The main types of non-impact printer are as follows:

Fig. 1.32: A daisywheel showing detail of the characters

(a) Thermal printer

(b) Laser printer

(c) Inkjet printer

Thermal printer: In thermal printer, characters are formed by heated elements being placed in contact with special heat sensitive paper forming darkened dots when the elements reach a critical temperature. Thermal printer paper tends to darken over time due to exposure to sunlight and heat The standard of print produced is poor. Thermal printers (Fig. 1.33) are widely used in battery powered equipment such as portable calculators.

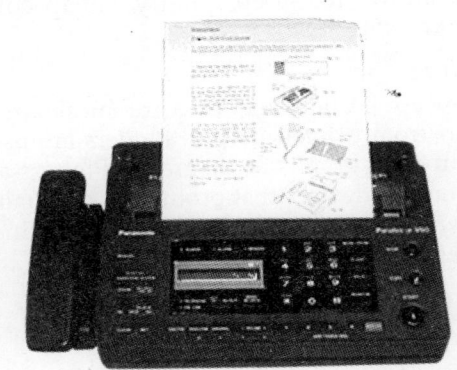

Fig. 1.33: A fax machine uses a thermal printer

Laser printer: Laser printers (Fig. 1.34) use a laser beam and dry powdered ink to produce a fine dot matrix pattern. This method of printing can generate about 4 pages of A4 paper per minute. The standard of print is very good and laser printers can produce very good quality printed graphic images too. A photoconductive drum is initially charged and then a high intensity laser beam is used to discharge selected areas on the drum. These discharged areas correspond to the white

Fig. 1.34: A laser printer

areas of the printed document. Toner is attracted to parts of the drum with a high charge. The drum rotates and transfers the toner to the paper which has an even greater electrical charge. Finally a heater fixes the toner onto the paper.

The working of a laser printer is graphically represented in Fig. 1.35.

Inkjet printer: In inkjet printer (Fig. 1.36), characters are formed as a result of electrically charged or heated ink being sprayed in fine jets onto the paper. Individual nozzles in the printing head produce high resolution (up to 400 dots per inch or 400 dpi) dot matrix characters. Ink jet printers use color cartridges which combine magenta, yellow and cyan inks to create color tones. A black cartridge is also used for crisp monochrome output. This method of printing can generate up to 200 cps and allows for good quality, cheap color printing (Fig. 1.36).

Plotters

Many applications require a graphical output apart from printed output. Plotters are used to produce graphs or diagrams. Plotters are the output devices that produce good quality drawings and graphs. Plotters are classified into two types as follows:

1. Pen plotters
2. Electrostatic plotters

Pen plotters

Pen plotters have an ink pen attached to draw the images and pen plotters are classified into two categories as follows:

(a) Flatbed plotter
(b) Drum plotters

(a) Flatbed plotter: This is a plotter (Fig. 1.37) where the paper is fixed on a flat surface and pens are moved to draw the image. This plotter can use several different color pens to draw with. The size of the plot is limited only by the size of the plotter's bed.

(b) Drum plotters: In drum plotters (Fig. 1.38) the pen is moved in a single axis track and the paper itself moves on a cylindrical drum to add the other axis or dimension. The size

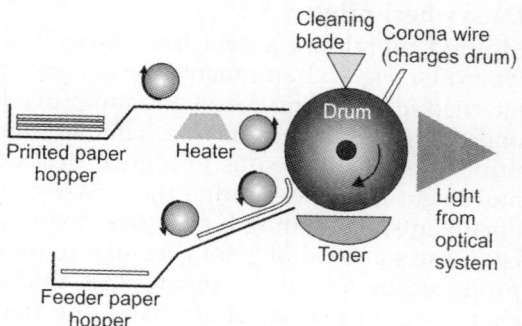

Fig. 1.35: How a laser printer works

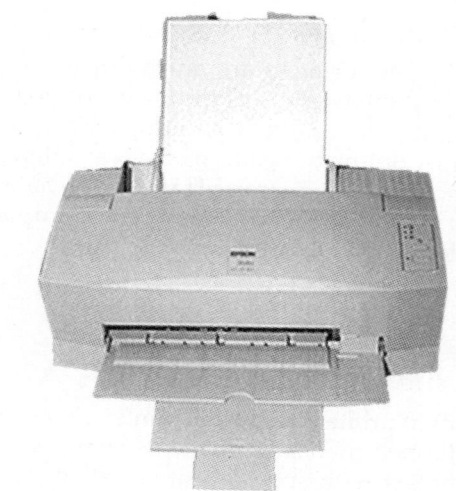

Fig. 1.36: An inkjet printer

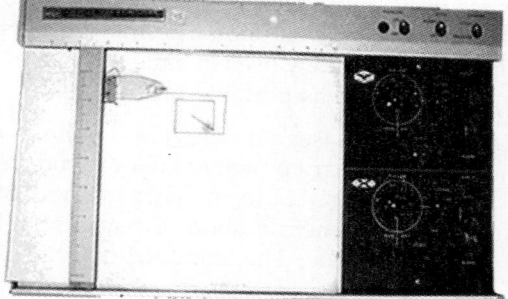

Fig. 1.37: Top view of a flatbed plotter

of the graph is therefore limited only by the width of the drum and can be of any length.

Fig. 1.38: A drum plotter

Electrostatic plotters

An electrostatic plotter produces a raster image by charging the paper with a high voltage. This voltage attracts toner which is then melted into the paper with heat. This type of plotter is fast, but the quality is generally considered to be poor when compared to pen plotters.

Storage Unit

Since the computer is an electronic device can understand only the electricity signals whatever input is given to it, is firstly converted into electric signals. An electric signal has just two states: ON or OFF. That means a computer can understand just two symbols: On (signified by 1) and Off (signified by 0). Therefore, computer's language is made up of just 1's and 0's and this language is also known as binary language? Why binary language? Well bi means two; so a language that has just two symbols is known as binary language.

The two binary digits 0 and 1 are known as bits. The bit is the smallest unit of storage. Bigger units are listed as follows:

- 1 byte = 8 bits
- 1 kilobyte (K/Kb) = 2^{10} bytes = 1,024 bytes
- 1 megabyte (M/MB) = 2^{10} kilobytes = 1,048,576 bytes
- 1 gigabyte (G/GB) = 2^{10} megabytes = 1,073,741,824 bytes
- 1 terabyte (T/TB) = 2^{10} gigabytes = 1,099,511,627,776 bytes

- 1 petabyte (P/PB) = 2^{10} terabytes = 1,125,899,906,842,624 bytes
- 1 exabyte (E/EB) = 2^{60} petabytes = 1,152,921,504,606,846,976 bytes.

The unit byte is very important as, any one character can be stored in a byte. That is, if you want to store 'SRMS', it will consume 4 bytes as there are 4 characters to be stored in it.

The Memory

Memory is the internal storage areas in the computer. The term memory identifies data storage that comes in the form of chips, and the word storage is used for memory that exists on tapes or disks. Moreover, the term memory is usually used as shorthand for physical memory, which refers to the actual chips capable of holding data. Some computers also use virtual memory, which expands physical memory onto a hard disk.

Every computer comes with a certain amount of physical memory, usually referred to as main memory or RAM. You can think of main memory as an array of boxes, each of which can hold a single byte of information. A computer that has 1 megabyte of memory, therefore, can hold about 1 million bytes (or characters) of information.

The memory of computer is often called main memory or primary memory. It is generally called the third component of CPU.

Types of Memory

The computer makes use of two types of memories:

1. Internal memory (also known as main memory or primary memory)
2. External memory (also known as auxiliary memory or secondary memory).

Internal memory

In a computer, all of the storage spaces that are accessible by a processor without the use of the computer input-output channels is known as internal memory. Internal memory usually includes several types of storage, such as main storage, cache memory, and special registers, all of which can be directly accessed by the processor.

The internal memory is the built-in memory where CPU holds the programs and data being manipulated. There are two types of internal memory

(a) RAM (Random Access Memory)

(b) ROM (Read-Only Memory).

(a) RAM (Random Access Memory): RAM is acronym for random access memory, a type of computer memory that can be accessed randomly; that is, any byte of memory can be accessed without touching the preceding bytes. RAM is the most common type of memory found in computers and other devices, such as printers. There are two basic types of RAM:

1. Dynamic RAM (DRAM)
2. Static RAM (SRAM)

These two types differ in the technology they use to hold data, dynamic RAM being the more common type. Dynamic RAM needs to be refreshed thousands of times per second. Static RAM does not need to be refreshed, which makes it faster; but it is also more expensive than dynamic RAM. Both types of RAM are *volatile,* meaning that they lose their contents when the power is turned off.

In common usage, the term RAM is synonymous with *main memory,* the memory available to programs. For example, a computer with 8M RAM has approximately 8 million bytes of memory that programs can use. In contrast, ROM (read-only memory) refers to special memory used to store programs that boot the computer and perform diagnostics. Most personal computers have a small amount of ROM (a few thousand bytes). In fact, both types of memory (ROM and RAM) allow random access. To be precise, therefore, RAM should be referred to as *read/write RAM* and ROM as *read-only* RAM. The limitations of RAM are:

• Limited storage capacity

• Volatile in nature, i.e. its contents get erased when power is turned off.

(b) ROM (Read-Only Memory): ROM is acronym for read-only memory, computer memory on which data has been prerecorded.

Once data has been written onto a ROM chip, it cannot be removed and can only be read. Unlike main memory (RAM), ROM retains its contents even when the computer is turned off. ROM is referred to as being nonvolatile, whereas RAM is volatile. Most personal computers contain a small amount of ROM that stores critical programs such as the program that boots the computer. In addition, ROMs are used extensively in calculators and peripheral devices such as laser printers, whose fonts are often stored in ROMs. A variation of a ROM is a PROM (programmable read-only memory). PROMs are manufactured as blank chips on which data can be written with a special device called a PROM programmer.

External Memory

External memory which is sometimes called *backing store* or *secondary memory,* allows the permanent storage of large quantities of data. Some method of magnetic recording on magnetic disks or tapes is most commonly used. More recently optical methods which rely upon marks etched by a laser beam on the surface of a disc (CD-ROM) have become popular, although they remain more expensive than magnetic media. The capacity of external memory is high, usually measured in hundreds of megabytes or even in **gigabytes** (thousand million bytes) at present. External memory has the important property that the information stored is not lost when the computer is switched off.

Most common external storage devices are: hard drives, floppy disks, magnetic tapes, CD ROM, video disks, etc.

Now we will discuss some external storage devices.

Floppy disk: A floppy disk is a thin magnetic-coated disk contained in a flexible or semi-rigid protective jacket. Data is stored in tracks and sectors. The floppy disks are usually 3.5" in size. However, older floppy disks may be in use; these would be 5.25" in size. Double sided high density 3.5" disks can hold 1.44 MB of data. Once data is stored on a floppy disk it can be 'write protected' by clicking a tab on the disk. This prevents any new data being stored or any old data being erased. The floppy disk has now nearly been replaced by the compact disk (CD).

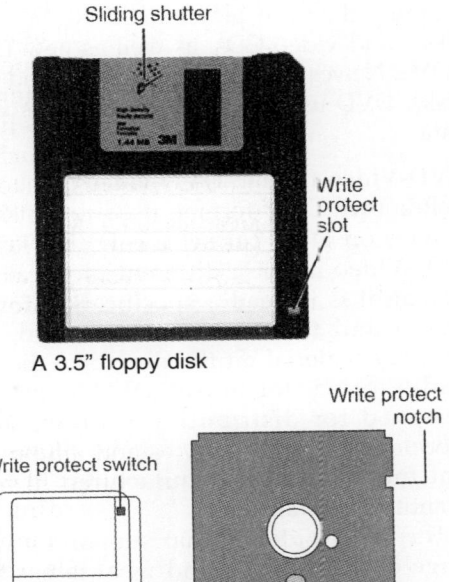

A 3.5" floppy disk

3½-inch   5¼-inch

Hard Disk

The hard disk is direct-access storage medium with a rigid hard magnetic disk. The data is stored as magnetized spots arranged in concentric circles (tracks) on the disk. Each track is divided into sectors. The number of tracks and sectors on a disk is known as its 'format'.

High data rates demand that the disk rotates at a high speed (about 3,600 rpm). As the disk rotates, read/write heads move to the correct track. The disk is sealed and lubricated and the head hovers on a cushion of air just above the disk to avoid damage. These are therefore called floating heads. The storage capacity of a hard disk can be several gigabytes (GB), i.e. thousands of megabytes (1000 MB), of information.

A single hard disk usually consists of several *platters*. Each platter requires two read/write heads, one for each side. All the read/write heads are attached to a single access arm so that they cannot move independently. Each platter has the same number of *tracks*, and a track location that cuts across all platters is called a cylinder. For example, a typical 84 megabyte hard disk for a PC might have two platters (four sides) and 1,053 cylinders. The picture of a hard disk showing internal mechanism is given in Fig. 1.39.

Magnetic tape: This recording medium consisting of a thin tape with a coating of a fine magnetic material, used for recording analogue or digital data. Data is stored in frames across the width of the tape. The frames are grouped into blocks or records which are separated from other blocks by gaps. Magnetic tape is a serial access medium, and so data cannot be quickly located. Like videotape, computer tape is made of flexible plastic with one side coated with a ferromagnetic material. Tapes were originally open reels, but were superseded by cartridges (Fig. 1.40) and cassettes of many sizes and shapes.

Fig. 1.39: Hard disk showing internal mechanisms

Fig. 1.40: A JAZ™ drive with cartridge

However, large amounts of information can be stored within magnetic tape. This characteristic has prompted its use in the regular backing up of hard disks.

Compact disks (CDs): A compact disk (CD) is a small, portable, round medium made of molded polymer (close in size to the floppy disk) for electronically recording, storing, and playing back audio, video, text, and other information in digital form (Fig. 1.41).

Initially, CDs were read-only, but newer technology allows users to record as well. CDs will probably continue to be popular for music recording and playback. A newer technology, the digital versatile disk (DVD), stores much more in the same space and is used for playing back movies.

Some variations of the CD include:

- CD-ROM
- CD-i
- CD-RW
- CD-ROM XA
- CD-W
- Photo CD
- Video CD

Digital Video Disk or Digital Versatile Disk (DVD)

DVD is a type of optical disk technology similar to the CD-ROM (Fig. 1.42). A DVD holds a minimum of 4.7 GB of data, enough for a full-length movie. DVDs are commonly used as a medium for digital representation of movies and other multimedia presentations that combine sound with graphics.

The DVD specification supports disks with capacities of from 4.7 GB to 17 GB and access rates of 600 kbps to 1.3 Mbps. One of the best features of DVD drives is that they are backward-compatible with CD-ROMs, meaning they can play old CD-ROMs, CD-I disks, and video CDs, as well as new DVD-ROMs. Newer DVD players can also read CDR disks. DVD uses MPEG-2 to compress video data.

DVD-Video format: DVD-video is the most well-known DVD format, used to distribute movies on DVD for set-top DVD players. DVD-Video is not a different physical disc format, it is instead a specification for the layout and format of video, audio, and ancillary material on the disc. Like the CD-Audio format for music, DVD-Video was designed for distributing professionally produced material (movies and albums) so that they could be manufactured in mass quantity for retail sale.

The DVD-Video format supports a wide range of features, beyond what is used for typical movies. For video, it supports widescreen and letterbox aspect ratio, and up to 9 user-selected camera angles. For audio, it supports

Fig. 1.41: A CD-ROM

Media type	Optical disc
Capacity	~4.7 GB (single-sided single-layer), ~8.54 GB (single-sided double-layer)
Read mechanism	1350 Kb/s (1×)
Write mechanism	1350 Kb/s (1×)
Usage	Data storage, audio, video, games

Fig. 1.42: DVD

CD-quality surround sound, up to 8 audio tracks for multiple languages or commentary, and up to 32 subtitle/karaoke tracks. It also provides for extensive user interaction and even limited programmability with menus and navigation features. And for the movie studios, it provides copy protection and region coding to prevent discs sold in one part of the world from being played in another.

DVD-ROM (DVD-Read-Only Memory) format: It is the base read-only DVD format. Like CD-ROM, it provides a cost-effective format for manufacturing a disc full of data that is designed to be read (but not written) on computers. Compared to CD, of course, DVD offers seven times the storage capacity, faster data rates, and the ability to store an entire two-hour movie on one disc.

TYPES OF COMPUTERS

Computers come in a variety of types designed for different purposes, with different capabilities and costs.

Microcomputer

A **microcomputer** is a computer that has a microprocessor chip as its CPU. They are often called **personal computers** because they are designed to be used by one person at a time. Personal computers are typically used at home, at school, or at a business. Popular uses for microcomputers include word processing, surfing the web, sending and receiving e-mail, spreadsheet calculations, database management, editing photographs, creating graphics, and playing music or games.

Personal computers come in two major varieties, desktop computers and laptop computers.

Desktop computers are larger and not meant to be portable. They usually sit in one place on a desk or table and are plugged into a wall outlet for power. The cabinet of the computer holds the motherboard, drives, power supply, and expansion cards. This cabinet may lay flat on the desk, or it may be a **tower** that stands vertically (Fig. 1.43). The computer has a separate monitor (either a

CRT or LCD). A separate keyboard and mouse allow the user to input data and commands.

Laptop or **notebook** computers are small and lightweight enough to be carried around with the user (Fig. 1.44). They run on battery power, but can also be plugged into a wall outlet. They typically have a built-in LCD display that folds down to protect the display when the computer is carried around. They also feature a built-in keyboard and some kind of built-in pointing device (such as a touch pad).

While some laptops are less powerful than typical desktop machines, this is not true in all cases. Laptops, however, cost more than desktop units of equivalent processing power because the smaller components needed to build laptops are more expensive.

Fig. 1.43: A desktop computer system

Fig. 1.44: A laptop computer

Workstations/Servers

A **workstation** is a powerful, high-end microcomputer (Fig. 1.45). They contain one or more microprocessor CPUs. They may be used by a single-user for applications requiring more power than a typical PC (rendering complex graphics, or performing intensive scientific calculations).

Alternately, workstation-class microcomputers may be used as **server** computers that supply files to **client** computers over a network. This class of powerful microcomputers can also be used to handle the processing for many users simultaneously who are connected via terminals; in this respect, high-end workstations have essentially supplanted the role of mini-computers (see below).

The term "workstation" also has an alternate meaning: In networking, any client computer connected to the network that accesses server resources may be called a **workstation**. Such a network client workstation could be a personal computer or even a "workstation" as defined here. Note: Dumb terminals are not considered to be network workstations (client workstations on the network are capable of running programs independently of the server, but a terminal is not capable of independent processing).

Minicomputers

A **minicomputer** is a multi-user computer that is less powerful than a mainframe. This class of computers became available in the 1960s when large scale integrated circuits made it possible to build a computer much cheaper than the then existing mainframes. The niche previously filled by the minicomputer has been largely taken over by high-end microcomputer workstations serving multiple users.

Mainframe Computer

A **mainframe** computer is a large, powerful computer that handles the processing for many users simultaneously (up to several hundred users). The name mainframe originated after minicomputers appeared in the 1960s to distinguish the larger systems from the smaller minicomputers.

Users connect to the mainframe using terminals and submit their tasks for processing by the mainframe. A **terminal** is a device that has a screen and keyboard for input and output, but it does not do its own processing (they are also called **dumb terminals** since they cannot process data on their own). The processing power of the mainframe is time-shared between all of the users. (Note that a personal computer may be used to "emulate" a dumb terminal to connect to a mainframe or minicomputer; you run a program on the PC that pretends to be a dumb terminal.)

Mainframes typically cost several hundred thousand dollars. They are used in situations where a company wants the processing power and information storage in a centralized location. Mainframes are also now being used as high-capacity server computers for networks with many client workstations (Fig. 1.46).

Supercomputer

A **supercomputer** is mainframe computer that has been optimized for speed and processing power (Fig. 1.47). The most famous series of supercomputers were designed by the company founded and named after Seymour Cray. The **Cray-1** was built in the 1976 and installed at Los Alamos National Laboratory. Supercomputers are used for extremely calculation-intensive tasks such simulating nuclear bomb detonations, aerodynamic flows, and global weather patterns. A supercomputer typically costs several million dollars.

Recently, some supercomputers have been constructed by connecting together large numbers of individual processing units (in some cases, these processing units are standard microcomputer hardware).

HISTORY OF COMPUTERS

The history of computers is as follows.

Mechanical Beginnings

Some mechanical control and computing devices preceded the development of the

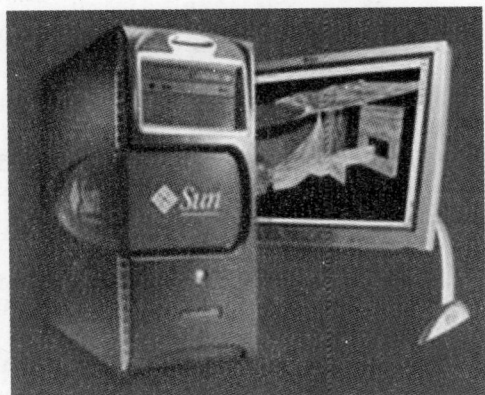

Fig. 1.45: A workstation

Fig. 1.46: A mainframe computer

Fig. 1.47: A supercomputer

modern computer. In the early 1800s, a French inventor named Joseph-Marie Jacquard produced a loom that could weave complex patterns into cloth. The loom was controlled automatically by reading instructions punched as holes in cards.

An American named Herman Hollerith invented machines using **punch cards** that were used to tabulate statistics for the 1890 US census. Hollerith eventually sold out to a company named CTR (Computer Tabulating Recording), which later became IBM (International Business Machines).

More interesting from a theoretical standpoint was the work in the 1800s by Englishman Charles Babbage. Babbage designed a mechanical computing device named the Analytical Engine in 1834. Although he was never able to build the device (he could not get funding to develop the precisely-machined gears, wheels, and lever systems of the machine), his ideas included many concepts that were later incorporated in modern computers.

ABC: The First Digital Electronic Computer

The first digital electronic computer was built by John Vincent Atanasoff and his assistant Clifford Berry at Iowa State University between 1937 and 1942. The **Atanasoff Berry Computer** (ABC) used punched cards for input and output, vacuum tube electronics to process data in binary format, and rotating drums of capacitors to store data.

The ABC, however, only performed one task: it was built to solve large systems of simultaneous equations (up to 29 equations with 29 unknowns), an onerous computing task commonly found in science and engineering. So, the ABC was not a general-purpose computer.

Similarly, another special-purpose electronic computer named Colossus was built in England starting in 1943 for the purpose of breaking German codes. The project was worked on by Alan Turing and Max Newman. The existence of this computer was kept secret until the 1970s.

ENIAC: The First General-purpose Electronic Computer

The first **general-purpose** digital electronic computer, one that could be programmed to perform a variety of calculational tasks, was the **ENIAC** (Electronic Numerical Integrator and Calculator). It was designed and built in the fall of 1945 by John Mauchly and J. Presber Eckert. ENIAC was originally built to calculate ballistic tables for the US military to aim their big guns.

ENIAC was a monster of a machine, filling a large room and weighing 30 tons. It included 18,000 vacuum tubes and used 200 kilowatts of electrical power (the lights dimmed in its Philadelphia neighborhood when it was first turned on). ENIAC was the first general-purpose computer because it could be **programmed** (given different sets of instructions to follow) by the cumbersome procedure of reconnecting cables and flipping switches.

Later computers were much more flexible because they incorporated the idea of **stored programs**, conceived in 1945 by mathematician John Von Neumann (who worked on the Manhattan Project in Los Alamos). In this scheme, both the data being manipulated and the program of instructions for the computer are stored in memory. Modern computers use this same method.

Mauchly and Eckert later went on to work for the Univac division of Remmington Rand corporation. The **UNIVAC I** (Universal Automatic Computer) was the first commercial computer, coming out in 1951.

IBM (International Business Machines) entered the computer market in 1953 with its 701 computer (Fig. 1.48). By 1960, IBM was the dominant force in the market of large mainframe computers. Smaller players in the mainframe market included Burroughs, Control Data, General Electric, Honeywell, NCR, RCA, and UNIVAC.

Transistors and Integrated Circuits

Vacuum tubes consume lots of electrical power and are prone to burning out, which caused problems for early computers that used thousands of them. By 1960, the **transistor** replaced the vacuum tube as the electrical switching device in computers. The transistor (developed at Bell Labs by William Shockley and others in the 1950s) is a solid-state semiconductor device typically made of silicon or germanium. It is much smaller, much more reliable, and consumes much less energy than a vacuum tube. A vacuum tube computer that previous filled a sizable portion of a room could be replaced by a transistorized computer system that filled a few cabinets. A good example of an early computer using transistors is the IBM 360, which dominated the mainframe computer market in the mid to late 1960s.

The early 1960s also saw the development of the **microchip**, or **integrated circuit (IC)**, invented by Jack Kirby and Robert Noyce (Fig. 1.49). An integrated circuit incorporates many transistors and other electrical components, all formed into a miniature circuit onto a single chip of silicon.

The invention of the integrated circuit allowed computers to become even smaller, with the whole central processing unit (CPU) of the computer fitting onto one circuit board. These **minicomputers** were cheaper and smaller than a mainframe (the computer was roughly the size of a drawer in a large filing cabinet) allowing many more businesses and universities to afford their own computer systems.

The most successful minicomputers were the PDP and Vax series made by Digital Equipment Corporation (DEC). Mini-computers were multi-user systems, in much the same way as mainframe computers, but on a smaller scale.

Minicomputers are now a mainly obsolete class of computer, having been largely replaced by high-end microprocessor workstations.

Microprocessors

As IC technology progressed, chip manufacturers could fit more and more circuitry onto the tiny silicon chips. By 1971, a company named **Intel** developed the first **microprocessor** (also called **MPU**) that could fit a whole CPU

Fig. 1.48: IBM 701 computer

Fig. 1.49: An integrated circuit

onto one microchip. The Intel 4004 processor contained 2300 transistors on a chip of silicon 1/8" * 1/16" in size.

By 1974, Intel introduced their 8080 chip, a general purpose microprocessor offering ten times the performance of the earlier MPU. It was not too long before electronics hobbyists began building small computer systems based on the rapidly improving microprocessor chips.

The First Microcomputers

The first commercially available microcomputer of note was the Altair 8800 computer sold by MITS (Micro Instrumentation and Telemetry Systems), a company founded by Dr. Ed Roberts that was based in Albuquerque, New Mexico. The computer was featured on the cover of the January 1975 issue of Popular Electronics, and was sold as a kit for $397 or assembled for $439. It used a 2 MHz Intel 8080 processor and had 256 bytes of RAM.

Remember that a computer cannot do anything without software, and some companies sprang into existence to fill this need. One small company of note was formed in Albuquerque by a Harvard dropout to provide software (a BASIC language) for the Altair computer. The founder's name was Bill Gates, and the company he form (along with his partner Paul Allen) was **Microsoft**.

Dozens of companies (most of which have long since vanished) began offering microcomputers for sale, most of them based on the Intel 8080 processor and running the CP/M operating system. Other companies had proprietary operating systems, such as Radio Shack, Atari, and Commodore. Of particular note is a company named **Apple** founded by Steve Jobs and Steve Wozniak on April 1, 1976. Their Apple II computer was a hit, especially in the home and education markets.

Two things caused the microcomputer market to really take off in the late 1970s and early 1980s: spreadsheet software, and the **IBM PC**. Spreadsheet software (the first was Visicalc for the Apple II, written by Dan Bricklin) finally convinced business people that there was a serious use for microcomputers. The IBM PC, released in 1981, gave a legitimacy to the microcomputer by virtue of the IBM name (remember, IBM was the maker of big mainframe computers; and it was said "Nobody ever gets fired for buying IBM"). It used a 4.77 MHz Intel 8088 processor.

Microsoft went to IBM about an operating system (OS) for their new PC. Bill Gates told them, "We have an OS that will run on this new machine you are planning," and made a deal. Microsoft did not, in fact, have such an OS, but they quickly bought one from a third party and converted it into PC-DOS. But what Gates did that was really clever was to make a deal with IBM that allowed Microsoft to also sell the OS to other companies as **MS-DOS**...and Microsoft's future was set.

IBM PC sales skyrocketed and IBM dominated the market within two years, releasing the PC XT (1983) and PC AT (1984) using the Intel 80286 processor.

But, almost as quickly, IBM lost it dominance in the PC market place when other companies (such as Compaq) began to release "**PC compatible**" computers (also called "**PC clones**"). By 1986, the clones owned most of the market, and IBM never regained its dominance. Microsoft, on the other hand, supplied their operating systems to all PCs, becoming a huge corporation.

Graphical User Interface

Computers were traditionally very difficult to use, requiring the user to memorize and type in the necessary commands (this is called a command line interface). To make computers more accessible, the **graphical user interface (GUI)** was developed. In a GUI, the user interacts with a graphical display on the screen containing icons and windows and controls. Commands are chosen from menus rather than typed in.

The GUI was developed at the Xerox Palo Alto Research Center, but the management at Xerox failed to see the usefulness of it. When Steve Jobs of Apple saw the GUI, however, he recognized its value. Apple licensed the concepts from Xerox, developed them further, and released the first successful GUI computer, the **Macintosh**, in 1984. Macintosh computers used the Motorola 68000 series of microprocessors (and later the PowerPC series of microprocessors).

Microsoft was also quick to realize the worth of a GUI, but its graphical user interface, **Windows**, was slow in displacing DOS on PCs (the first versions of Windows left much to be desired).

GENERATION OF COMPUTERS

The history of computer development is often referred to in reference to the different generations of computing devices. Each generation of computer is characterized by a major technological development that fundamentally changed the way computers operate, resulting in increasingly smaller, cheaper, and more powerful and more efficient and reliable devices.

First Generation (1940–1956): Vacuum Tubes

The first computers used vacuum tubes for circuitry and magnetic drums for memory, and were often enormous, taking up entire rooms. They were very expensive to operate and in addition to using a great deal of electricity, generated a lot of heat, which was often the cause of malfunctions. First generation computers relied on machine language to perform operations, and they could only solve one problem at a time. Input was based on punched cards and paper tape, and output was displayed on printouts.

The UNIVAC and ENIAC computers are examples of first generation computing devices. The UNIVAC was the first commercial computer delivered to a business client, the US Census Bureau in 1951.

Second Generation (1956–1963): Transistors

Transistors replaced vacuum tubes and ushered in the second generation of computers. The transistor was invented in 1947 but did not see widespread use in computers until the late 1950s. The transistor was far superior to the vacuum tube, allowing computers to become smaller, faster, cheaper, more energy-efficient and more reliable than their first generation predecessors. Though the transistor still generated a great deal of heat that subjected the computer to damage, it was a vast improvement over the vacuum tube. The second-generation computers still relied on punched cards for input and printouts for output.

The second generation computers moved from cryptic binary machine language to symbolic, or assembly, languages, which allowed programmers to specify instructions in words. High-level programming languages were also being developed at this time, such as early versions of COBOL and FORTRAN. These were also the first computers that stored their instructions in their memory, which moved from a magnetic drum to magnetic core technology. The first computers of this generation were developed for the atomic energy industry.

Third Generation (1964–1971): Integrated Circuits

The development of the integrated circuit was the hallmark of the third generation of computers. Transistors were miniaturized and placed on silicon chips, called integrated circuits, which drastically increased the speed and efficiency of computers.

Instead of punched cards and printouts, users interacted with the third generation computers through keyboards and monitors and interfaced with an operating system, which allowed the device to run many different applications at one time with a central program that monitored the memory. Computers for the first time became accessible to a mass audience because they were smaller and cheaper than their predecessors.

Fourth Generation (1971–Present): Microprocessors

The microprocessor brought the fourth generation of computers, as thousands of integrated circuits were built onto a single silicon chip. What in the first generation filled an entire room could now fit in the palm of the hand. The Intel 4004 chip, developed in 1971, located all the components of the computer from the central processing unit and memory to input/output controls—on a single chip.

In 1981, IBM introduced its first computer for the home user, and in 1984, Apple introduced the Macintosh. Microprocessors also moved out of the realm of desktop computers and into many areas of life as more and more everyday products began to use microprocessors.

As these small computers became more powerful, they could be linked together to form networks, which eventually led to the development of the Internet. Fourth generation computers also saw the development of GUIs, the mouse and handheld devices.

Fifth Generation (Present and Beyond): Artificial Intelligence

The fifth generation computing devices, based on artificial intelligence, are still in development, though there are some applications, such as voice recognition, that are being used today. The use of parallel processing and superconductors is helping to make artificial intelligence a reality. Quantum computation and molecular and nanotechnology will radically change the face of computers in years to come. The goal of fifth generation computing is to develop devices that respond to natural language input and are capable of learning and self-organization.

COMPUTER HARDWARE

The term **hardware** refers to the physical components of the computer system (as opposed to the software). Computer hardware consists the devices within the case of the computer itself, and any **peripheral devices** that are connected to the computer (such as the mouse and keyboard).

The primary component of the computer is the **motherboard** (also called the **main circuit board**, main logic board, mainboard, or systemboard). The motherboard is a large printed circuit board with microchips, connectors, and other components mounted on it, and with copper circuitry traces that connect the components together.

A motherboard typically holds the following items:

- **CPU (central processing unit)** where the actual processing of data takes place.
- **System clock** circuitry (that keeps all of the digital chips in lockstep)
- Other controller chips that act as traffic cops directing data flow along the system busses (the circuitry connecting the chips to the CPU) and **I/O ports**.
- **RAM** (the main memory, plus additional slots for adding more memory)

- **ROM** (containing the **BIOS**)
- "**CMOS** memory"
- **Expansion slots** (for adding expansion cards such as video cards and sound cards).

Along with the motherboard, the case of your computer typically contains a **power supply** (to convert the AC line current from the wall outlet to the low-voltage DC current used by the computer) and several storage devices located in the **expansion bays** of the case (such as: hard drives, floppy drives, zip drives, and CD drives, and DVD drives).

Configurations and Specifications

Each **configuration** includes a particular set of parts or components (both hardware and software) in a specific arrangement. A similar term, **architecture**, also describes the layout and interactions of the components of a computer system.

Each system configuration will have a **specification** that lists the details about the components included in that particular system. Below you will see a typical computer system specification; look it over carefully. Typical specification for a desktop computer system is as follows:

1. Processor: Intel® Pentium® 4 processor, 3 GHz, with 800 MHz front side bus, 512 KB Level 2 Cache.
2. Microsoft® Windows® XP Professional.
3. Memory: 512 MB DDR SDRAM at 400 MHz (expandable to 4 GB).
4. Hard drive: 250 GB Serial ATA, 7 ms seek time, 7200 RPM, 512 Kb cache.
5. Floppy drive: 3.5" 1.44 MB.
6. Optical drive: 12x DVD-ROM / 48x CD-RW combo drive.
7. Expansion slots: 1 AGP and 5 PCI.
8. External ports: Six USB 2.0 (two on front panel), one Parallel, one Serial, two PS/2, and one IEEE 1394.
9. Modem: 56 K PCI FAX/modem.
10. Video card: 256 MB *RADEON™ 9800* AGP graphics card.
11. Monitor: 17" CRT (16" viewable), 1,024 × 768, 27 dp.
12. Sound card: *Sound Blaster® Audigy™2* card w/*Dolby 5.1* stereo.
13. Speakers: *Bose® B775* surrounds sound speaker system with subwoofer.
14. Networking: Ethernet 10/100
15. Keyboard: 101-key multifunction keyboard
16. Mouse: *Logitech® MX™ 500* optical mouse with scroll wheel.
17. Case: Tower case with 6 expansion bays (two for internal-only drives).
18. Application Software: Microsoft® Office Professional 2003.

COMPUTER SOFTWARE

Computer software means computer instructions or data. Anything that can be stored electronically is software. The storage devices and display devices are hardware.

The terms *software* and *hardware* are used as both nouns and adjectives. For example, you can say: "The problem lies in the software," meaning that there is a problem with the program or data, not with the computer itself. You can also say: "It is a software problem."

The distinction between software and hardware is sometimes confusing because they are so integrally linked. Clearly, when you purchase a program, you are buying software. But to buy the software, you need to buy the disk (hardware) on which the software is recorded.

We know software is a set of instructions that are used to carry out a task. Software can be grouped into two categories namely application software and system software. The application software is one, which is application oriented, for example inventory program and payroll program. Similarly, system software is used for system oriented tasks. Examples are compilers, assemblers, and loaders.

Classification of Software

Softwares are classified into two major categories:
(a) Application software
(b) System software

Application Software

Application software is developed for application of the computer to common problems and tasks. They are available for business applications, science and engineering applications and so on. Personal productivity programs are categorized based on the nature of their use in word processing, generating spreadsheet, presenting graphics and maintaining databases. Application software is also available as packages and usually with a user manual. Some of the application software's are described below:

Word Processors

A word processor is used to prepare a report, a personal or business letter, in desktop publishing and so on. These offer formatting features such as using different character styles, line spacing, and page numbering and so on. Documents prepared using a word processor can be easily printed in any type of printer.

Electronic Spreadsheets

An electronic spreadsheet software is used to prepare documents containing information or data in the form of numbers or characters. The information is arranged in rows and columns for further processing and analysis, preparing reports and generating charts. It is also capable of performing arithmetic operations and using functions.

Database Software

Databases are records related to a person or an organization. Database software has capability to edit and update data in a file. The data are processed to prepare and print salary details of employees, annual sales details and so on. One of the major applications of a computer is database management.

System Software

System software is designed for a specific type of hardware. For example, the disk operating system is used to co-ordinate the peripherals of a computer. The system software controls the activities of a computer, application programs, flow of data in and out of memory and disk storage. Our operating system, compilers, assemblers, linker and loaders are the example of system software.

System software also handles data in communication applications and within the computer systems in a computer network. The communication software transfers data from one computer to another. These programs also provide data security and error checking along with the transfer of data between the computer systems.

Examples of system software include operating system, compiler, interpreter and assembler that we will discuss later in this book.

SUMMARY

- A computer is an electronic device which takes information and process information according to the program and produces the output.
- A computer system has five basic functional units.
- The central processing unit (CPU) is the brain of the computer.
- The arithmetic logic unit (ALU) is the unit of the computer that performs arithmetic and logical operations on the data.
- The control unit controls the overall activities of the components of the computer.
- The memory unit is the unit where all the input data and results are stored.
- Stored program concept uses the memory unit to store both instruction or operation code and data or operands.
- Random access memory (RAM) is a temporary storage medium in a computer. All data to be processed by the computer are transferred from a storage device or keyboard to RAM
- Read-only memory (ROM) is a permanent storage medium which stores the start up programs
- A floppy disk is used to store data permanently.
- Hard disk is a reliable and permanent storage disk. It has a set of metal disks

coated with magnetic material and mounted on a central spindle which rotates at 7200 rpm

- Compact disk (CD) is an optical disk used to store data permanently.
- A keyboard is an input device used to enter data into a computer.
- The keyboard contains function keys, numeric keys and toggle keys (caps lock, num lock, and scroll lock) and so on.
- A mouse is an input device used to select a command by moving it in any direction on a flat surface
- The software developed with graphical interface requires the mouse.
- A peripheral input device used to assist in the entry of data into a computer system.
- A printer is an output device used to print text or graphics on paper or on any other hardcopy medium such as microfilm.
- Printers are of two basic types impact and non-impact.
- The most popular kind of printer for small computers is the dot matrix printer, which forms characters as arrays of dots.
- Laser printers are the fastest type of non-impact electrostatic printers. They produce high quality prints at high speeds.
- An inkjet is a non-impact printer. It sprays tiny drops of ink to form character and graphic images on paper.
- A plotter is an output device used to print engineering drawing or graphics on large size sheets.

Exercises

Section A

Multiple Choice Questions

1. Which of these is not a microcomputer?
   (a) Desktop
   (b) Laptop
   (c) Mainframe
   (d) Handheld PC

2. Another name for the main circuit board of a personal computer is...
   (a) RAM board
   (b) Peripheral board
   (c) Motherboard
   (d) Central processing chip

3. The first general purpose computer that could be programmed was the
   (a) IBM PC   (b) Altair
   (c) HOOVERVAC   (d) ENIAC

4. The first microprocessor chip was developed by
   (a) IBM   (b) Apple
   (c) Motorola   (d) Intel

5. A personal computer is a
   (a) Microcomputer
   (b) terminal
   (c) Personal Digital Assistant
   (d) mainframe

6. Which of these is not designed to be portable?
   (a) Notebook computer
   (b) PDA
   (c) Laptop computer
   (d) Minicomputer

7. Which of these would NOT be classified as system software?
   (a) Microsoft Word
   (b) Microsoft Windows
   (c) Device drivers
   (d) Hard disk recovery utility

8. Which of these would NOT be classified as application software?
   (a) Microsoft Excel
   (b) Microsoft Access
   (c) Macintosh OSX
   (d) Adobe Photoshop

9. If your computer reports that a file has a size of one kilobyte, then how much space it would be
   (a) 8 bytes
   (b) 1,024 bytes
   (c) 1,048,576 bytes
   (d) Exactly 1,000 bytes

10. The code used to store letters in the form of numbers in a computer is called
    (a) ROM code   (b) ASCII
    (c) PCMCIA   (d) SCSI

11. Which component are you unlikely to find as part of a motherboard?
    (a) ROM   (b) CPU
    (c) CMOS memory (d) CD-ROM

12. A computer must be capable of doing all of these things *except...*
    (a) Electrolyze data (b) output data
    (c) Process data   (d) store data

13. Which of these is not an output device?
    (a) Monitor   (b) Microphone
    (c) Laser printer (d) LCD display

14. Which of these is a pointing device?
    (a) Laser printer (b) Touch pad
    (c) Data bus   (d) Decimal point

15. You could measure a microprocessor's clock speed in...
    (a) Gigabytes   (b) Avishertz
    (c) Bits per second (d) gigahertz

16. CPU stands for:
    (a) Computation procedure unit
    (b) Chip production utility
    (c) Computer programmers union
    (d) Central processing unit

17. Which of these is volatile?
    (a) ROM   (b) RAM
    (c) PRAM   (d) BIOS

18. Which type of memory holds data that does not change?
    (a) ROM   (b) RAM
    (c) ASCII   (d) CMOS

19. A CD-ROM can hold approximately
    (a) 1.44 megabytes
    (b) 250 gigabytes
    (c) 670 megabytes
    (d) 1.44 gigabytes

20. This device stores data by magnetizing tiny spots on a metal disk.
    (a) CD-ROM disk drive
    (b) Hard disk drive
    (c) USB flash drive
    (d) DVD disc drive

21. Which of these can a personal computer NOT write data on?
    (a) CD-ROM   (b) Hard disk
    (c) Zip disk   (d) CD-RW

22. These are small expansion cards that can be inserted into many laptop computers.
    (a) SCSIS cards
    (b) ASCII cards
    (c) PCMCIA cards
    (d) CPU cards

23. This is a common networking technology.
    (a) IEEE 1492   (b) Ethernet
    (c) PCI   (d) SCSI

24. Which of these ports would you most likely use to connect a keyboard to a personal computer?
    (a) Parallel port (b) SCSI
    (c) USB   (d) PCMCIA port

25. A device that converts digital computer data to analog signals for telecommunications.
    (a) Modem   (b) Ethernet card
    (c) Web browser   (d) I/O port

26. A device with a display and a keyboard, but no independent processing power is called...
    (a) Dumb PC   (b) LAN
    (c) Workstation   (d) Terminal

27. A computer consists of _____ units.
    (a) 3   (b) 4
    (c) 5   (d) 6

28. Keyboard is an example of _____ unit.
    (a) Memory   (b) Input
    (c) Output   (d) ALU

29. ALU stands for _____
    (a) Arithmetic logic unit
    (b) Arithmetic lower unit
    (c) Add logical unit
    (d) None of the above

30. RAM is considered as a _____
    (a) Volatile memory
    (b) Non-volatile memory
    (c) Permanent
    (d) None of the above

31. _____ is programmed during the manufacturing itself.
    (a) RAM
    (b) ROM
    (c) Both a and b
    (d) None of the above

32. _____ unit is used to store information.
    (a) Input   (b) Output
    (c) Control   (d) Memory

33. In stored program concept _____ and _____ are stored in the same memory.
    (a) Data and instruction
    (b) Data and operands
    (c) Instruction and operation code
    (d) None of the above

34. Microprocessor is the heart of _____ computer.
    (a) Digital
    (b) Analog
    (c) Both a and b
    (d) None of the above

Fill in the Blanks

1. _____ and _____ are examples of input device.
2. Printout of a program is considered as _____.
3. Payroll program stored on CD is considered as _____.
4. Dot matrix printer is an example of _____.
5. Laser printer is an example of _____.
6. The speed of dot matrix printer is expressed as _____.
7. The speed of laser printer is expressed as _____.
8. RAM is a _____ memory.
9. Floppy disk storage capacity is _____.
10. Concentric circles in a floppy disk is known as _____.
11. ROM is also known as _____.
12. ROM is a _____ memory.
13. Hard disk is ____ than primary memory.
14. CD-ROM storage capacity is _____.
15. Information stored on a CD is accessed through _____.

Section B

Short Answer Type Questions

1. Mention the basic functional units of a computer.
2. With a neat diagram explain the working organization of a computer.
3. What is stored program concept or John Von Neumann concept?
4. What is microprocessor?
5. What are the differences between RAM and ROM?
6. Mention the basic functional units of a computer.
7. With a neat diagram explain the working organization of a computer.
8. Mention the storage devices used in a personal computer.
9. Explain primary memory its properties and its types.
10. What is the need for secondary storage? Briefly describe secondary storage devices like (i) floppy disk, (ii) hard disk, and (iii) CD ROM.
11. Mention the components of a personal computer.
12. What is the difference between volatile memory and non-volatile memory?

Section C

Long Answer Type Questions

1. Define the concept of computer system and define the working of various units of a computer system in detail.
2. Define the term keyboard and define the complete working of keyboard.
3. Define the term mouse and define the complete working of mouse.
4. Define the term laser printer and define the complete working of laser printer.
5. Differentiate between primary memory and secondary memory with suitable example.
6. Differentiate between system software and application software with suitable example.
7. Define the various functional components of a computer system.
8. Define the applications of computer system in detail.
9. Describe the history of computer with proper explanation.
10. Describe the generations of computer system with proper explanation.

2 Introduction to Operating System

Computer software can be divided into two main categories: application software and system software. According to Brookshear (1997), "application software consists of the programs for performing tasks particular to the machine's utilization. Examples of application software include spreadsheets, database systems, desktop publishing systems, program development software, and games." Application software is generally what we think of when someone speaks of computer programs. This software is designed to solve a particular problem for users.

On the other hand, system software is more transparent and less noticed by the typical computer·user. This software "provides a general programming environment in which programmers can create specific applications to suit their needs. This environment provides new functions that are not available at the hardware level and performs tasks related to executing the application program" (Nutt 1997). System software acts as an interface between the hardware of the computer and the application software that users need to run on the computer.

Figure 2.1 illustrates the relationship between application software and system software.

OPERATING SYSTEM

Operating system is the most important program that runs on a computer. Every general-purpose computer must have an operating system to run other programs. Operating systems perform basic tasks, such as recognizing input from the keyboard, sending output to the display screen, keeping track of files and directories on the disk, and controlling peripheral devices such as disk drives and printers.

For large systems, the operating system has even greater responsibilities and powers. It is like a traffic cop—it makes sure that different program and users running at the same time do not interfere with each other. The operating system is also responsible for *security*, ensuring that unauthorized users do not access the system.

Operating systems provide a software platform on top of which other programs, called *application programs,* can run. The application programs must be written to run on top of a particular operating system.

We know that an operating system is a collection of programs and it is the interface between user and the computer. An operating system is a program which connects the user and the electronic hardware in a computer. It is a set of programs which supervises the activities of a computer and activates the

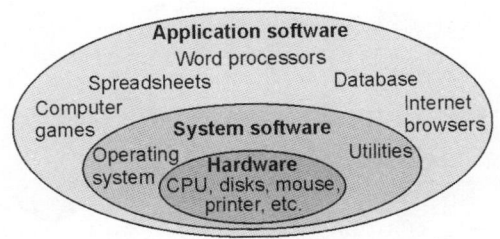

Fig. 2.1: Relationship between application software and system software

35

operations of the hardware components such as CPU, main memory, disk drives, keyboard, monitor and printer and so on. Some of the startup programs initially loaded to RAM are stored in ROM, mainly the BIOS programs which are recorded by the manufacturers of the computer system. Service programs available in operating system for operations like copying a file, deleting a file, formatting a disk, printing a file and so on are usually stored in the disk. Error messages are displayed on the screen if there is any malfunctioning of hardware.

Figure 2.2 represents the relationship between the operating system and the other parts of the computer system.

FUNCTIONS OF OPERATING SYSTEM

In any computer, the operating system:
- Controls peripherals such as disk drives and printers
- Controls the loading and running of programs
- Organizes the use of memory between programs
- Organizes the processing time between programs and users
- Organizes priorities between program and users
- Maintains security and access rights of users
- Deals with errors and user instructions.

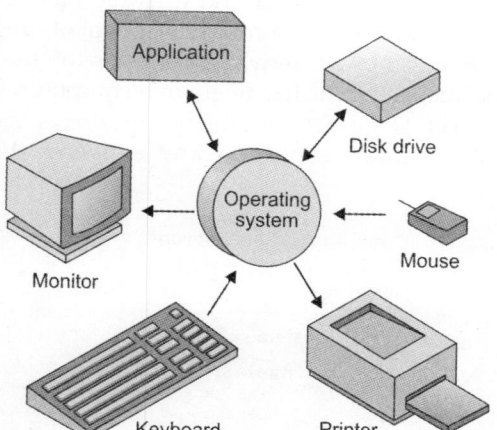

Fig. 2.2: Relationship between operating system and computer hardware

On a personal computer the **operating system** will:
- Deal with the transfer of programs in and out of memory
- Allow the user to save files to a backing store
- Control the transfer of **data** to peripherals such as printers
- Provide the **interface** between the user and the computer, for example, Windows XP and OSX.
- Issue simple error messages.

ROLES OF OPERATING SYSTEM

The operating system performs various roles as follows:

Management of the Processor

The operating system is responsible for managing the allocation of the processor between the different programmes using a **scheduling algorithm**. The type of scheduler is totally dependent on the operating system, according to the desired objective.

Management of the Random Access Memory

The operating system is responsible for managing the memory space allocated to each application and, where relevant, to each user. If there is insufficient physical memory, the operating system can create a memory zone on the hard drive, known as "**virtual memory**". The virtual memory lets you run applications requiring more memory than that available in RAM on the system. However, this memory is a great deal slower.

Management of Input/Output

The operating system allows the unification and control of access of programs to material resources via drivers (also known as peripheral administrators or input/output administrators).

Management of Execution of Applications

The operating system is responsible for smooth execution of applications by allocating the resources required for them to operate. This means that an application which is not responding correctly can be "killed".

Management of Authorizations

The operating system is responsible for security relating to execution of programs by guaranteeing that the resources are used only by programs and users with the relevant authorizations.

File Management

The operating system manages reading and writing in the file system and the user and application file access authorizations.

Information Management

The operating system provides a certain number of indicators that can be used to diagnose the correct operation of the machine.

COMPONENTS OF THE OPERATING SYSTEM

The operating system comprises a set of software packages that can be used to manage interactions with the hardware. The following elements are generally included in this set of software:

- The **kernel**, which represents the operating system's basic functions such as management of memory, processes, files, main inputs/outputs and communication functionalities.
- The **shell**, allowing communication with the operating system via a control language, letting the user control the peripherals without knowing the characteristics of the hardware used, management of physical addresses, etc.
- The **file system**, allowing files to be recorded in a tree structure.

CLASSIFICATION OF OPERATING SYSTEM

Operating system can be classified into single user and multi-user categories.

Single User Operating System

A single user operating system allows one single user to login at a time. There is no user account database which makes the level of

security low and so users cannot protect their files from being viewed, copied or deleted.

Examples of this type of operating system are DOS and Windows vista.

Multi-user Operating System

The multi-user operating system has a user database account which states the right that users have on certain resources. It is more secure than the single user operating system since the access is limited.

An example of this type of operating system is UNIX.

TYPES OF OPERATING SYSTEM

Several different types of operating system are described below.

Multi-threaded Operating System

An operating system is known as **multi-threaded** when several "**tasks**" (also known as *processes*) may be run at the same time. The applications consist of a sequence of instructions known as "**threads**". These threads will be alternately active, on standby, suspended or destroyed, according to the priority accorded to them or may be run simultaneously.

A system is known as **pre-emptive** when it has a **scheduler** (also called *planner*), which, according to priority criteria, allocates the machine time between the various processes requesting it.

The system is called a **shared time** system when a time quota is allocated to each process by the scheduler. This is the case of multi-user systems which allow several users to use different or similar applications on the same machine at the same time. The system is then referred to as a "**transactional system**". To do this, the system allocates a period of time to each user.

Multiprocessor Operating System

Multiprocessing is a technique that involves operating several processors in parallel to obtain a higher calculation power than that obtained using a high-end processor or to increase the availability of the system (in the event of processor breakdown).

The term **SMP** (*symmetric multiprocessing* or *Symmetric multiprocessor*) refers to an architecture in which all processors access the same shared memory.

A multiprocessor system must not only be able to manage memory sharing between several processors but also be able to distribute the workload.

Embedded System

Embedded systems are operating systems designed to operate on small machines, such as PDAs (*personal digital assistants*) or autonomous electronic devices (spatial probes, robot, on-board vehicle computer, etc.) with reduced autonomy. Thus, an essential feature of embedded systems is their advanced energy management and ability to operate with limited resources.

The main "general use" embedded systems for PDAs are as follows:

- PalmOS
- Windows CE/Windows Mobile/Window Smartphone

Realtime Systems

Realtime systems, used mainly in industry, are the systems designed to operate in a time-constrained environment. A realtime system must also operate reliably according to specific time constraints; in other words, it must be able to properly process information received at clearly-defined intervals (regular or otherwise). Here are some examples of the realtime operating systems:

- OS-9
- RTLinux (Real Time Linux)
- QNX
- VxWorks.

Batch Processing Operating System

In batch processing operating systems, the tasks are performed in batches, the operating system waits for a batch of tasks to arrive and then performs the operation on that batch.

Time Sharing Operating System

A time sharing operating system shares the time between the tasks, i.e. if job A takes 6 minutes to complete and the predefined time

to the operating system is 2 minutes it first performs the job A for 2 minutes, then takes up the job B (the next one) and this way shares its time between all the jobs arriving.

Distributed Operating System

In distributed operating systems, different processors that are loosely coupled operate on different jobs and communicate with each other.

DISK OPERATING SYSTEM (DOS)

MS-DOS is a single user operating system developed by Microsoft Corporation. An operating system has a collection of program. When the computer is switched on, the file COMMAND.COM is loaded to the RAM and after the successful start of the computer, the DOS prompt or command prompt will be displayed. The DOS prompt displays the letter associated with the disk drive followed by a > symbol. For floppy disk drive, A> or A:> is displayed and for hard disk drive C> or C:\> is displayed. It indicates the operating system is ready to take commands from the user. DOS operating system consists of three parts, namely, the resident part, the initialization part and the transient part. Most of the command programs are located in the resident part. While booting, the number of files and buffers to open is contained in the initialization part and the transient part is the flexible part of the operating system. The commands are not case sensitive.

Specific Features of DOS

The following are the specific features of the disk operating system.

File

A file is a collection of related information. For example, like the contents of a file folder in a desk drawer. Files on the disk can contain letters, memos and executable programs.

Program

Programs are special types of files. These are series of instructions written in computer languages. These programs instruct the computer to perform the task.

Directory

DOS uses a filing system to store its files. The filing system uses storage areas called directories. A directory is nothing more than an expandable file folder that can hold other expandable file folders. These file folders hold the data files. A directory is a table of contents for a disk. It contains the names of files, their sizes, and the dates they were last modified. All of the different directories are stored under one master directory. This directory is called the root directory.

In addition to directories, DOS uses an area on a disk called the file allocation table (FAT). The FAT is similar to our contents page in our book. It holds the information as to where the file is stored in the disk.

Multilevel Directories

When there are two or more users who share a computer, when you are working on several different projects, the number of files in the directory can become a large and unwieldy. Using directories is one way of dividing our files into convenient groups. Any one directory can contain many files. This directory may also contain other directories or sub directories. This organized file structure is called a hierarchical directory system.

Preliminary Commands of DOS

The preliminary commands of the disk operating system are described below.

DIR Command

DIR command is an internal command which is used to display the contents in disk directory. To locate data files and programs on a specific disk, DOS uses the directory along with a file allocation table (FAT).

Examples

C:\>DIR

This command will display the disk directory in the default drive.

C:\>DIR/W

This command will display the disk directory in the default drive in a wide format.

C:\>DIR/P

This command will display the disk directory in the default drive but page wise. This command is useful when the disk contains numerous files.

C:\>DIR A:

This command will display the disk contents in A drive.

C:\>DIR *.C

This command will display the disk contents in the default drive with only the files having the extension .C. Here * is known as wild card character. It means all matching characters are represented by *.

CLS Command

CLS command is used to clear the screen. When this command is entered, all the previously displayed text or messages are removed from the screen.

The general syntax of cls command is as follows:

C:\>CLS

REN or RENAME Command

REN or RENAME command is used to rename an existing file. Consider the following example to rename a file in the current directory.

Example

C:\>REN A.BAK A.C

Or

C:\>RENAME A.BAK A.C

When this command is entered, the file A.BAK is renamed as A.C.

DEL Command

DEL command is used to delete files in a directory.

Consider the following example to delete files in the current directory.

C:\>DEL A.BAK

This command will delete all the files in the directory. The message "Bad command or file name" is displayed when the file is not as available in the directory.

C:\>DEL *.BAK

This command will delete all the files in the directory with the extension. BAK The message "Bad command or file name" is displayed when the file is not as available in the directory.

C:\>DEL *.*

This command will delete all the files in the directory. When this command is used, the message "Are you sure to delete all files (y/n)?" is displayed. Press y to confirm deletion.

ERASE Command

ERASE command is used to erase or remove files in the directory.

Consider the following examples to erase files in the directory.

C:\> ERASE A.BAK

This command will erase the file A.BAK in the directory. The message "Bad command or file name" is displayed when the file is not as available in the directory.

C:>\ ERASE *.*

This command will erase all the files in the directory. When this command is entered, the message "Are you sure to delete all files (y/n)?", is displayed. Press y to confirm deletion.

DATE Command

DATE command is used to display the current system date. The computer also maintains a calendar. This command will display the current system date in mm-dd-yyyy (month-date-year) format and the user may enter the new date. Consider the following example to display the current date.

C:\>DATE

Current date is Sun 09-25-2005
Enter new date (mm-dd-yyyy):

TIME Command

TIME command is used to display the current system time. The computer also maintains a clock. This command will display the current system time in (hours: minutes: seconds) format and the user may enter the new time.

Consider the following example to display the current time.

C:\> TIME

Current time is 11:37:25.34p
Enter new time:

CD Command

CD (change directory) command is used to change the directory to another specified directory/location in the disk. A message "invalid directory" is displayed if the directory mentioned is not available.

Consider the following examples:

(I) C:\>CD ABC:

This command will change the current directory to the specified directory ABC in the disk. Now the prompt is displayed as follows.
C:\ABC>

(II) C:\>CD TC

C:\TC>CD ABCP

These commands will change the current directory to the specified directory TC and then to the directory ABCP in the disk. Now the prompt is displayed as follows.
C:\TC\ABCP>

(III) C:\>CD\TC\ABCP

This command will also change the directory to the specified directory TC and then to the directory ABCP in the disk. Now the prompt is displayed as follows.
C:\TC\ABCP>

(IV) C> cd furniture:

Moves you to the directory called 'FURNITURE'

(V) C> cd \furniture\chairs:

Moves you to the directory called 'CHAIRS' under the directory called 'FURNITURE'.

(VI) C> cd. . :

Moves you up one level in the path.

(VII) C> cd \ :

Takes you back to the root directory (c: in this case).

MD Command

MD (make directory) command is used to create a new directory in the storage device to store programs. Consider the following example to create a new directory:

C:\>MD ABCP

This command will create a new directory in the current directory. To transfer the control to the new directory a CD command is used.

C:\>CD ABCP
CD:\ABCP>

RD Command

RD (remove directory) command is used to remove a directory permanently from the disk. Note that all the files in that directory should be removed before the RD command is used. Also know that you should quit from the directory being removed.

Consider the following example:

If we want to remove the ABCP directory which is the part of TC directory then we have to follow the following steps:

1. Firstly we delete all the contents of the ABCP directory

   C:\TC\ABCP>DEL *.*

   All files in directory will be deleted!

   Are you sure (y/n) ? Press y to confirm the delete option.

2. Then we go to the parent directory of the ABCP directory and execute the RD command.

   C:\TC\ABCP>CD

   C:\TC>RD ABCP

   One of these commands can be used to remove the directory ABCP from the disk.

COPY Command

COPY command is used to copy a file to a new location or directory in the disk. A file cannot be copied to itself in the same directory in the same name.

Consider the following example:

(I) C:\>COPY AB.TXT AB.BAK

This command will copy the file AB.TXT in the same directory as AB.BAK. The first file name in the command AB.TXT is the source file and the file AB.BAK is the target file which is a copy of AB.TXT.

(II) C:\>COPY AB.TXT A:

This command will copy the file AB.TXT from the C drive to A drive and is copied in the same name.

(III) C:\>COPY AB.TXT A: AB.BAK

This command will copy the file AB.TXT from the C: drive to A: drive but the target file is named as AB.BAK.

(IV) C:\>COPY *.* A:

This command will copy all the files in the current directory of C: drive to A: drive in the respective file names

(V) C:\>COPY *.* A:

This command will copy all the C program files in the current directory of C: drive to A: drive in the respective file names.

The COPY command can also be used to create a file in the console using the keyword CON along with COPY command.

(VI) C:\>COPY CON sample.txt

This command is used to create a simple.txt file and write the matter and press F6 then file is automatically saved.

TYPE Command

This command is used to display the contents of a file on the monitor screen. TYPE command can also be used to send the contents of the file to the printer in the console using the keyword PRN.

Consider the following examples:

(I) C:\>TYPE AB.TXT:

This command will display the contents of the file AB.TXT on the screen.

(II) C:\>TYPE AB.TXT >PRN

This command will print the contents of the file AB.TXT

FORMAT Command

FORMAT command is used to format a new disk. It is a utility program that is available on hard disk. Only the formatted disks can be used by the operating system to store files or programs. Consider the following example to format a new floppy disk. Insert the new floppy disk in A drive and give the following command.

C:\>FORMAT A:

Nowadays new formatted floppies are available in packs which need not be for- matted again. If any old floppy is formatted then the contents will be erased automatically.

DISKCOPY Command

DISKCOPY command is used to copy all the contents of various directories of a disk in another disk. It is faster and useful to take backup copies.

Consider the following example:

C:\>DISKCOPY A: C:

Now the contents of floppy disk in A drive is copied to the hard disk drive C.

CHKDSK Command

CHKDSK command is used to get the report about a disk such as disk directories, files, storage space available, etc. Consider the following example to check the floppy disk in drive A.

C:\>CHKDSK A:

Now the details of floppy disk in A drive is displayed. It is also possible to display details using the SCANDISK command.

UNIX OPERATING SYSTEM

UNIX is an operating system which was first developed in the 1960s, and has been under constant development ever since. By operating system, we mean the suite of programs which make the computer work. It is a stable, multi-user, multi-tasking system for servers, desktops and laptops.

UNIX systems also have a graphical user interface (GUI) similar to Microsoft Windows which provides an easy to use environment. However, knowledge of UNIX is required for operations which are not covered by a graphical program, or for when there is no windows interface available, for example, in a telnet session.

Types of UNIX

There are many different versions of UNIX, although they share common similarities. The most popular varieties of UNIX are Sun Solaris, GNU/Linux, and MacOSX.

Parts of UNIX

The UNIX operating system is made up of three parts; the kernel, the shell and the files and processes.

The Kernel

The kernel of UNIX is the hub of the operating system: it allocates time and memory to programs and handles the filestore and communications in response to system calls.

As an illustration of the way that the shell and the kernel work together, suppose a user types **rm myfile** (which has the effect of removing the file **myfile**). The shell searches the filestore for the file containing the program **rm**, and then requests the kernel, through system calls, to execute the program **rm** on **myfile**. When the process **rm myfile** has finished running, the shell then returns the UNIX prompt % to the user, indicating that it is waiting for further commands.

The Shell

The shell acts as an interface between the user and the kernel. When a user logs in, the login program checks the username and password, and then starts another program called the shell. The shell is a command line interpreter (CLI). It interprets the commands the user types in and arranges for them to be carried out. The commands are themselves programs: when they terminate, the shell gives the user another prompt (% on our systems).

The adept user can customize his/her own shell, and users can use different shells on the same machine.

History—The shell keeps a list of the commands you have typed in. If you need to repeat a command, use the cursor keys to scroll up and down the list or type history for a list of previous commands.

Files and Processes

Everything in UNIX is either a file or a process. A process is an executing program identified by a unique PID (process identifier). A file is a collection of data. They are created by users using text editors, running compilers, etc.

Examples of files:

- A document (report, essay, etc.)
- The text of a program written in some high-level programming language
- Instructions comprehensible directly to the machine and incomprehensible to a casual user, for example, a collection of binary digits (an executable or binary file)
- A directory, containing information about its contents, which may be a mixture of other directories (sub-directories) and ordinary files.

Specific Features of the UNIX

The following are the specific features of the UNIX operating system:

Case Sensitiveness

The commands in UNIX are case sensitive. It means that, lower case a and uppercase A are considered differently.

Multitasking

It refers to performing a number of tasks simultaneously. For example, when a document is printed, you may run another program to sort large data and at the same time you may edit a document in the foreground screen. UNIX switches between the tasks and executes them one by one at small interval of time. This process of sharing the CPU to perform various tasks simultaneously is called time-sharing. If more number of the tasks are submitted then we end up with a slower response from the computer.

Multi-user Capability

UNIX allows the computer to be used by several users through several terminals connected to a powerful computer. A terminal will have a keyboard and a monitor. The computer to which terminals are connected is called as the host computer or server. Any user on the terminal can run various programs, read file information or print a document at the same time. Multi-user computers, are economical and efficient compared to stand-alone computers.

Portability

One of the outstanding features of UNIX is its ability to port itself to another installation. For example, an application program developed in UNIX environment can also be used on a different platform.

Security

UNIX provides a good security for users. The users are required to authenticate themselves before they use the system. The password is encrypted.

File System

UNIX identifies three types of users, owner, group and others. For each group, it provides permission on the files like to read, write and execute operation.

UNIX Commands

The following are the UNIX important commands:

Who Command

Who command is used to list users who are currently logged to the system. The username together with the terminal, date and time the user last logged will be displayed.

Example
```
$who
abhi tty1 sep 29 13:01
reva tty2 sep 29 14:15
user1 tty3 sep 29 15:45
$
$who –u
```
This option –u (means unused) will list the user with the unused time shown below
```
$who –u
abhi tty1 sep 29 13:01 00:05
reva tty2 sep 29 14:15 01:20
user1 tty3 sep 29 15:45
$
```
Note that the users abhi and reva were idle for 5 minutes and 1 hour 20 minutes respectively.

To display user name, terminal line, date and time of login, the who command is given as follows:
```
$who am i
user1 tty3 sep 29 15:45
```
To confirm login name, type the following command
```
$ logname
user1
$
```

PWD Command

pwd (print working directory) command will display the current working directory.
```
$pwd
/usr/user1
$
```
Note that/(slash) represents the root directory. In MS DOS, the root directory is represented by \(back slash).

Echo Command

echo command will display the text typed from the keyboard

Example
```
$ echo learning UNIX is fun
Learning UNIX is fun
$
```

CAT Command

The "cat" command lets you view text files. "cat" is short for concatenate. Note that the cat command echoes each line as soon as it has been typed in.

Examples

(I) cat /etc/passwd

This command displays the "/etc/passwd" file on your screen.

(II) cat /etc/profile

This command displays the "/etc/profile" file on your screen. Notice that some of the contents of this file may scroll off of your screen.

(III) cat file1 file2 file3 > file4

This command combines the contents of the first three files into the fourth file.

SORT Command

SORT command is used to sort the contents of a file and name it students_list

Example

```
$cat > students_list
Abhi       20101
Revathi    20125
Preethi    20104
Ravi       20121
ctrl -d
$
```

Using sort command, the above list may be sorted alphabetically as shown below:

```
$sort students_list
Abhi       20101
Preethi    20104
ravi       20121
Revathi    20125
```

WC Command

WC command is used to count the number of lines, words and characters in a file.

Example

```
$wc students_list
5 10 53 students_list
```

Note that there are 5 lines, 10 words and 53 characters in the file. Also note that every line is terminated by pressing Enter key which is represented by an invisible new line character. These characters are also accounted to get the number of characters as 53 instead of the actual number of characters as 48.

The wc command has the following options:
(I) $wc –l will display the number of lines in a file.

Example

```
$wc–l students_list
```

The output of the above command is the number of lines in student_list file.

(II) $wc–w will display the number of words in a file.

Example

```
$wc–w students_list
```

The output of the above command is the number of words in student_list file.

(III) $wc–c will display the number of characters in a file.

Example

```
$wc–c students_list
```

The output of the above command is the number of characters in student_list file.

GREP Command

GREP command is used to search and display a line for a given word or pattern in a given file name.

Consider the following example to display the register number of a student.

```
$grep Abhi students_list
Abhi 20101
```

Filters

These refer to any command that can take input from standard input, perform some operations and write the results to standard output.

Consider the following example:

```
$who
cseabhi tty1 sep 29 13:01
itreva tty2 sep 29 14:15
ituser1 tty3 sep 29 15:45
$
```

The above list can be short listed to specific users and display the list in alphabetical order as shown below.

```
$who | grep it | sort
itreva tty2 sep 29 14:15
ituser1 tty3 sep 29 15:45
```

$

The output of who is fed into the input of grep which will filter and display those users containing the pattern it.

LS Command

LS command is used to list the files stored in the directory.

Consider the following example

$ls

jp.c   test1   sample1   sample2

sam.txt   ex1.c   letter.doc

The following options are available with ls command.

The ls –l option is used for long listing of files in the current directory.

$ls -l

Total 24

dwxr-xr-x2 user1 group 480 Sep 05 02:15 first.dir

-rw-r—r— 1 abhi group 80 Sep 05 02:13 myfile.c

-rw-r—r— 1 reva group 80 Sep 05 02:34 jp.c

$ls –r will display files in reverse alphabetically order.

$ls* will display all the files and directories in the current directory.

MKDIR Command

MKDIR command is used to create a new directory.

Consider the following example

$mkdir jpdir

$

will create a directory named jpdir

RMDIR Command

RMDIR command is used to remove a directory.

Consider the following example.

$rmdir jpdir

$

Note that all fields in the directory should be deleted before removing that directory.

CD Command

CD command is used to switch from the current directory to another directory.

Consider the following example.

$ cd tjp

$

Note that the current directory will now be tjp. The following options are available with cd command.

$cd or $cd. . is used to switch to the home directory.

$cd/usr/user1 is used to switch to the directory user1.

CP Command

CP command is used to copy a file.

Consider the following example.

$ cp jp.prg jp.copy

$

Note that jp.prg is an existing file which will be copied as jp.copy. if the file jp.copy already exists in the same directory, it will be over written without any warning.

The following options are available with cp command.

$cp jp.prg/usr/usr1 is used to copy the file to the specified directory.

$cp jp.prg/usr/user1/jp.bak is used to copy and change the name of the file.

RM Command

RM command is used to remove or delete a file.

Consider the following example:

$rm jp.bak

$

The file jp.bak will be permanently removed without any warning.

The following options are available with rm command

$rm –i jp.bak will remove a file after confirmation from the user.

RM: remove 'jp.bak' ? y press y then enter to delete the file.

$

MV Command

MV command is used to rename a file.

Consider the following example.

$mv jp jp.c

$

Note that the file jp is renamed as jp.c

man command

man command is used to display the help manual for UNIX commands.

LINUX OPERATING SYSTEM

Linux is a free open-source operating system based on UNIX. Linux was originally created by Linus Torvalds with the assistance of developers from around the globe. Linux is free to download, edit and distribute. Linux is a very powerful operating system and it is gradually becoming popular throughout the world.

Advantages of Linux

The following are the advantages of LINUX operating System:

Low Cost

There is no need to spend time and huge amount money to obtain licenses since Linux and much of it's software come with the GNU General Public License. There is no need to worry about any software's that you use in Linux.

Stability

Linux has high stability compared with other operating systems. There is no need to reboot the Linux system to maintain performance levels. Rarely it freeze up or slow down. It has a continuous up-time of hundreds of days or more.

Performance

Linux provides high performance on various networks. It has the ability to handle large numbers of users simultaneously.

Networking

Linux provides a strong support for network functionality; client and server systems can be easily set up on any computer running Linux. It can perform tasks like network backup much faster than other operating systems.

Flexibility

Linux is very flexible. Linux can be used for high performance server applications, desktop applications, and embedded systems. You can install only the needed components for a particular use. You can also restrict the use of specific computers.

Compatibility

It runs all common Unix software packages and can process all common file formats.

Wider Choice

There is a large number of Linux distributions which gives you a wider choice. Each organization develop and support different distribution. You can pick the one you like best; the core function's are the same.

Fast and Easy Installation

Linux distributions come with user-friendly installation.

Better use of Hard Disk

Linux uses its resources well enough even when the hard disk is almost full.

Multitasking

Linux is a multitasking operating system. It can handle many things at the same time.

Security

Linux is one of the most secure operating systems. File ownership and permissions make Linux more secure.

Open Source

Linux is an open source operating system. You can easily get the source code for Linux and edit it to develop your personal operating system.

Today, Linux is widely used for both basic home and office uses. It is the main operating system used for high performance business and in web servers. Linux has made a high impact in this world.

Linux vs Windows

Linux is an open-source operating system. People can change codes and add programs to Linux OS which will help use your computer better. Linux evolved as a reaction to the monopoly position of windows. you can not change any code for windows OS. You can not even see which processes do what and build your own extension. Linux wants the programmers to extend and redesign it's OS. Linux user's can edit its OS and design new OS.

All flavors of Windows come from Microsoft. Linux comes from different companies like LIndows, Lycoris, Red Hat, SuSe, Mandrake, Knopping, Slackware.

Linux is customizable but Windows is not. For example, NASlite is a version of Linux that runs off a single floppy disk and converts an old computer into a file server. This ultra small edition of Linux is capable of networking, file sharing and being a web server.

Linux is freely available for desktop or home use but Windows is expensive. For server use, Linux is cheap compared to Windows. Microsoft allows a single copy of Windows to be used on one computer. You can run Linux on any number of computers.

Linux has hign security. You have to log on to Linux with a userid and password. You can login as root or as normal user. The root has full previlege.

Linux has a reputation for fewer bugs than Windows.

Windows must boot from a primary partition. Linux can boot from either a primary partition or a logical partition inside an extended partition. Windows must boot from the first hard disk. Linux can boot from any hard disk in the computer.

Windows uses a hidden file for its swap file. Typically this file resides in the same partition as the OS (advanced users can opt to put the file in another partition). Linux uses a dedicated partition for its swap file.

Windows separates directories with a back slash while Linux uses a normal forward slash.

Windows file names are not case sensitive. Linux file names are case sensitive. For example, "abc" and "aBC" are different files in Linux, whereas in Windows it would refer to the same file.

Windows and Linux have different concepts for their file hierarchy. Windows uses a volume-based file hierarchy while Linux uses a unified scheme. Windows uses letters of the alphabet to represent different devices and different hard disk partitions. For example, c:, d:, e:, etc. while in Linux " / " is the main directory.

Linux and windows support the concept of hidden files. In Linux hidden files begin with ". ", For example, . filename

In Linux each user will have a home directory and all his files will be save under it, while in windows the user saves his files anywhere in the drive. This makes difficult to have backup for his contents. In Linux it is easy to have backup's.

WINDOWS OPERATING SYSTEM

Windows is a personal computer operating system from Microsoft that, together with some commonly used business applications such as Microsoft Word and Excel, has become a de facto "standard" for individual users in most corporations as well as in most homes.

The original 1985 version of Windows introduced to home and business PC users many of the graphical user interface (GUI) ideas that were developed at an experimental lab at Xerox and introduced commercially by Apple's Lisa and Macintosh computers. Some of the well-known versions of Windows have included:

- Windows 286
- Windows 386
- Windows 3.0 and 3.11
- Windows 95
- Windows 98
- Windows NT
- Windows 2000
- Windows CE for use in small mobile computers
- Windows Me
- Windows XP
- Windows Vista
- Windows 7
- Windows 8

With the advent of the Internet, Microsoft has repositioned Windows as a kind of "window to the world," and its efforts to take the lead in Web browsers have made Internet Explorer the most popular browser. Microsoft's. NET initiative represents an attempt to become industry-dominant in furnishing products and services that facilitate the use of remote application services on the Web.

Introduction to the Desktop

After you have started the computer, the area you are looking at is called the desktop. The desktop is usually different from one computer to another. This is because some items get added as new programs are installed on a computer, and other items get deleted at will. You will learn how to change the way your desktop looks.

The center and right empty area you are looking at is actually the desktop. Whenever you are asked to use the desktop, the request refers to the whole area you are looking at. But if you are asked to right-click on the desktop, it refers to the empty area of the desktop. From now on, unless specified otherwise, the word "desktop" will always refer to the whole area of the monitor screen.

The Microsoft Windows desktop is made of various parts. From the upper left to the lower left side of the screen, there are small pictures or images called icons. Each one is used to make the computer do something.

Graphically the desktop area is represented as follows:

Desktop

My Computer

My Computer is used to explore the content of your computer and to do other routine things.

Network Neighborhood

You use Network Neighborhood to communicate with other computers if yours is part of a network.

Recycle Bin

When you get rid of (delete) some things on your computer (folders or files), they go to an area called the Recycle Bin where you still have a chance of recalling (retrieving or restoring) them.

Folder

A folder is one of the containers you will be using to store or locate your work. You will eventually learn how to create and manipulate folders.
The My Computer, Network Neighborhood, and Recycle Bin programs are a kind of folder referred to as System Folder (s), they come standard with your computer. They have fancy (artistic) icons. The folders you create are in yellow color with a small tab and are considered "user created" folders.

Taskbar

In the bottom section of the screen, there is (or there may be) a long object. It is called the Taskbar. On the left side of the taskbar, there is an area with the word Start

or . The appearance of this depends on the version of Windows you are using but it plays exactly the same role in any version.

On the right side of Start, there is a wide area that is empty when the computer starts. This is actually referred to as the taskbar. As you keep using the computer, this area would be filled with some objects. In some versions (Windows 95, 98, Windows Server 2003), the color of the taskbar may be gray (or Silver). In some other versions (Windows Millennium), the taskbar may be yellowish. In Windows XP, it may be blue.

On the far right side of the taskbar is a section called the tray area. One of the things that this area displays is the current time (as set on your computer).

Graphically the task bar is represented as follows:

Window

A window is a (usually) rectangular portion of the display on a computer monitor that presents its contents (e.g. the contents of a directory, a text file or an image) seemingly independently of the rest of the screen. Windows are one of the elements that comprise a graphical user interface (GUI).

A GUI is a type of human-computer interface (i.e. a system for people to interact with a computer) that uses windows, icons, pull-down menus and a pointer and that can be manipulated by a mouse (and usually to some extent by a keyboard as well). An icon is a small picture or symbol that represents a program (or command), file, directory (also called a folder) or device (such as a hard disk or floppy disk).

The GUI represents a major advantage over the command line interface (CLI) of the console, which displays only text (i.e. no images) and is accessed solely by a keyboard. It has made computers much easier to learn and work with, and it has also led to the development of major new applications for them, including desktop publishing and CAD (computer-aided design).

Windows Elements

The following are the basic elements of the window:

The Application Window

Application windows are the main part of almost all programs. Common elements of application windows include the control menu, menu bar, and border. Graphically it is represented as follows:

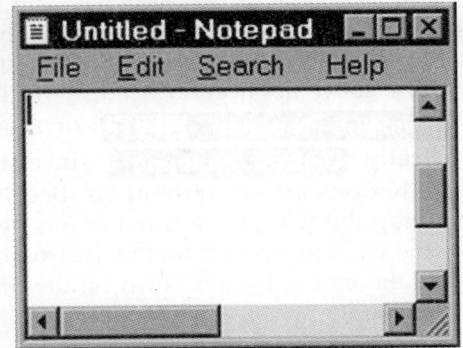

Dialog Box

It is also, technically, a window. Dialogs are usually tied to an application and help perform a specific task or give details for the application. Most dialog boxes lack several of the control buttons and a border, and will have other buttons inside the window to complete a request such as "OK" and "CANCEL".

Graphically it is represented as follows:

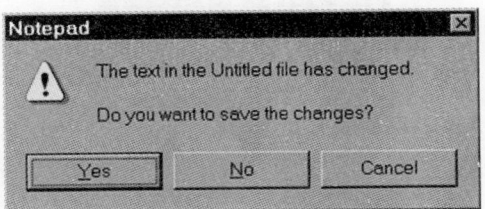

Border

A border is "a part that forms the outer edge of something." As you can see in the above examples, the application window has a thin border around the edges (it is the part that looks like this: |___). The border not only defines where the window is on the desktop, but it can also be used to change the size of most windows. If you point to the border with the mouse, the pointer will change to a double-headed arrow, like so: ↖. If you hold down the mouse button at this time, you can drag the border around and thus alter the size of the window.

Title Bar

A title is "a general or descriptive heading." A heading is "the title, subtitle, or topic that stands at the top or beginning." So, the title bar of the above application window would be: **Untitled – Notepad**; or more specifically, **Untitled – Notepad**, since the control buttons are separate items. Besides identifying the window, a title bar lets you move the window around on the desktop by holding the mouse button down on the title and moving it.

Control Buttons

Control buttons are all the little buttons that are on the title bar but are not actually part of the title. These include:

Control Menu

In Windows 3.x, the button that activates the control menu always looks like this: ▬. In

Windows 95, the button is a miniature version of the program icon, such as: ▤. In every case, the button is located at the left hand side of the title bar. Pressing it will give you a menu with several standard window manipulation commands, such as:

Note that all of these commands can be done with the mouse using other windows elements; their primary usefulness is in when you have to do any of these functions with the keyboard.

Minimize Button

In Windows 3.x, this button looks like: ▾. Pressing it will remove the window and replace it with a program icon somewhere on the desktop. Clicking on that icon will bring up the control menu, so that you can open the window up again. In Windows 95, the button looks like: ▬. Pressing it will remove the window, but leave the program's button on the taskbar. If you click that button with the right mouse button, it will bring up the control menu just like Windows 3. But if you click the program's button with the left mouse button, the program window will immediately be brought back.

Maximize/Restore Button

In Windows 3.x, the Maximize Button looks like: ▴. Pressing it will make the window as large as it can possibly go—usually as large as the screen. The button will then change to the Restore Button: ▴▾, which you can press to change the window back to its previous size. In Windows 95, the Maximize Button looks like: ▢. It does the same thing as its Windows 3 counterpart, except that the window will (usually) only go down as far as the Taskbar. This lets you see all the other programs that have buttons on the Taskbar so that you can easily switch between them using the mouse. The Restore Button in Windows 95 looks like: ▣.

Close Button

Windows 3.x does not have one of these. In Windows 95, it looks like: ☒. Pressing this button is just one way of closing the window, and it is the easiest. But take caution on dialog

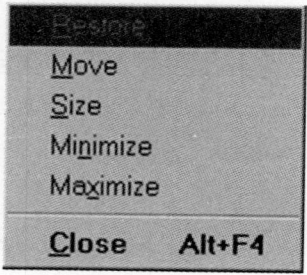

boxes: when this button is active, it usually has the same effect as pressing the "CANCEL" button, so be sure you do not need to save any changes you made in the dialog box. Other ways of closing the window include double-clicking the control menu or clicking on the File menu and then Exit if it is an application window, and clicking on the "OK" button if it's a dialog box.

Help Button

A special addition to some Windows 95 title bars is this button: [?]. If you press it, a question mark will be attached to the mouse pointer. Then when you click on something else in that window, you will see a little box describing the purpose of what you clicked on and/or how to use it:

Displays the Help Index. Type a topic you want to find, or scroll through the list of index entries. Click the index entry you want, and then click Display.

Resize Handle

A new addition to some Windows 95 windows, the resize handle: is actually an extension to the border, found in the lower right corner of the window. It is especially useful when you

want to change the size of the window but for some reason the border is too thin.

Menu Bar

A menu bar is present in merely every application window directly below the title bar. Each word on the bar is a separate menu. If you click on the word, the corresponding menu will appear, like so:

Some programs have cascading menus, which means that an item inside the menu will bring you to another related menu. A very common example in Windows 95 is the Start menu (which, by the way, is not part of a menu bar but rather a button on the Taskbar):

Note that you can identify a cascading menu by the right-pointing arrow at the side.

Popup Menu

Another type of menu is the popup menu. Unlike other menus, the popup does not have

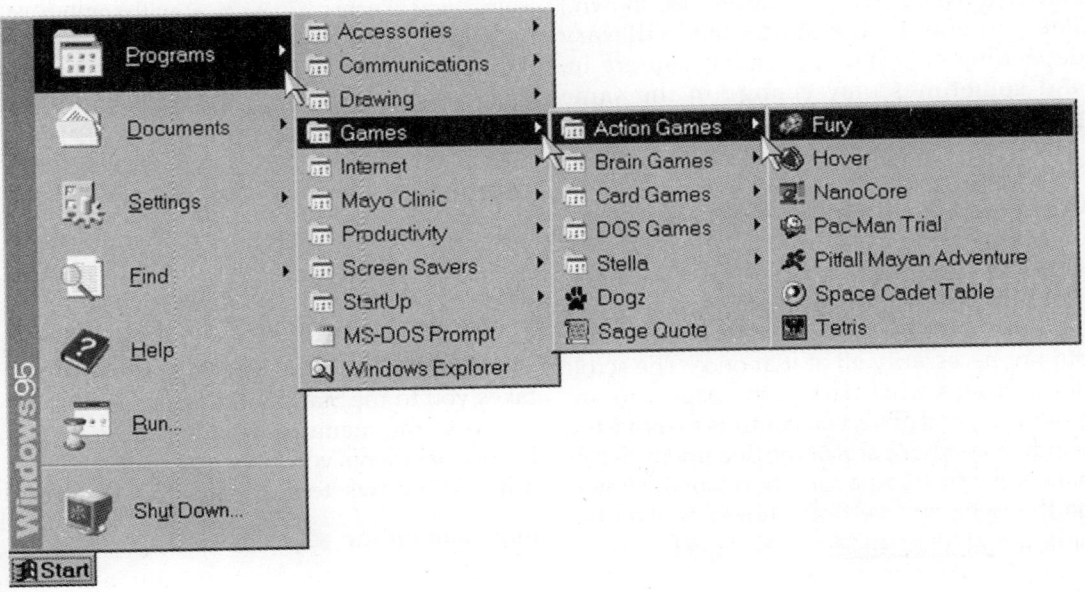

a menu heading and is not attached to a particular spot on a window. Popup menus are usually activated by clicking the right mouse button on something, and the menu you get will depend on where exactly you clicked. For example, a typical popup menu for text box would look like this:

Status Bar

Some windows have a status bar to indicate the current state of various parts of the application. This bar is usually located at the bottom of the window, and is a more-or-less solid gray line. For example, the status bar on the bottom of WordPad looks like this:

Quits the application; prompts you to save documents   CAP NUM

On the left is a one-liner to give you hints about what you are doing. To the right are a couple of boxes that indicate whether the Caps Lock and Num Lock keys are on. (Part of the border and the Resize Handle are also shown.) The contents of the status bar will vary depending on what application you are in, and sometimes may change in the same program depending on what you are doing.

Scroll Bar

Often, an application will show you a document or a list that is too big to fit inside the window. This is what really defines the 'window' concept: just like a real window, these rectangles show you a view of the world, but not necessarily all of it at once. The scroll bar indicates what part of the page you are looking at, and gives you controls to move the window to other parts of the document. Scroll bars come in two flavors: horizontal, shown on the right, and vertical, shown here on the bottom:

The middle part of the bar, called the scroll box: , indicates which part of the document or list you are looking at. Therefore, if the box is at the top of the scroll bar, then you are looking at the top of the list. In Windows 95, this box has the added benefit of indicating how much of the list is being displayed; if the box covers one-third of the length of the scroll bar, then you are seeing one-third of the list. If this box is missing, then you are viewing the entire list and there is nothing else to scroll to. If you press the mouse button down on this box and drag it, you will move the window over the document so that you can see other parts of it.

The arrows on either side of the bar: ▲ and ▼, can be used to move the window one line at a time. This is useful when the scroll box is very small and you need to fine-tune the window position.

If you want to jump from one 'page' to the next, click in the blank area on either side of the scroll box. This will move the window one window's length at a time.

Button

A button is one of the most basic control elements in Windows. It usually looks something like this: , with something written in it, such as "OK", "CANCEL", "YES", "NO", etc. To use the button, just click on it with the mouse. It's the easiest thing there is.

Taskbar

This is a nifty feature new to Windows 95. The taskbar is located at one side of the Desktop (it does not matter which side) and is logically divided into three parts. First is the Start button. Pressing this button takes you to the Start menu, which gives you access to the menu of installed programs, a list of documents you have used recently, and other Windows tools, including **the very**

important online  Help.

The middle part of the Taskbar holds buttons for each application window which you currently have open. This makes it really easy to do many things at once. To switch from one program to another, just click on its button on the taskbar. This is also, of course, a way to tell if you have any programs open and what they are.

At the other end of the taskbar is the "System Tray". This tray holds the clock, plus miniature icons for miscellaneous mini-applications. Simply pointing to one of these icons will pop up a brief description of the mini-app or its status. For example, the yellow speaker at the far left of the system tray lets you control the volume. If you point to the clock, it will briefly show you the date.

The taskbar is also a window. Like application windows, you can change its size; however, the border is only available on one side of the taskbar (the side that is not against the edge of the display.) You can only enlarge the taskbar to half the size of the Desktop, but windows will happily let you shrink it down to nothing but the border!

You can also move the taskbar, but unlike other windows, it must be stuck to one side of the display and run the entire length of that side. To move the taskbar, just hold the mouse button down on a part of the Taskbar that is not covered by buttons: ![Start taskbar 12:45 PM], and drag it to whichever side of the screen you want it to be.

Icon

An icon is "an image; a representation." It is used to represent a window ![window icon], program ![skull icon], file ![folder icon], or any other object. Icons behave differently depending on what they represent, but usually you need to double-click on the icon to activate it. In Windows 95, clicking the right mouse button on most icons will display a menu that gives you additional things you can do with that icon.

EXPLORE YOUR COMPUTER

Once your computer is booted up, you are ready to explore your computer, i.e. you can now see what all things (files, application, etc.) are there on your computer.

There are several ways that let you navigate around your computer. But the most commonly used one are as follows:

1. Start Button and Taskbar
2. My Computer
3. Windows Explorer

Now we will discuss these issues one by one.

The Start Button and Taskbar

You can use the taskbar and Start button to easily navigate through Windows 98. Both features are always available on your desktop, no matter how many open windows you have.

The Start Button

The start menu is the main navigational tool for launching programs. It also provides access to some system settings. The start menu is located in the bottom left hand corner of the screen.

- **Programs:** A submenu where all of the programs installed on a computer are located.
- **Documents:** A submenu of the recently opened **documents**.
- **Settings:** A submenu of system settings and **control panels**.
- **Search:** A submenu for different types of searches. A particular search is to **find files and folders**.
- **Help:** A help wizard with answers to common questions on particular areas of Windows 2000.
- **Run:** A command line interface for advanced users.
- **Shutdown:** Gives a user the option to turn off the computer (shutdown), restart or logoff.

Graphically the start menu is represented in the graphic (p. 54 top).

Most beginners have no idea how many programs they have, or what the names of those programs are. If you fall into that category, you may want to take a minute now to look around. That is, click the Start button then click on All Programs. Then, any time you see a menu option with a % on it, point to

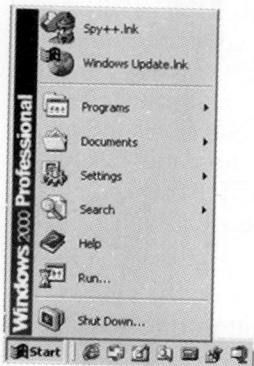

it or click on it to see the name of programs in that group. Read some of the program names, and look at their icons, just to start getting familiar with what you have on your computer.

Taskbar

If you think of your computers hard disk as a filing cabinet where all your programs and documents (files) are stored, then you can think of your Windows desktop as being like your real "wooden" desktop. Only the things you are working with right now are on your Windows desktop, as well as your real desktop.

At any given time, you might have several programs, folders, and documents open on your desktop. The short name for "anything that is currently open on your desktop" is *task*. That is, we can refer to each open item on your desktop—no matter what that item is, as a "task", short for "task-in-progress". The Windows taskbar, which is roughly centered across the bottom of your screen, as in following figure:

moment. Here are some things you can do with the taskbar along those lines:

- If a program window is buried in the mess, click its taskbar button to instantly bring it to the top of the stack.
- You can also click a task's taskbar button to make it invisible (so that it is not taking up any space on the desktop), then click that same button again to make it visible again.
- You can also *close* any open task (thereby removing it from the desktop and putting it back in the filing cabinet) by right-clicking its taskbar button and choosing Close.

My Computer

When the My computer **icon** on the **desktop** is double clicked, the my computer explorer window opens and displays a list of storage media on the current computer. These can include floppy disk drives, CD-ROM, DVD-ROM, removable disks such as ZIP disks, and hard disk drives.

- **3½ Floppy:** The floppy disk drive. To access files on a floppy disk, double click the floppy disk **icon**.
- **Local Disk:** The hard disk drive. Typically the C: drive is the main drive where the files for Windows 2000 are stored. It is possible to have multiple local disks. To access the files and folders, double click the local disk **icon**.
- **Compact Disk:** The CD-ROM drive. Double clicking on the compact disk **icon** will give access to the files and folders on the CD. CDs are **read only** if a **CDR/RW drive** is present.

Each taskbar button represents an open program (a "task in progress")

When you have lots of program windows open, they pile up on your Windows desktop, just like sheets of paper can pile up on your real desktop. You can use the taskbar to sort of "shuffle things around", so you are in control of what is, and is not visible at the

- **Removable Disk:** A removable disk can be any number of portable media from ZIP disks to USB thumb drives.

Search for Files and Folders

This utility allows a user to search for files on the computer. To open the search window, go to start, then search, then files and folders.

- **Search For:** The name of the file a user is trying to locate goes here.
- **Look In:** The location the file may be in. The utility will also search all subfolders in a given location.
- **Start Over:** Clears the search results.
- **Search Now:** Begins the search.
- **Also Search By:** Adds more filtering to the search results. For example, if the date when the file was created or last modified is known, then a user can check the date box and input a date or date range.

There are some wildcard characters available to help narrow a search down.

- **The Asterisk (*)** can be included in a file name as a substitution for zero or more characters. For example, if a user typed tax_return_* then the results would be tax_return_1998, tax_return_1999, tax_return_2000, and so on.
- **The Question Mark (?)** can be included in a file name as a substitution for a single character. For example, the search tax_return_200? Would return the files tax_return_2000, tax_return_2001, tax_return_2002, and so on.

System tray

The system tray is located in the bottom right hand corner of the screen. The system tray holds a clock and other **icons** for programs currently running on the system. Typically right or left clicking on these **icons** displays some options for interacting with the given program

Graphically the system tray is represented as follows:

The System Tray

Control Panel

The control panel is the main configuration of settings for a computer. To get to the control panel, go to start, then settings, then control panels. There are a number of **icons** for each control. Some of the controls adjust sensitive settings in the system and others adjust how input devices like the mouse and keyboard work. Here is a list of some of the basic settings.

- **Add/Remove Programs:** This is the control panel to use in order to properly remove software from a computer. It is important not to just delete a program because there are other system entries that are made when a program is installed and do not get removed when a program is simply deleted.
- **Date/Time:** The control panel to set the date and time of the computer.
- **Keyboard:** The control panel for keyboard settings such as the rate the cursor blinks when typing.
- **Mouse:** The control panel for mouse settings such as double click speed. This control panel may be different for each machine depending on the type of mouse.
- **Regional Options:** This control panel has settings such as what type of currency to use, or how numbers should be displayed.

Graphically the control panel is represented as follows:

Windows Explorer

Windows Explorer is a file management tool available in Microsoft Windows. You can use Windows Explorer to see all the resources available on your computer.

To Start Explorer

Click the **Start** button, point to **Programs** then **Accessories**, and then click **Windows Explorer**.

Other Methods of Starting Explorer

The following are the other methods to start the Explorer:

- Right-click the Start button and then click Explore.
- Right-click the My Computer icon, then click Explorer.
- Hold the "windows" key + e

Windows Management Tools and Helpful Hints

The Explorer Windows is graphically represented as follows:

The Menu Bar

The **Menu Bar** of the Explorer window gives access to all the commands available with Explorer. To see a description of what each menu command does, point and click on a command then rest your mouse pointer over it. Information about the command appears in the status bar at the bottom of the window. (Note: If the status bar is not displayed, click the **View** menu, and then click the **Status Bar** command.) Slide the pointer over other commands and sub-commands to see more descriptions.

The main area of the Explorer window is divided into two panes.

- The left pane is labeled *All Folders*. You will see all drives available to you, and the folders within the drives.
- The right pane of Windows Explorer shows the *Contents of* the drive or folder that is selected in the left pane. Your documents and programs are stored in

The Explorer Window

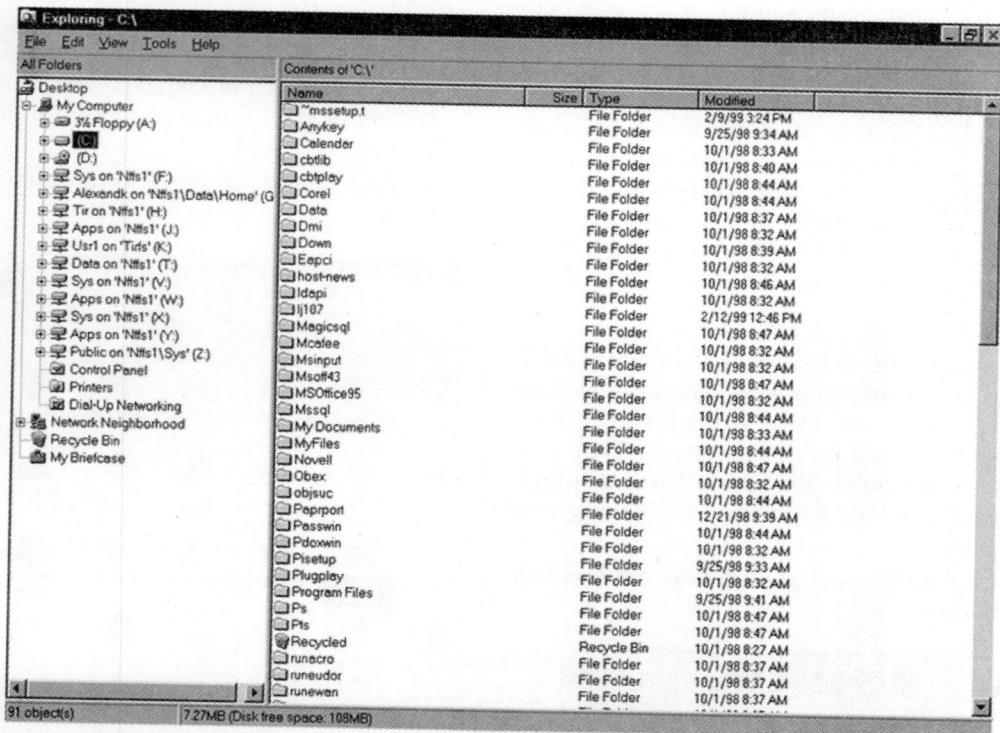

folders, which are indicated with a yellow folder icon.

Folders are sometimes referred to as directories.

- The + (plus sign) and – (minus sign) can be used to open and close folders.

The Recycle Bin

The **Recycle Bin** is a storage area for deleted files and folders. Deleted items can be recovered from the Recycle Bin. Once the Recycle Bin has been emptied, the files are gone forever.

- To restore a file from the Recycle Bin, select the file, then choose File, then Restore. (Or, select the file, Right-Click on the file, and choose Restore.)
- Shift+Delete and file will not go to recycle bin—it will be deleted from your hard drive permanently.

You may change the way files and folders are displayed by using the commands on the **View** menu.

To Change the Appearance of Items in a Folder

If you want to change the appearance of items in a folder then perform the following options:

Click the **View** menu, and then click one of the following commands:

- Large Icons
- Small Icons
- List
- Details

You can also sort the items by name, size, date, and type, depending on the view. Experiment with the **Arrange Icons** command on the **View** menu. Another method you can use to sort is to click the column heading in the contents frame.

To Change the Size of the Explorer Window Panels

Position the mouse pointer over the bar that separates the two panels. When you see the two-headed arrow (√) left-click and drag the bar right or left to resize the panels.

To Display Contents of a Folder

If you want to display the contents of a folder then perform the following operation:

- Double-click the folder name

OR

- Click the plus sign (+) to the left of the folder.

Opening a Program

If the program you want to open is on the desktop, simply double-click the icon and it will open. However, if it is not located on the desktop do the following steps:

1. Click on the **Start at bottom left corner of screen**
2. Click on **Programs**
3. Select the Program you want to open. Graphically the process is represented as follows.

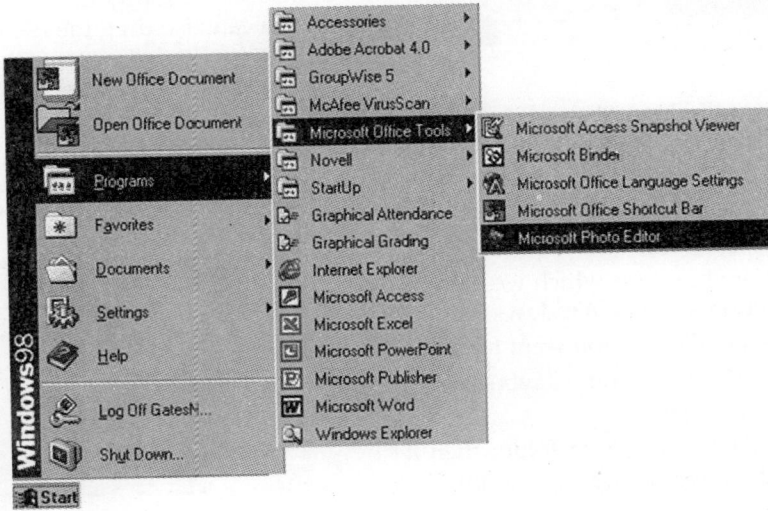

Managing File and Folders

In Windows 98, you can organize your documents and programs to suit your preferences. You can store these files in folders, and you can move, copy, and even search for files and folders.

Creating Folders

The following are the steps which we have to follow for creating a New Folder in Windows 98.

1. Select the Drive or Folder you want to create your Folder in
2. Right click in the right pane in the white area
3. Select **New**
4. Select **Folder**
5. Type in name for that Folder

Graphically the process is represented as follows:

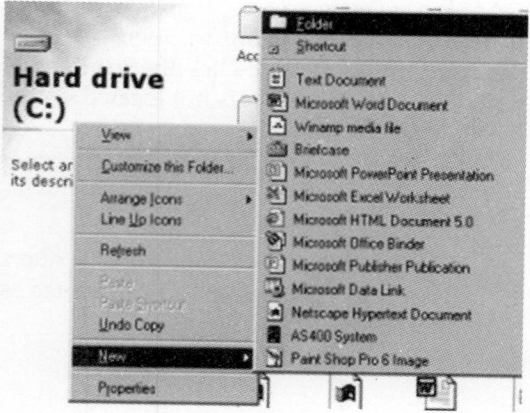

File names in Windows 98 can be up to 255 characters including spaces. However, file names cannot contain any of these characters: \ / : * ? " <> |.

Deleting a File or Folder

The following are the steps which we have to follow for deleting a file in Windows 98.

1. Select the File/Folder you want to delete
2. Hit the Delete Key on the keyboard
3. Confirm Deletion

When you delete any file or folder then it is automatically stores in the Recycle Bin.

To Restore a File from Recycle Bin

When we want to delete any file then the file automatically goes to Recycle Bin and if we want to restore a file from Recycle Bin then we have to follow the following steps:

1. Double click Recycle Bin icon on the Desktop.
2. Select the desired file in recycle Bin
3. Click Restore command in the file menu.

If you delete a file at the command prompt from a floppy disk (A: or B: drive), it does not go into the Recycle Bin.

Moving/Copying Files to Different Folders

The following are the steps which we have to follow for Moving/Copying files to different folders.

1. Select the Folder you want to move or copy
2. **To Move File**—Simply drag it to the destination you want it to go to
3. **To Copy File**—Click on **Edit** —> **Copy** from the top of the screen. Then go to the destination folder and click **Edit** —> **Paste**

Copying a File/Folder to a Floppy Disk

The following are the steps which we have to follow for copying a file/folder to a floppy disk.

1. Select the Folder you want to move or copy
2. Right click on the file or folder in the right pane
3. Select **Send To** —>
4. Select **3.5 Floppy (A)**
5. You can also drag the file to 3.5" Floppy (A:\) in the left pane

Graphically the process is represented as follows:

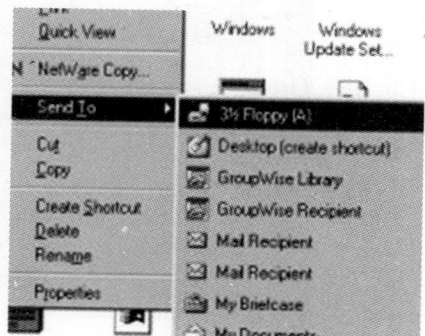

Copying a File/Folder from a Floppy Disk to Other Drive

The following are the steps which we have to follow for copying a file/folder from a floppy disk to other drive:

1. Select 3.5 Floppy (A:) in the left pane
2. Select the File/Folder you want to copy from in the floppy drive
3. Drag it to your destination

Renaming a File or Folder

The following are the steps which we have to follow for renaming a file or folder.

1. Select the File or Folder you want to rename
2. Right click on it
3. Select **Rename**
4. Rename File

Graphically the process is represented as follows:

Finding Files on Your Computer

The following are the steps which we have to follow for finding files on our computer system.

1. Click on the **Start**
2. Click on **Find**
3. Select the **Find Files or Folders**
4. Enter your search criteria. You do not need to complete every field. Only enter the criteria you want to use for your search. Switch through the different tabs (Date, Advanced, Name and Location) to further your search
5. Click the **Find Now** button

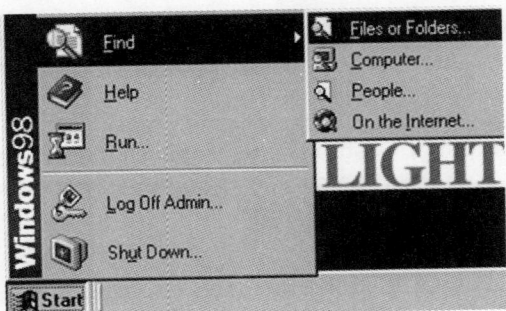

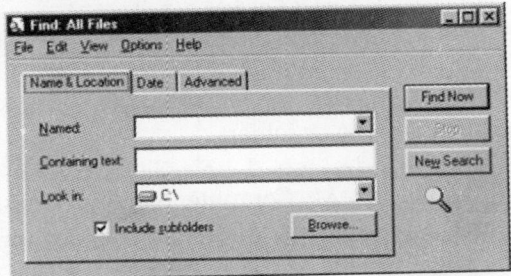

Searching Using Wildcard Characters

A wildcard character is a keyboard character such as an asterisk (*) or a question mark (?) that is used to represent one or more characters when you are searching for files, folders, printers, computers, or people. Wildcard characters are often used in place of one or more characters when you do not know what the real character is or you do not want to type the entire name.

The following table shows the working of (*) and (?) Wildcard characters:

Wildcard character	Uses
Asterisk (*)	Use the asterisk as a substitute for zero or more characters. If you are looking for a file that you know starts with "gloss" but you cannot remember the rest of the file name, type the following: **gloss***This locates all files of any file type that begin with "gloss" including Glossary.txt, Glossary .doc, and Glossy.doc. To narrow the search to a specific type of file, type: **gloss*.doc**This locates all files that begin with "gloss"

but have the file name extension .doc, such as Glossary.doc and Glossy.doc.

Question mark (?) Use the question mark as a substitute for a single character in a name. For example, if you type **gloss?.doc**, you will locate the file Glossy.doc or Gloss1.doc but not Glossary.doc.

Opening Files and Folders

After you have located the file you want, you can double-click to open it.

Now we have to follow the following steps to open any file or folder:

1. On the desktop, double-click **My Computer**. The My Computer window opens.

2. Double-click the drive that contains the file or folder you want to open.

3. Double-click the file or folder.

Working with Frequently Used Files

You can quickly open documents and programs that you use often. The Start menu lists the documents used most recently, so that you can quickly reopen them. The My Documents folder on your desktop is a convenient place for you to store frequently used files and folders and for easy access to a file that you use frequently, you can also create a *shortcut* to it. A shortcut does not change the location of a file—the shortcut is just a pointer that lets you open the file quickly. If you delete the shortcut, the original file is not deleted.

To open recently used documents we have to follow the following steps:

1. Click the **Start** button, and then point to **Documents**. A list of your recently opened documents appears.

2. Click a document on the list. The document opens.

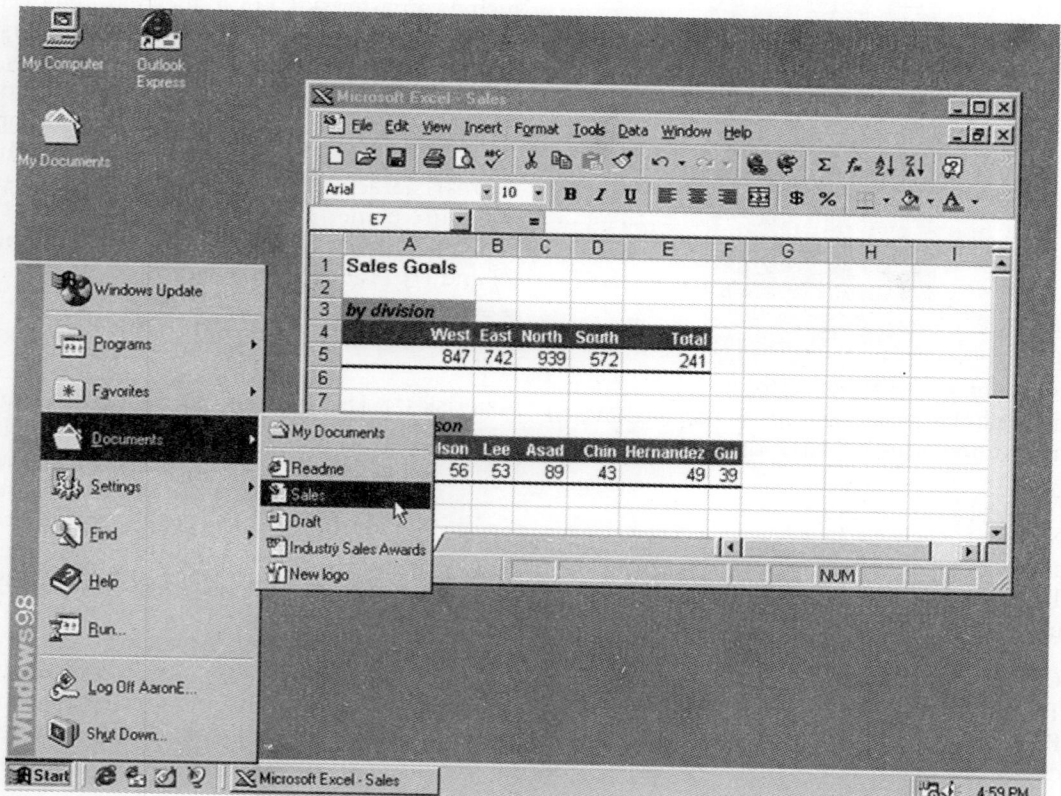

Create A Shortcut

The following steps are to be followed for creating a shortcut:

1. While in Explorer, **Right-click** the file you wish to make into a shortcut.
2. Choose **Send To** from the popup menu and then select **Desktop (create shortcut).**
3. Click **OK** to confirm.
4. Minimize the Explorer window to verify that the shortcut has been created.

Another method for creating a shortcut is as follows:

1. **Right-click** anywhere on the desktop
2. Select **New**, then **Shortcut**
3. Click the **Browse** button
4. In the Browse Window, find the **Windows folder** and **double-click** it. Then find the **tour.exe** file and click once to select it.
5. Click the **Open** button
6. Click the **Next** button then click **Finish.**

Delete A Shortcut

The following steps are to be followed for creating a shortcut

1. **Right-click** the shortcut you wish to delete.
2. Choose **Delete** from the popup menu and click **Yes** to confirm.

Basic Window Accessories

The windows operating system comes equipped with basic accessories like calculator, paint, notepad, WordPad, etc. Let us learn something more about using these accessories.

Notepad

Notepad is a basic text editor you can use for simple documents or for creating Web pages. To create or edit files that require formatting, use WordPad.

Opening Notepad

To open Notepad, click **Start**, point to **All Programs**, point to **Accessories**, and then click **Notepad**. Graphically the process is represented as follows:

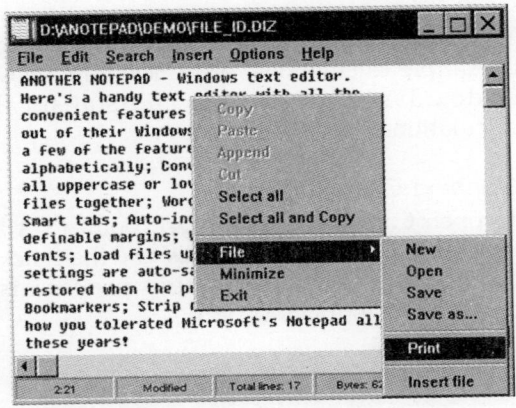

Paint

Paint is a drawing tool you can use to create simple or elaborate drawings. These drawings can be either black-and-white or color, and can be saved as bitmap files. You can print your drawing, use it for your desktop background, or paste it into another document. You can even use Paint to view and edit scanned photos.

You can also use Paint to work with pictures, such as .jpg, .gif, or .bmp files. You can paste a Paint picture into another document you have created, or use it as your desktop background.

Opening Paint

To open **Paint**, click **Start**, point to **All Programs**, point to **Accessories**, and then click **Paint**. Graphically the process is represented as follows:

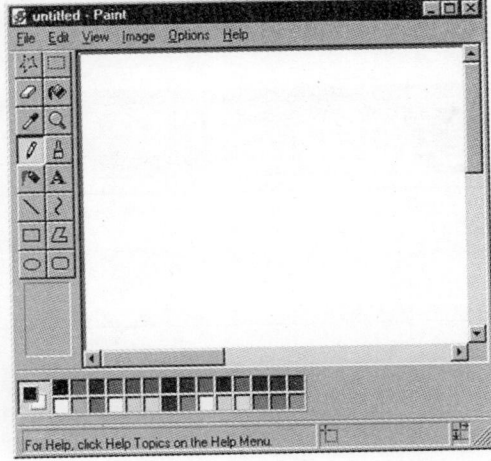

Calculator

You can use Calculator in Standard view to do simple calculations, or in Scientific view to do advanced scientific and statistical calculations.

Opening Calculator

To open Calculator, click **Start**, point to **All Programs**, point to **Accessories**, and then click **Calculator.** Graphically this process is represented as follows:

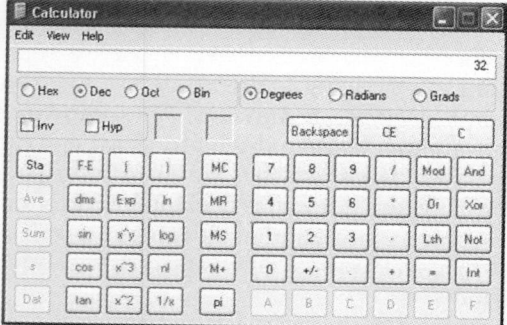

Wordpad

You can use Wordpad to create or edit text files that contain formatting or graphics. Use Notepad for basic text editing or for creating Web pages.

Opening Wordpad

To open Wordpad, click **Start**, point to **All Programs**, point to **Accessories**, and then click **WordPad**. Graphically this process is represented as follows:

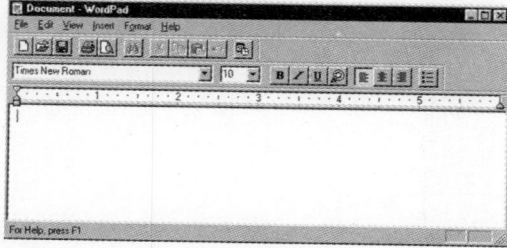

Clipboard

The Clipboard is the storage area for items that have been cut or copied. Each time you execute Cut or Copy, you replace the old information on the Clipboard with whatever you just cut or copied. You can paste Clipboard information as often as you like, until you replace it with something else.

SHUTTING DOWN THE COMPUTER

The following are the steps that we have to follow to shut down the computer:

1. Click on the **Start**
2. Click on **Shut Down** (A Shut Down Windows dialog box will appear)
3. Click on **Shut Down**
4. **Click on OK**

Graphically the process is represented as follows:

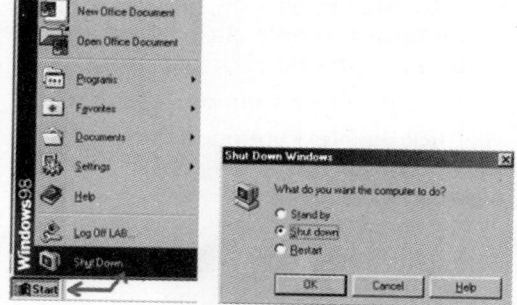

SUMMARY

- An operating system is the system software which is used to operate the computer system.
- Operating systems can be classified into the two categories: (a) Single user operating system, and (b) Multiple user operating system
- Single user Operating system means that only one user can work at the one time
- A multiple user operating system means that more than one user can work simultaneously at the same time.
- DOS, Windows 95 are the examples of single user operating systems.
- UNIX, LINUX, WINDOWS NT are the examples of multiple user operating system.

Exercises

Section A

Multiple Choice Questions

1. Which command displays all the files having the same name but different extensions?
   (a) Dir filename.*
   (b) Dir filename.ext
   (c) Dir *.sys
   (d) Dir *.ext

2. Which command displays only file and directory names without size, date and time information?
   (a) Dir/w   (b) Dir a:
   (c) Dir /b   (d) Dir /s

3. Which software is used to activate the computer system?
   (a) Commands
   (b) User Interface
   (c) Operating System
   (d) Utilities

4. Booting means
   (a) Carrying out a self test
   (b) Checking Computer's Memory
   (c) Loading Operating System in Computer's Memory
   (d) Executing a command

5. Which of the following management function is not performed by Operating System?
   (a) Software Management
   (b) Information Management
   (c) Device Management
   (d) Memory Management

6. Which of the following is a single user multitasking operating system?
   (a) MS-DOS   (b) Windows-XP
   (c) Windows 2000 (d) UNIX

7. A GUI is
   (a) Hardware
   (b) Language interpreter
   (c) Software interface
   (d) An operating system

8. Multiprogramming refers to
   (a) Having several programs in RAM at the same time

9. Multiprogramming is a prerequisite for
   (a) Multitasking
   (b) An operating system
   (c) To run more than one program at the same time
   (d) None of the above

10. Which of the following is not a GUI component?
    (a) Radio Button   (b) Menu
    (c) Command   (d) Text box

11. Which type of User Interface does MS-DOS provide?
    (a) GUI   (b) CUI
    (c) ABI   (d) MGI

12. In which year was the first operating system was developed
    (a) 1910   (b) 1940
    (c) 1950   (d) 1980

13. MS-DOS was developed in
    (a) 1991   (b) 1984
    (c) 1971   (d) 1961

14. CHKDSK command is used to
    (a) Analyze the hard disk error
    (b) Diagnose the hard disk error
    (c) Report the status of files on disk
    (d) All of the above

15. Internal commands in DOS are
    (a) Cls, rd label
    (b) Dir, ren, sys
    (c) Time, type, dir
    (d) Del, disk copy, label

16. How many maximum options can be selected from a Check Box List?
    (a) As many as required
    (b) All the options presented in the list
    (c) 1   (d) 0

17. Which command is used to see the version?
    (a) Version   (b) Ver
    (c) Verson   (d) None of the above

18. Which command can be used to create the disk's tracks and sectors?
    (a) Fdisk   (b) Format
    (c) Chkdsk   (d) Attrib

- (b) Multitasking
- (c) Writing programs in multiple languages
- (d) None of the above

19. Which command is used to copy all files from drive A with extension .txt to the currently logged drive and directory?
    (a) Copy a. *.txt   (b) Copy *.txt a:
    (c) Copy *.txt c: (d) Copy *.txt all.txt

20. Which command is used to create root directory and FAT on disk?
    (a) CHKDSK   (b) Command.com
    (c) Format   (d) Fat

Fill in the Blanks

1. Edit is an example of _____.
2. MS-word is an example of _____.
3. To display files in the MS-DOS, ____ _____ command is used.
4. To display copy a file in UNIX_____ _____ command is used.
5. In UNIX, commands are _____ sensitive.
6. GUI is acronym for _____.
7. _____ is an example of a network operating system.
8. _____ is an example of a multi user operating system.
9. _____ Command of MS-DOS will be used to view the names of all the files present in current directory.
10. DOS is example of _____.

Section B

Short Answer Type Questions

1. What do you mean by software and what are the various classification of software?
2. What do you mean by hardware and also explain the various items of hardware used in computer system?
3. Differentiate between Application Software and System Software with suitable examples?
4. Define the concept of UNIX operating System with suitable examples.
5. Define the concept of LINUX operating system with suitable examples.
6. Define the concept of WINDOWS Operating System with suitable examples.
7. Define the concept of DESKTOP related to Windows operating system.
8. Define the concept of taskbar related to Windows operating system.
9. Define the concept of Menu Bar related to Windows operating system.
10. Define the concept of MY COMPUTER related to Windows operating system.

Section C

Long Answer Type Questions

1. Define the concept of Windows Explorer and also define the working of Windows Explorer.
2. Define the basic Windows accessories with a suitable example.
3. How to open NOTEPAD in Windows Environment? Explain with a suitable example.
4. How to open WORDPAD in Windows Environment? Explain with a suitable example.
5. How to open PAINT in Windows Environment? Explain with a suitable example.
6. How to open Calculator in Windows Environment? Explain with a suitable example.
7. How to open MSWORD in Windows Environment? Explain with a suitable example.
8. How to open MSEXCEL in Windows Environment? Explain with a suitable example.
9. How to open MSPOWERPOINT in windows Environment? Explain with a suitable example?
10. Define the term operating system and explain what is the purpose of operating system. Name any four popular operating system.

3 Fundamental of Programming

Computer programming is the procedure of writing, testing, debugging/troubleshooting, and maintaining the source code of programs. This source code is written in a programming language. The code may be a change of an existent source or something completely new. The purpose of programming is to produce a program that exhibits a certain desired behavior (customization). The process of writing source code needs expertise in many different subjects, including knowledge of the application domain, specialized algorithms and formal logic.

PROGRAMMING ENVIRONMENT

A programming environment incorporated into a software application that furnishes a GUI builder, a text or code editor, a compiler and/or interpreter and a debugger.

Graphically the programming environment may be represented as shown in Fig. 3.1.

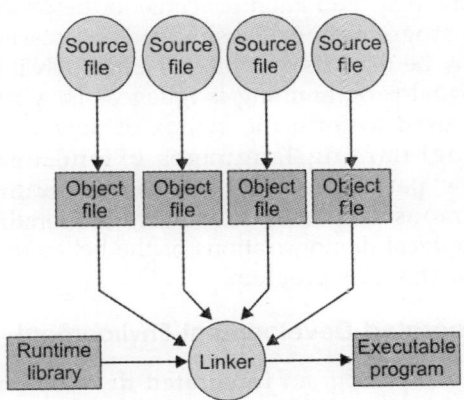

Fig. 3.1: Programming environment

In the context of programming environment we will discuss the following terms:
- Assembler
- Compiler
- Interpreter
- Linker
- Loader

Assembler

Assembler is a program that translates programs from assembly language to machine language.

Compiler

A program that transforms *source code* into *object code*. The compiler gains its name from the way it works, looking at the entire piece of source code and collecting and reorganizing the instructions. Thus, a compiler differs from an *interpreter*, which examines and fulfills each line of source code in succession, without looking at the entire program. The advantage of interpreters is that they can accomplish a program immediately. Compilers need some time before an executable program emerges. However, programs produced by compilers run much quicker than the same programs executed by an interpreter.

Every high-level programming language (except strictly interpretive languages) accompanies a compiler. In effect, the compiler is the language, because it defines which instructions are satisfactory.

Because compilers transform source code into object code, which is unique for each type of computer, many compilers are usable for the same language. For example, there is a

FORTRAN compiler for Personal computer and another for Apple Macintosh computers. In addition, the compiler industry is quite aggressive, so there are actually many compilers for each language on each type of computer. More than a dozen companies develop and sell C compilers for the Personal computer.

Interpreter

An interpreter is a program that translates the high level language code into an equivalent machine level code and the execution is done line by line.

An interpreter interprets high-level commands into an intermediate form, which it then executes. In contrast, a compiler transforms high-level instructions directly into machine language. Compiled programs broadly run faster than interpreted programs. The advantage of an interpreter, however, is that it does not need to go through the compilation stage throughout which machine commands are generated. This process can be long if the program is long. The interpreter, on the other hand, can now execute high-level programs. For this reason, interpreters are sometimes used throughout the development of a program, when a programmer wants to add small sections at a time and test them rapidly. In addition, interpreters are frequently used in education because they allow students to program interactively.

Both interpreters and compilers are usable for most high-level languages. However, Visual Basic and Prolog are especially designed to be executed by an interpreter. In addition, page description programming languages, such as PostScript, use an interpreter. Every PostScript printing machine, for example, has a built-in interpreter that fulfills PostScript instructions.

Linker

Linker is also called *link editor* and *binder*, a linker is a program that aggregates object modules to form an executable program (Fig. 3.1). Many programming languages allow for you to write different pieces of code, called *modules*, separately. In addition to combining modules, a linker also substitutes symbolic addresses with real addresses.

Loader

Loader is a program routine that copies a program into computer memory for execution. In a computer operating system, a loader is a part that locates a given computer program (which can be an application or, in some cases, part of the operating system itself) in offline storage (such as a hard disk), loads it into main storage (in a personal computer, it is called random access memory), and gives that computer program control of the computing machine (allows it to execute its instructions).

A program that is loaded may itself contain elements that are not initially loaded into main storage, but can be loaded if and when their logic is needed. In a multitasking operating system, a program that is sometimes called a *dispatcher* juggles the computer processor's time among different tasks and calls the loader when a program associated with a task is not already in main storage. (By program here, we mean a binary file that is the result of a programming language compilation, linkage editing, or some other program formulation process.)

Visual Programming Environment

It is the computer software which allows the use of visual expressions (such as graphics, drawings, animation or icons) in the process of programming. These visual expressions may be used as graphical user interface for textual programming languages. They may be used to form the syntax of new visual programming languages extending to new paradigms such as programming by demonstration or they may be used in graphical demonstrations of the behavior or structure of a program.

Integrated Development Environment

In computing, an **integrated development environment (IDE)** is a computer software application that provides comprehensive facilities to computer programmers for software development. An IDE normally comprises a source code editor, a compiler and/or interpreter, build automation tools, and (usually) a debugger. Sometimes a version control system and various tools are incorporated to simplify the construction of a GUI. Many modern IDEs also have a class web browser, an object inspector, and a class hierarchy diagram, for use with object oriented software development.

IDEs are designed to maximize programmer productiveness by providing tightly-knit components with similar user interfaces. This should mean that the computer programmer has much less mode changing to do than when using discrete development programs.

Typically an IDE is committed to a specific programming language, so as to provide a feature set which most closely matches the programming paradigms of the language. However, some multiple-language IDEs are in use, such as Eclipse, Active State Komodo, recent versions of Net Beans, and Microsoft Visual Studio.

SOFTWARE (PROGRAM) DEVELOPMENT LIFE CYCLE

Software development life cycle or SDLC is a model of a detailed plan on how to create, develop, implement and eventually fold the software. It is a complete plan outlining how the software will be born, raised and eventually be retired from its function. Although some of the models do not explicitly say how the program will be folded, it is already common knowledge that software will eventually have it is ending in a never ending world of changing web, software and programming technology.

Stages in SDLC

There are several ways of developing a software. Almost each technology based company has its own way of developing software. Although different companies to follow different theories of development, they certainly follow the following general stages:

Planning

Before a program is created, a programmer has to know what to create. Software development companies therefore use this stage to determine the need of the present market. Surveys and project proposal are usual in this stage. Management is often involved in this stage to determine what the developers need to do and how their work will impact the market.

Design

After the management has approved the plan and the budget allocated, the nest step is to create the architecture of the program. Developers work together and discuss the ways of developing the program. The workflow of the software is outlined in this stage. Some software development models approach this stage in a rather simple manner for example the iterative development model often treats this stage as the beginning of a software's actual creation. The initial programs are created in this part of development model.

Testing

It is necessary to make sure that the created programs work well in different environments. The normal way of testing requires testers to expolite the program under different conditions. The other way of testing get normal and intended users to test programs is to in a restricted manner under actual conditions so that the necessary modification can be made before the software is released for general use. The testing stage may include implementing the software in beta testing just to make sure that it can withstand multiple users at the same time.

Implementation

Once testing is completed and the software declare fit for implementation, it is released for sell to the public or removed from beta version. Initially a serious challenge of fixing bugs may be faced as discovered by different users.

Maintenance

With the software properly implemented, developers role in this software does not end there. Rather, they will have to work reactively for this software. Instead of looking for the problems in their created software, computer programmer will only be providing answers to their problems.

Termination

No software will last forever. Updates are undertaken and there will be occasions when a software would need to be scrapped altogether. The evolution of a new coding language or the adaptation of the same software to a different platform may also warrant termination of the original software.

WRITE AND EXECUTE THE PROGRAM

The following are the steps usually followed in any programming environment for writing and executing the program:

1. The program is first written in the editor of the programming language.

2. The program is then compiled or interpreted. Then the object code of that program is automatically created.

3. The program is then linked and the direct executable file is automatically created. The program is executed and the direct executable file created.

In the C programming language the write and execute the program is represented as follows:

1. First the source file in the Editor of C language is created, i.e. first.c

2. Then first .c file is compiled with the help of ALT+F9 then the object file is automatically created, i.e. first.obj

3. The program is then linked with the help of CTRL+F9 and the direct executable file of that particular file is automatically created and program automatically executed.

DIGITAL COMPUTER

We know that information processing, plays a very important role in decision making. In this context, computing machine play a significant role in bulk of information processing. Here, we study what a computer is and what the organization of a computer is. The computer operates a program or set of statements. We will discuss the important contribution made by the John Von Neumann. The objective is to understand the definition of computer, working concepts of the computer, the stored program concept and microprocessor.

BASIC FUNCTIONAL UNITS OF A DIGITAL COMPUTER

A computer is an electronic device which accepts information and processes the information according to the program and produces the output. Computer programs may be written in a high level languages like Pascal, FORTRAN, COBOL and so on. Some programers also write programs in assembly language to carry out the desired task.

A computer system comprises hardware and software. A hardware refers to any physical, electrical, electromechanically component of the computer. For example, keyboard and mouse are considered hardware terms. A software denotes to a program or set of instructions that is written to achieve a specified task.

A computer system has five basic functional units as listed below.

(a) Input Unit

(b) Output Unit

(c) Control Unit ⎫ Central

(d) Memory Unit ⎬ Processing

(e) Arithmetic Logic Unit ⎭ Unit

Graphically a computer system may be represented as follows:

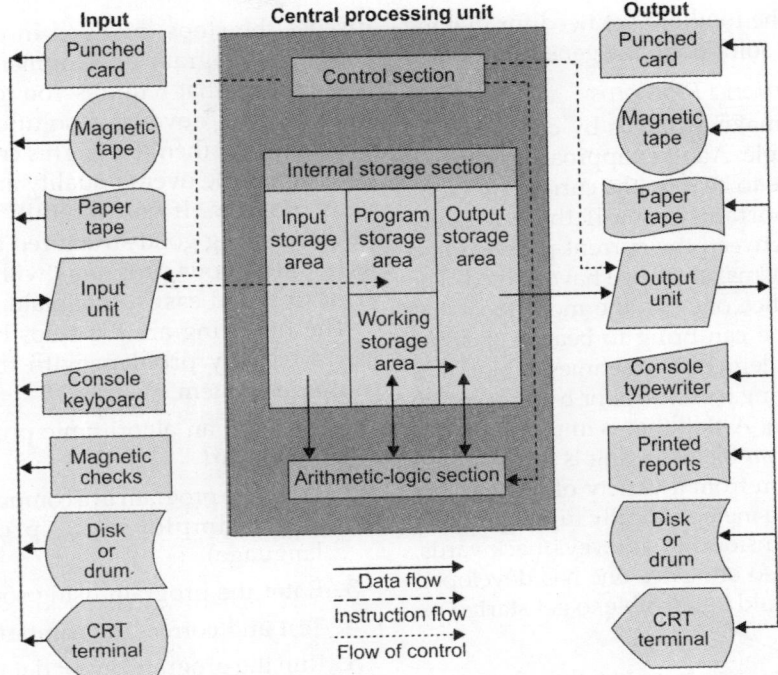

PROBLEM-SOLVING TECHNIQUES

Problem-solving is a creative process which defines systematization and mechanization. There are a number of steps that can be taken to raise the level of one's performance in problem solving.

Steps for Problem-Solving

A problem-solving technique usually comprises the following steps to find the solution.

Problem Definition Phase

The success in solving any problem lies in first fully understanding the problem. We cannot hope to solve a problem, which we do not understand. So, the problem understanding is the first step towards the solution of the problem. It is *what must be done* rather than *how is it to be done*. We have to define the set of task required to be performed to solve the problem.

Getting Started on a Problem

There are many ways of solving a problem and there may be several solutions. So, it is difficult to recognize immediately which path could be more productive. Sometimes you do not have any idea where to begin solving a problem, even if the problem has been defined. Such block sometimes occurs because you are overly concerned with the details of the implementation even before you have completely understood or worked out a solution. The best advice is not to get concerned with the details. Those can come later when the intricacies of the problem have been understood.

The Use of Specific Examples

To get started on a problem, we can make use of heuristics i.e., the guideline. This approach will allow us to start on the problem by picking a specific problem we wish to solve and try to work out the mechanism that will allow solving this particular problem. It is usually much easier to work out the details of a solution to a specific problem because the relationship between the mechanism and the problem is more clearly defined. This approach of focusing on a particular problem can give us the foothold we need for making a start on the solution to the general problem.

Similarities among Problems

One way to make a start is by considering a specific example. Another approach is to bring the experience to bear on the current problem. So, it is important to see if there are any similarities between the current problem and the past problems which we have solved. The more experience one has the more tools and techniques one can bring to bear in tackling the given problem. But sometimes, it blocks us from discovering a desirable or better solution to the problem. A skill that is important to try to develop in problem-solving is the ability to view a problem from a variety of angles. One must be able to metaphorically turn a problem upside down, inside out, sideways, backwards, forwards and so on. Once one has developed this skill it should be possible to get started on any problem.

Working Backwards from the Solution

In some cases we can assume that we already have the solution to the problem and then try to work backwards to the starting point. Even a guess at the solution to the problem may be enough to give us a foothold to start on the problem. We can systematize the investigations and avoid duplicate efforts by writing down the various steps taken and explorations made. Another practice that helps to develop the problem solving skills is, once we have solved a problem, to consciously reflect back on the way we went about discovering the solution.

Using Computer as a Problem-Solving Tool

The computer is a resource—a versatile tool—that can help you solve some of the problems that you encounter. A computer is a very commanding general-purpose tool. Computers can solve or help to solve many types of problems. There are also many ways in which a computing device can enhance the effectiveness of the time and effort that you are willing to devote to solving a problem. Thus, it will prove to be well worth the time and cause you spend to learn how to make effective use of this tool. In this section, we

discuss the steps involved in developing a program. Program development is a multi-step process that requires you to understand the problem, develop a solution, write the program, and then test it. This critical process determines the overall quality and success of your program. If you carefully design each program using good structured development techniques, your programs will be efficient, error free, and easy to maintain.

The following are the steps by which we can solve any problem with the help of a computer system.

1. Develop an algorithmic program and a flow chart
2. Write the program in a computer language (for example, say C programming language)
3. Enter the program using some editor
4. Test and correct the computer program
5. Run the program, input data, and get the outcomes.

DESIGN OF ALGORITHM

The first step in the program development is to devise and describe a precise plan of what you want the computer to do. This plan, showed as a sequence of operations, is called an algorithm. An algorithm is just an outline or idea behind a program something resembling C, but with some statements in english rather than the programming language. It is then expected that each pseudo code ( ) statement would be translated to a small number of lines of actual code, easily and mechanically.

Algorithm Definition

An **algorithm** is a finite set of steps defining the solution of a particular problem. An algorithm is expressed in pseudo code—something resembling C language, but with some statements in english rather than in the programming language. Developing an efficient algorithm needs a lot of practice and skill. It must be noted that an efficient algorithm is one which is capable of giving the solution to the problem by using minimum resources of the system such as memory and processor's time. Algorithm is a language independent, well structured and detailed. Algorithm enables the programmer to translate the problem into a computer program using any high-level language.

Criteria to be Followed by an Algorithm

Criteria need to be satisfied by an algorithm:

(a) **Input:** There should be zero or more values which are to be supplied.

(b) **Output:** At least one result is to be produced.

(c) **Definiteness:** Each step must be clear and unambiguous.

(d) **Finiteness:** If we trace the steps of an algorithm, then for all cases, the algorithm must terminate after a finite number of steps.

(e) **Effectiveness:** Each step must be sufficiently basic that a person using only paper and pencil can in principle carry it out. Each step should not only be definite, but also be feasible.

Features of Algorithm

The following features must be present in any algorithm:

Proper Understanding of a Problem

Proper understanding of a problem is must for designing an efficient algorithm. This is normally the outcome of the problem definition phase.

Use of Procedures/Functions to Emphasize Modularity

To assist the development, implementation and readability of the program, it is usually helpful to modularize (section) the program. Independent functions perform specific and well-defined tasks. In applying modularization, it is important to watch that the process is not taken so far to a point at which the implementation becomes difficult to read because of fragmentation. The program then can be implemented as calls to the various procedures that will be needed in the final implementations.

Choice of Variable Names

Proper variable names and constant names can make the program more meaningful and easier to understand. This practice tends to make the program more self documenting. A clear definition of all variables and constants at the start of the procedure/algorithm can also be helpful. For example, it is better to use variable *day* for the day of the weeks, instead of the variable *a* or something else.

Documentation of a Program

Brief information about the segment of the code can be included in the program to facilitate debugging and provide information. A related part of the documentation is the information that the programmer presents to the user during the execution of the program. Since the program is often used by persons who are unfamiliar with the working and input requirements of the program, proper documentation must be provided. That is, the program must specify what responses are required from the user. Care should also be taken to avoid ambiguities in these specifications. Also the program should "catch" incorrect responses to its requests and inform the user in an appropriate manner.

Writing Algorithms to Depict Logic

We have now understood what an algorithm is and what its characteristics are. So write some algorithms for the following problems:

Problem 1

Design an algorithm to input two numbers and then print the sum of both the numbers.

Algorithm

1. Start
2. Input a and b
3. Set c = a + b
4. Print c
5. End

Problem 2

Design an algorithm to input two numbers and then print the greater of the two numbers.

Algorithm

1. Start
2. Input a and b
3. If a > b then
4. Print a
5. Else
6. Print b
7. End

Problem 3

Design an algorithm to compute and print the average of a set of data values.

Algorithm

1. Start
2. Set the sum of the data values and the count to zero.
3. As long as the data values survive, add the next data value to the sum and add 1 to the count.
4. To compute the mean, divide the sum by the count.
5. Display the average.
6. Stop

Problem 4

Design an algorithm to calculate the factorial of a given number.

Algorithm

1. Start
2. Read the number n
3. [Initialize] $i \leftarrow 1$, fact $\leftarrow 1$
4. Repeat steps 4 through 6 until i = n
5. Fact $\leftarrow$ fact * i
6. $i \leftarrow i + 1$
7. Print fact
8. Stop

Problem 5

Design an algorithm to check whether the given number is prime or not.

Algorithm

1. Start
2. Read the number num
3. [Initialize] $i \leftarrow 2$, flag $\leftarrow 1$
4. Repeat steps 4 through 6 until i < num or flag = 0
5. Rem $\leftarrow$ num mod i
6. If rem = 0 then

   flag $\leftarrow 0$

   else

   $i \leftarrow i + 1$
7. If flag = 0 then

   Print Number is not prime

   Else

   Print Number is prime
8. Stop

ALGORITHM CORRECTNESS AND TERMINATION

We say that an algorithm is correct if for every input data that satisfies some condition, which is called the precondition of the algorithm, the output data satisfies a certain predefined condition, which is called the postcondition of the algorithm.

Example

DIV (A, B)
1. R: = A/B
2. return R

For this algorithm we have:

Precondition: A, B are real numbers and B<>0.

Postcondition: R = A/B.

The proof of correctness of an algorithm has two parts:

1. Partial correctness. If the algorithm terminates then it will give the right result (the result will satisfy the post condition).
2. Termination. Proof that the algorithm terminates.

Termination of an Algorithm

Termination of an algorithm means at what condition the algorithm terminates. This concept is very useful in the case of iteration and recursion.

FLOW CHART

The next step after the algorithm development is the flow charting. Flow charts are used in programming to diagram the path in which information is treated through a computer to obtain the desired results. Flow chart is a pictorial illustration of an algorithm. It makes use of symbols which are connected among them to indicate the flow of information and processing. It will show the general outline of how to solve a problem or perform a task. It is prepared for better understanding of the algorithm.

Basic Symbols Used in Flow Chart Design

The following basic symbols are used in the design of flow charts

	Start/End
	Question, decision (use in branching)

	Input/output
	Lines or arrows represent the direction of the flow of control
	Connector (connect one part of the flow chart to another)
	Process, instruction
	Comments, explanations, definitions

The following are the additional symbols followed for advanced programming.

	Preparation (may be used with "do Loops")
	Refers to a separate flowchart

Examples of Flow Chart

Problem 1

Draw a flow chart to input two numbers and print the sum of both the numbers.

Solution

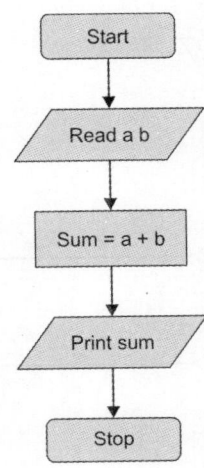

Problem 2

Draw a flow chart to input a number and print the factorial of that number

Solution

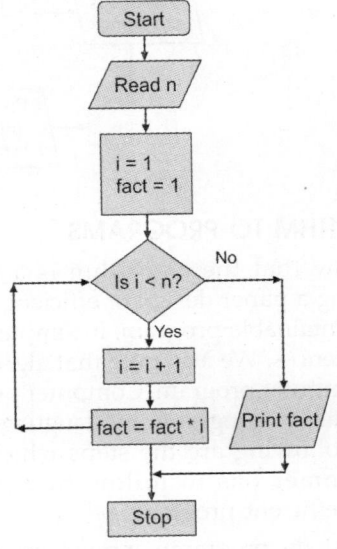

Problem 3

Draw a flow chart to input a number and check whether the given number is prime or not?

Solution

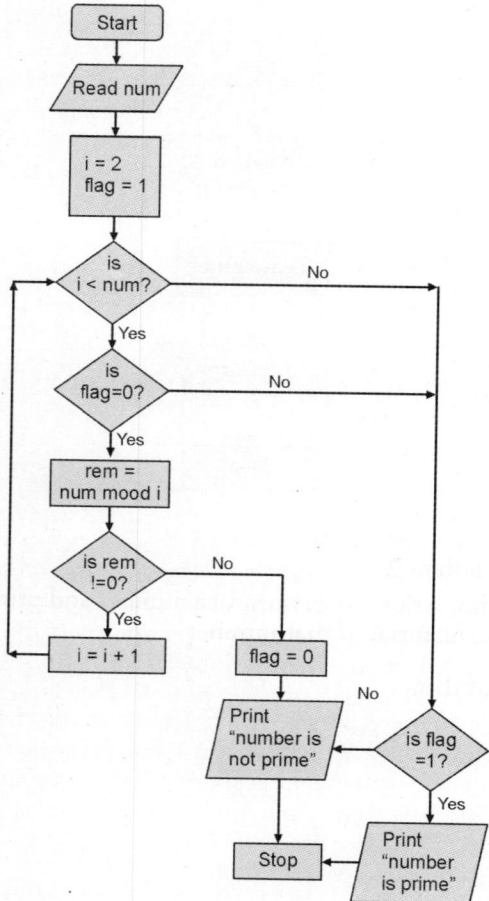

ALGORITHM TO PROGRAMS

We know that the algorithm is a tool for preparing a paper design of efficient, correct, and maintainable program, in simple english like sentences. We also note that algorithm is not a computer program. Computer programs are written in programming languages.

The following are the steps which every programmer has to follow to write and execute efficient programs.

1. First, the programmer must understands the problem

2. After understanding the problem the programmer write the algorithm for that problem

3. Finally, the programmer writes the code in any programming language based on that algorithm.

The above rules can be summarized as follows:

- Define and analyze the problem
- Formulate an algorithm
- Transform the algorithm into a program
- Compile, test, and debug the computer program
- Document and maintain the computer program.

Top-down Design or Stepwise Refinement

Once we have defined the problem and have an idea of how to solve it, we can then use the available techniques for designing algorithms. Most of the problems are complex or large and to solve them we have to focus on, a very limited span of logic or instructions at one time. A technique for algorithm design that tries to accommodate this human limitation is known as **top-down design or stepwise refinement.**

Top-down design provides the way of handling the logical complexity and detail encountered in computer algorithm. It allows building solutions to problems in systematically. In this way, specific and complex details of the implementation are encountered only at the stage when sufficient ground-work on the overall structure including the relationships among the various parts of the problem has been carried out.

Before the top-down design can be applied to any problem, there must be available at least the outlines of a solution. Sometimes this might demand a lengthy and creative investigation into the problem while at another time the problem description may in itself provide the necessary starting point for the top-down design.

Top-down design is all about taking the general statements about the solution one at a time, and then breaking them down into a more precise subtask/sub-problem. These sub-problems should more accurately describe how the final goal can be reached. The process of repeatedly breaking a task down into a subtask and then each subtask into smaller subtasks must continue until the sub-problem can be implemented as the program statement. With each splitting, it is essential to define how sub-problems interact with each other. In this way, the overall structure of the solution to the problem can be maintained. Preservation of the overall structure is important for making the algorithm comprehensible and also for making it possible to prove the correctness of the solution.

Graphically the topdown design is represented as follows:

Advantages of Top-down Design
The following are the advantages of the top-down approach.

- Separating the low-level work from the higher-level objects leads to a modular design
- Modular design means that the development can be self-contained
- Having the "skeleton" code illustrates clearly how the low level modules integrate
- Less operations errors (To reduce errors, because each module has to be processed separately, so programmers get more time for processing)
- Much less time consuming (each programmer is only involved in a part of the big project)

- Very optimized way of processing (each programmer has to apply his/her own knowledge and experience to his/her parts (modules), so that the project will become an optimized one)
- Easy to maintain (if an error occurs in the output, it is easy to identify the module from which the error might have been generated).

Disadvantages of Top-down Programming
The disadvantage of top-down programming is that the functionality needs to be inserted into low level objects, or otherwise functionality will be lacking until the development of low level objects is complete.

DATA ORGANIZATION

In pure mathematics, a value may take an arbitrary number of bits. Computers, on the other hand, broadly work with some specific number of bits. Common collections are single bits, groups of four bits (called nibbles), groups of eight bits called bytes, groups of 16 bits called words, groups of 32 bits called double words or dwords, groups of 64 bits called quad words or qword, and more. Thus, the sizes are not arbitrary. There is a good reason for these especial values. This section will describe the bit groups commonly used in computers.

Bits
The smallest "unit" of data on a binary computer is a single bit. Since a single bit is

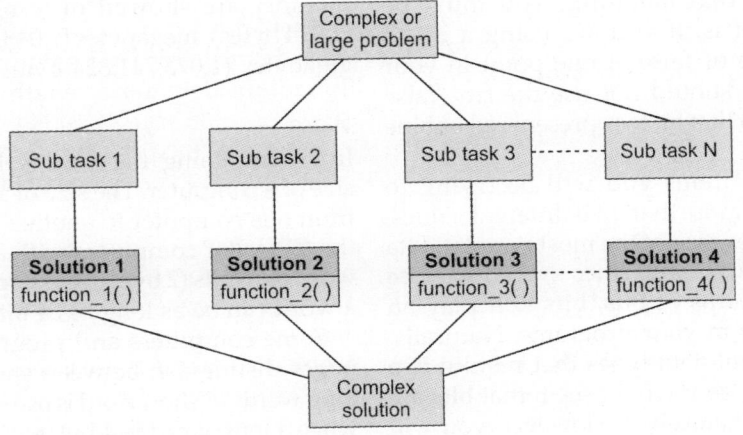

capable of constituting only two different values (typically zero or one) you may get the impression that there are a very small number of items you can represent with a single bit. Not true! There are an infinite number of items you can constitute with a single bit.

With a single bit, you can constitute any two distinct items. Examples include zero or one, true or false, on or off, male or female, and correct or wrong. However, you are not limited to constituting binary data types (that is, those objects which have only two distinct values). You could use a single bit to constitute the numbers 723 and 1,245. Or may be 6,254 and 5. You could also use a single bit to constitute the colors red and blue. You could even constitute two unrelated objects with a single bit. For example, you could constitute the color red and the number 3,256 with a single bit. You can constitute any two different values with a single bit. Still, you can represent only two different values with a single bit.

To confuse things even more, different bits can constitute different things. For example, one bit might be used to constitute the values zero and one, while an adjacent bit might be used to represent the values true and false. How can you tell by viewing the bits? The result, of course, is that you can't. But this exemplifies the whole idea behind computer data structures: data is what you define it to be. If you use a bit to constitute a Boolean (true/false) value then that bit (by your definition) represents true or false. For the bit to have any real meaning, you must be uniform. That is, if you are using a bit to represent true or false at one point in your program, you should not use the true/false value stored in that bit to represent red or blue later.

Since most items you will be trying to model require more than two different values, single bit values are not the most popular data type you will use. Still, since everything else consists of groups of bits, bits will play an important role in your programs. Naturally, there are several data types that require two distinct values, so it would seem that bits are important by themselves. However, you will

soon see that individual bits are difficult to manipulate, so we will often use other data types to represent boolean values.

Nibbles

A nibble is a collection of four bits. It would not be a particularly interesting data structure except for two items: BCD (binary coded decimal) numbers and hexadecimal numbers. It takes four bits to constitute a single BCD or hexadecimal digit. With a nibble, we can represent up to 16 distinct values since there are 16 unique combinations of a string of four bits:

0000	0100	1000	1100
0001	0101	1001	1101
0010	0110	1010	1110
0011	0111	1011	1111

In the case of hexadecimal numbers, the values 0, 1, 2, 3, 4, 5, 6, 7, 8, 9, A, B, C, D, E, and F are represented with four bits). BCD uses ten different digits (0, 1, 2, 3, 4, 5, 6, 7, 8, 9) and needs four bits (since you can only represent eight different values with three bits). In fact, any sixteen distinct values can be constituted with a nibble, but hexadecimal and BCD digits are the primary items we can represent with a single nibble.

Bytes

Byte is an abbreviation for binary term, a unit of storage capable of holding a single character. On almost all modern computing device, a byte is equal to 8 bits. Large amounts of memory are showed in terms of kilobytes (1,024 bytes), megabytes (1,048,576 bytes), and gigabytes (1,073,741,824 bytes).

Word

In programming, the word is the natural data size of a computer. The size of a word changes from one computer to another, depending on the CPU. For computers with a 16-bit CPU, a word is 16 bits (2 bytes). On large mainframes, a word can be as long as 64 bits (8 bytes).

Some computers and programming languages distinguish between short words and long words. A short word is usually 2 bytes long, while a long word is 4 bytes.

Double Word

A double word is exactly what its name implies, a pair of words. Therefore, a double word quantity is 32 bits long.

Kilo Byte

In decimal systems, kilo stands for 1,000, but in binary systems, a kilo is 1,024 (2^{10}). Technically, so, a kilobyte is 1,024 bytes, but it is often used loosely as a synonym for 1,000 bytes. For example, a computer that has 256 K main memory can store about 256,000 bytes (or characters) in memory at one time.

A megabyte is 2^{20} (approximately 1 million) and a gigabyte is 2^{30} power (approximately 1 billion).

In computer literature, kilobyte is usually shortened as K or Kb. To distinguish between a decimal K (1,000) and a binary K (1,024), the IEEE has proposed following the convention of using a small k for a decimal kilo and a capital K for a binary kilo, but this convention is by no means strictly followed.

Mega Byte

In decimal systems, the prefix mega means one million, but in binary systems, mega stands for 2^{20} or 1,048,576. One megabyte, therefore, is either 1,000,000 or 1,048,576 bytes (this is equivalent to 1,024 K), depending on the context.

Giga Byte

1024 megabyte contains one gigabyte.

Tera Byte

A terabyte is 2^{40} or 1,099,511,627,776 bytes. It can be guessed as 10 to the 12th power, or 1,000,000,000,000 bytes. A terabyte is 1,024 gigabytes and comes before the petabyte unit of measurement. While today's consumer hard drives are typically measured in gigabytes, Web servers and file servers may have several terabytes of space. A single 500 GB hard drive can also be predicted a half-terabyte drive.

INTRODUCTION TO PROGRAMMING

We will now discuss the various types of programming languages.

Programming Language

A vocabulary and set of grammatical rules for instructing a computer to perform specific tasks. The term *programming language* usually refers to high-level languages, such as BASIC, C, C++, COBOL, FORTRAN, Ada, and Pascal. Each language has a unique set of keywords (words that it understands) and a special syntax for organizing program instructions.

High-level programming languages, while simple compared to human languages, are more complex than the languages the computer actually understands, called *machine languages*. Each different type of CPU has its own unique machine language.

Lying between machine languages and high-level languages are languages called assembly languages. Assembly languages are similar to machine languages, but they are much easier to program in because they allow a programmer to substitute names for numbers. Machine languages consist of numbers only.

Lying above high-level languages are languages called *fourth generation languages* (usually abbreviated *4GL*). 4GLs are far removed from machine languages and represent the class of computer languages closest to human languages.

Regardless of what language you use, you eventually need to convert your program into machine language so that the computer can understand it. There are two ways to behave this:

1. Compile the program
2. Interpret the program

Generations of Programming Languages

There are five generation of programming languages.

First Generation Languages (1GLs): Machine Language

The first generation language is the lowest level computer language. Information is conveyed to the computer by the programmer as binary instructions. Binary instructions are the equivalent of the on/off signals used by computers to carry out operations. The language consists of zeros and ones. In the 1940s and 1950s, computers were programmed by scientists sitting before control panels equipped with toggle switches so that they could input instructions as strings of zeros and ones.

The lowest-level programming language (except for computers that utilize pro-grammable microcode) Machine languages are the only languages understood by computers. While easily understood by computers, machine languages are almost impossible for humans to use because they consist entirely of numbers. Programmers, therefore, use either a high-level programming language or an assembly language. An assembly language contains the same instructions as a machine language, but the instructions and variables have names instead of being just numbers. Programs written in high-level languages are translated into assembly language or machine language by a compiler. Assembly language programs are translated into machine language by a program called an assembler. Every CPU has its own unique machine language. Programs must be rewritten or recompiled, therefore, to run on different types of computers.

Graphically the generation of language may be represented as follows:

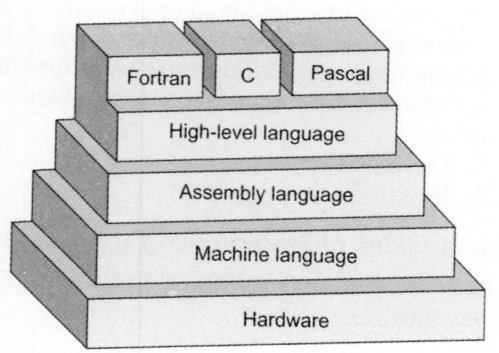

Second Generation Languages (2GLs): Assembly Language

The second generation languages are also known as assembly languages.

Assembly language is a low level pro-gramming language using the human readable instructions of the CPU. On PCs, the assembly language looks like this

```
mov ebx, eax
mov esi, 66
mov [edx+ebx*4+4], ecx
mov [ebx], ah
```

To compile this into machine code, you need an assembler. It is just a compiler for assembly language.

Many compilers can output the assembly language equivalent of code in C or C++.

The problem with assembly language is that it requires a high level of technical knowledge, and it is slow to write. In the same time that you take to write ten lines of assembly language—that is ten instructions, you could write ten lines of C, perhaps the equivalent of 500 instructions!

Advantages of assembly language
The advantages of assembly languages are

1. The symbolic programming of Assembly Language is easier to understand and saves a lot of time and effort of the pro-grammer.

2. It is easier to correct errors and modify program instructions.

3. Assembly language has the same efficiency of execution as the machine level language. Being one-to-one translator between assembly language program and its corresponding machine language program.

Disadvantages of assembly language
The disadvantages of assembly language are:

1. One of the major disadvantages is that assembly language is machine dependent. A program written for one computer might not run in other computers with different hardware configuration.

2. It is much more difficult to program and interpret the assembly language. Also assembly language is platform specific where as many of the new higher level languages are not because they have virtual machines built for different platforms

The machine language and assembly language together are known as low level languages.

Third Generation Languages (3GLs): High-level Language

A third generation language (3GL) is a programming language designed to be easier for a human to understand, including things like named variables, abstract data types, and algebraic expression syntax. Another crucial difference compared to second generation programming languages was abstraction away from the underlying processor. A fragment might be:

Let b = c + 2 * d

First introduced in the late 1950s, FORTRAN, ALGOL and COBOL are early examples of this sort of language. Most "modern" languages (BASIC, C, C++, Delphi, and Java) are also third-generation languages. Most 3GLs support structured programming.

A programming language such as C, FORTRAN, or Pascal that enables a programmer to write programs that is more or less independent of a particular type of computer. Such languages are considered high-level because they are closer to human languages and farther from machine languages. In contrast, assembly languages are considered low-level because they are very close to machine languages.

The main advantage of high-level languages over low-level languages is that they are easier to read, write, and maintain. Ultimately, programs written in a high-level language must be translated into machine language by a compiler or interpreter.

The first high-level programming languages were designed in the 1950s. Now there are dozens of different languages, including Ada, Algol, BASIC, COBOL, C, C++, FORTRAN, LISP, Pascal, and Prolog.

C is not a pure high-level language it is generally known as middle level language because in C language we can do the higher level programming as well as lower level programming.

Advantages of high-level language

The advantages of high level language are:

1. The main advantage of high-level languages over low-level languages is that they are easier to read, write, modify and maintain.
2. The documentation is simple.
3. The high-level languages resemble the English language, therefore, they are easy to learn and use.
4. The learning of these languages does not require any background of computer hardware; therefore these languages can be learnt by non-specialist as well.
5. The programs written in high-level languages are independent of the structure of the computer. The high-level languages are problem oriented rather than machine oriented.
6. The writing of program in high-level languages is easier and faster than assembly languages and easier to understand also. A high level language is ten times faster than assembly language.
7. High level programs are much easier to understand than assembly language ones.

Disadvantages of high-level language

The disadvantages of high level language are:

1. High-level language programs normally take up more space and execute more slowly than the equivalent assembly language programs.
2. The speed of operating the high-level language is slow therefore the computation time is more compared with machine language.
3. Low efficiency of memory utilization. A program in high-level language occupies a large memory.
4. One has to learn the special rules for writing programs in a particular high-level language.

Irrespective of all these disadvantages, the high-level languages are universally used because they are easy to learn and use.

Fourth Generation Languages (4GLs)

The fourth generation languages, or 4GL, are languages that consist of statements similar to statements in a human language. Fourth generation languages are commonly used in database programming and scripts. The term "application specific" language was coined by Jim Martin to refer to non-procedural high-level languages built around database systems. The first three generations were developed fairly quickly, but it was still frustrating, slow, and error prone to program computers, leading to the first "programming crisis", in which the amount of work that might be assigned to programmers greatly exceeded the amount of programmer time available to do it. Meanwhile, a lot of experience was gathered in certain areas, and it became clear that certain applications could be generalized by adding limited programming languages to them. Thus were born the report-generator languages, which were fed a description of the data format and the report to be generated and turned that into a COBOL (or other language) program which actually contained the commands to read and process the data and place the results on the page.

Some other successful fourth generation languages are: database query languages, for example SQL; Focus, Metafont, PostScript, RPG-II, S, IDL-PV/WAVE, Gauss, Mathematica and data-stream languages such as AVS, APE, and Iris Explorer.

Advantages of fourth generation language

The advantages of fourth generation language are:

1. They possess friendly interfaces.
2. They are easier to use than the previously used high level languages.
3. The programming language contained within a 4GL is closely linked to the english language structure.

Disadvantages of Fourth Generation Language

The disadvantages of fourth generation language are:

1. The programs run slower than those of the earlier language generations because their machine code equivalent is considerably longer and more complicated to execute
2. The recent popularity of 4GLs is closely linked to the development of fast microprocessors

Fifth Generation Languages (5GLs)

The fifth generation languages, or 5GL, are the programming languages that contain visual tools to help develop a program. A good example of a fifth generation language is Visual Basic. The Fifth generation languages are normally associated with the field of artificial intelligence. Artificial intelligence, built up through logic programming, models a real world environment or situation. Artificial intelligence aims to support flexible and informed patterns of behavior within a computer.

The main contenders within the fifth generation language category are as follows:

- Prolog 2
- Expert systems
- Knowledge based systems

Objective of Programming Language

Besides the main purpose of programming languages to provide instructions to a computer, there have been attempts to design one "universal" computer language that serves all purposes, although all of them have failed to be accepted in this role. The need for diverse computer languages arises from the diversity of contexts in which languages are used:

- Programs range from tiny scripts written by individual hobbyists to huge systems written by hundreds of programmers.
- Programmers range in expertise from novices who need simplicity above all else, to experts who may be comfortable with considerable complexity.
- Programs must balance speed, size, and simplicity on systems ranging from microcontrollers to supercomputers.
- Programs may be written once and not changed for generations, or they may undergo nearly constant modifications.

- Finally, programmers may simply differ in their tastes: they may be accustomed to discussing problems and expressing them in a particular language.

One common trend in the development of programming languages has been to add more ability to solve problems using a higher level of abstraction. The earliest programming languages were tied very closely to the underlying hardware of the computer. As new programming languages have developed, features have been added that let programmers express ideas that are more remote from simple translation into underlying hardware instructions. Because programmers are less tied to the complexity of the computer, their programs can do more computing with less effort from the programmer. This lets them write more functionality per time unit.

Usage of Programming Language

Programming languages differ from most other forms of human expression in that they require a greater degree of precision and completeness. When using a natural language to communicate with other people, people can be ambiguous and make small errors, and still expect their intent to be understood. However, figuratively speaking, computers "do exactly what they are told to do", and cannot "understand" what code the programmer intended to write. The combination of the language definition, a program, and the program's inputs must fully specify the external behavior that occurs when the program is executed, within the domain of control of that program.

Programs for a computer can be executed in a batch process without human interaction, or the programmer can use the programming language to write commands in an interactive session of an interpreter. "Commands" are simply single-line programs, whose execution may be chained together. When a language is used to give commands to a software application (such as a shell) it is called a scripting language.

Quality Requirements of Programming Language

Whatever the approach to software development may be, the final program must satisfy some important fundamental properties such as.

- **Efficiency:** The amount of system resources a program consumes (processor time, memory space, slow devices, network bandwidth and to some extent even user interaction), the more efficient it is.

- **Reliability:** How often are the results of a program are correct? This depends on prevention of error propagation resulting from data conversion and prevention of errors resulting from buffer overflows, underflows, and zero division.

- **Robustness:** How well does a program anticipates situations of data type conflict and other incompatibilities that result in run time errors and program halts? The focus is mainly on user interaction and the handling of exceptions.

- **Usability:** The clarity and intuitiveness of a programs output can make or break its success. This involves a wide range of textual and graphical elements that makes a program easy and comfortable to use.

- **Portability:** The range of hardware and OS platforms on which the source code of a program can be compiled and run. This depends mainly on the range of platform specific compilers for the language of the source code rather than anything having to do with the program directly.

USE OF HIGH LEVEL PROGRAMMING LANGUAGE FOR THE SYSTEMATIC DEVELOPMENT OF PROGRAMS

We know that a programmer generally use a high level programming languages because of its simplicity. Now in this section we will discuss the various high level programming languages and their uses.

LOGO (Logic Oriented Graphic Oriented)

This was meant to be a translator and mainly developed for children. It helps them learn basic mathematical and geometric skills easily.

FORTRAN (FORmula TRANslation)

One of the oldest programming languages, the FORTRAN was developed by a team of programmers at IBM led by John Backus, and was first published in 1957. The name FORTRAN is an acronym for FORmula TRANslation, because it was designed to allow easy translation of math formulas into codes.

Often referred to as a *scientific language*, FORTRAN was the first *high level* language, using the first compiler ever developed. Prior to the development of FORTRAN computer programmers were required to program in machine/assembly code, which was an extremely difficult and time-consuming task, not to mention the dreadful chore of debugging the code. The objective during its design was to create a programming language that would be: simple to learn, suitable for a wide variety of applications, *machine independent*, and would allow complex mathematical expressions to be stated similarly to regular algebraic notation. While still being almost as efficient in execution as assembly language. Since FORTRAN was so much easier to code, programmers were able to write programs 500% faster than before, while execution efficiency was only reduced by 20%, this allowed them to focus more on the problem solving aspects of a problem, and less on coding.

FORTRAN was so innovative not only because it was the first high-level language, but also because of its compiler, which is credited as giving rise to the branch of computer science now known as *compiler theory*. Several years after its release FORTRAN had developed many different *dialects* (due to special *tweaking* by programmers trying to make it better suit their personal needs) making it very difficult to transfer programs from one machine to another.

COBOL (Common Business Oriented Language)

COBOL is a high level programming language first developed by the CODASYL Committee (**Co**nference on **Da**ta **Sy**stems **L**anguages) in 1960. Since then the responsibility for developing new COBOL standards has been assumed by the American National Standards Institute (ANSI).

Three ANSI standards for COBOL have been produced: in 1968, 1974 and 1985. A new COBOL standard introducing object-oriented programming to COBOL is due within the next few years.

The word **COBOL** is an acronym that stands for **CO**mmon **B**usiness **O**riented **L**anguage. As the expanded acronym indicates, COBOL was designed for developing business, typically file-oriented, applications. It is not designed for writing systems programs. For instance you would not develop an operating system or a compiler using COBOL.

PASCAL

Pronounced *pass-kal*. A high-level programming language developed by Niklaus Wirth in the late 1960s. The language is named after Blaise Pascal, a seventeenth-century French mathematician who constructed one of the first mechanical adding machines.

Pascal is best known for its affinity to structured programming techniques. The nature of the language forces programmers to design programs methodically and carefully. For this reason, it is a popular teaching language not today.

Despite its success in academia, Pascal has had only modest success in the business world. Part of the resistance to Pascal by professional programmers stems from its inflexibility and lack of tools for developing large applications.

To address some of these criticisms, Wirth designed a new language called Modula-2. Modula-2 is similar to Pascal in many respects, but it contains additional features.

PL-1

PL/1 was developed as an IBM product in the mid 1960s, and was originally named NPL

(New Programming Language). The name was changed to PL/1 to avoid confusion of NPL with the National Physical Laboratory in England. If the compiler had been developed outside of the United Kingdom, the name may have remained PL/1. Until the time this new language was developed, all previous languages had focused on one particular area of application, such as science, artificial intelligence, or business. PL/1 was not designed to be used in the same way. It was the first large-scale attempt to design a language that could be used in a variety of application areas.

C

C is a general-purpose programming language with high-level and low-level capabilities. It is a statically typed, free-form, multi-paradigm, usually compiled language supporting procedural programming, data abstraction, and generic programming.

C is regarded as a *mid-level* language. This indicates that C comprises a combination of both high-level and low-level language features. Dr. Bjarne Stroustrup developed C in 1979 at Bell Labs as an enhancement to the C programming language and named it "C with Classes". In 1983, it was renamed to C. Enhancements started with the addition of classes, followed by, among other features, virtual functions, operator overloading, multiple inheritance, templates, and exception handling.

JAVA

JAVA is a high-level programming language developed by Sun Microsystems. Java was originally called *Oak*, and was designed for handheld devices and set-top boxes. Oak was unsuccessful so in 1995 Sun changed the name to Java and modified the language to take advantage of the burgeoning World Wide Web.

Java is an object-oriented language similar to C++, but simplified to eliminate language features that cause common programming errors. Java source code files (files with a *.java* extension) are compiled into a format called

byte code (files with a *.class* extension), which can then be executed by a Java interpreter. Compiled Java code can run on most computers because Java interpreters and runtime environments, known as *Java Virtual Machines (VMs)*, exist for most operating systems, including UNIX, the Macintosh OS, and Windows. Byte code can also be converted directly into machine language instructions by a just-in-time (JIT) compiler.

Java is a general purpose programming language with a number of features that make the language well suited for use on the World Wide Web. Small Java applications are called Java applets and can be downloaded from a Web server and run on your computer by a Java-compatible Web browser, such as Netscape Navigator or Microsoft Internet Explorer.

DESIGN AND IMPLEMENTATION OF CORRECT, EFFICIENT AND MAINTAINABLE PROGRAM

Writing a computer program (instructions) is not a big deal, if you are familiar with at least one computer language. But program acceptability is definitely a big issue. The program that we write should be acceptable to the users.

The following issues are closely related to the acceptability of computer programs.

Correctness of an Algorithm

In theoretical computer science, **correctness** of an algorithm is asserted when it is said that the algorithm is correct with respect to a specification. *Functional* correctness refers to the input-output behaviour of the algorithm (i.e. for each input it produces the correct output). A distinction is made between **total correctness**, which additionally requires that the algorithm terminates, and **partial correctness**, which simply requires that *if* an answer is returned it will be correct. Since there is no general solution to the halting problem, a total correctness assertion may lie much deeper.

For example, if we are successively searching through integers 1, 2, 3, ... to see if we can find an example of some phenomenon—say an odd perfect number—it is quite easy to write a partially correct program (use integer factorization to check n as perfect or not). But to say that this program is totally correct would be to assert something currently not known in number theory.

A proof would have to be a mathematical proof, assuming both the algorithm and specification are given formally. In particular it is not expected to be a correctness assertion for a given program implementing the algorithm on a given machine. That would involve such considerations as limitations on memory.

Efficiency of an Algorithm

In computer science, **efficiency** is used to describe the properties of an algorithm relating to how much of various types of resources it consumes. The two most frequently parameters encountered are:

1. Speed or running time—the time it takes for an algorithm to complete.
2. Space—the memory or 'non-volatile storage' used by the algorithm during its operation.

But efficiency also might apply to:

- transmission size or external memory such as required bandwidth or disk space.

The process of making code as efficient as possible is known as optimization and in the case of automatic optimization (i.e. compiler optimization)—performed by compilers (on request or by default)—usually focus on space at the cost of speed, or vice versa. There are also quite simple programming techniques and 'avoidance strategies' that can actually improve both at the same time, usually irrespective of hardware, software or language. Even the re-ordering of nested conditional statements to put the least frequently occurring condition first (example: test patients for blood type ='AB-', before testing age > 18, since this type of blood occurs in only about 1 in 100 of the population—thereby eliminating

the second test at run-time in 99% of instances), can reduce actual instruction path length, something an optimizing compiler would almost certainly not be aware of but which a programmer can research relatively easily even without specialist medical knowledge.

Analysis of Algorithm Efficiency

Every algorithm uses some of the computer's resources like central processing time and internal memory to complete its task. Because of high cost of computing resources, it is desirable to design algorithms that are economical in the use of CPU time and memory. Efficiency considerations for algorithms are tied in with the design, implementation and analysis of algorithm. Analysis of algorithms is less obviously necessary, but has several purposes:

- Analysis can be more reliable than experimentation. If we experiment, we only know the behavior of a program on certain specific test cases, while analysis can give us guarantees about the performance on all inputs.

- It helps one choose among different solutions to problems. As we will see, there can be many different solutions to the same problem. A careful analysis and comparison can help us decide which one would be the best for our purpose, without requiring that all be implemented and tested.

- We can predict the performance of a program before we take the time to write code. In a large project, if we waited until after all the code was written to discover that something runs very slowly, it could be a major disaster, but if we do the analysis first we have time to discover speed problems and work around them.

- By analyzing an algorithm, we gain a better understanding of where the fast and slow parts are, and what to work on or work around in order to speed it up. There is no simpler way of designing efficient algorithm, but a few suggestions as shown below can sometimes be useful in designing an efficient algorithm.

Analysis of Algorithm Complexity

Algorithms usually possess the following qualities and capabilities:

- Easily modifiable if necessary.
- They are correct for clearly defined solution.
- Require less computer time, storage and peripherals, i.e. they are more economical.
- They are documented well enough to be used by others who do not have a detailed knowledge of the inner working.
- The solution is pleasing and satisfying to its designer and user.
- They are able to be used as a sub-procedure for other problems.

Two or more algorithms can solve the same problem in different ways. So, quantitative measures are valuable in that they provide a way of comparing the performance of two or more algorithms that are intended to solve the same problem. This is an important step because the use of an algorithm that is more efficient in terms of time, resources required, can save time and money.

Computational Algorithm Complexity

We can characterize an algorithm's performance in terms of the size (usually n) of the problem being solved. More computing resources are needed to solve larger problems in the same class. Table 3.1 illustrates the comparative cost of solving the problem for a range of n values.

Table 3.1 shows that only very small problems can be solved with an algorithm that exhibit exponential behavior. An exponential problem with $n = 100$ would take immeasurably longer time. At the other extreme, for an algorithm with logarithmic dependency would merely take much less time (13 steps in case of $\log_2 n$ in Table 3.1). These examples

Table 3.1

$Log_2 n$	n	$nlog_2 n$	n^2	n^3	2^n
1	2	2	4	8	4
3.322	10	33.22	10^2	10^3	$>10^3$
6.644	10^2	664.4	10^4	10^6	$>>10^{25}$
9.966	10^3	9966.0	10^6	10^9	$>>10^{250}$
13.287	10^4	132877	10^8	10^{12}	$>>10^{2500}$

emphasize the importance of the way in which algorithms behave as a function of the problem size. Analysis of an algorithm also provides the theoretical model of the inherent computational complexity of a particular problem.

To decide how to characterize the behavior of an algorithm as a function of size of the problem n, we must study the mechanism very carefully to decide just what constitutes the dominant mechanism. It may be the number of times a particular expression is evaluated, or the number of comparisons or exchanges that must be made as n grows. For example, comparisons, exchanges, and moves count most in sorting algorithm. The number of comparisons usually dominates so we use comparisons in the computational model for sorting algorithms.

The Order of Notation

The O-notation gives an upper bound to a function within a constant factor. For a given function $g(n)$, we denote by $O(g(n))$ the set of functions. $O(g(n)) = \{f(n) :$ there exist positive constants c and $n0$, such that $0 <= f(n) <= cg(n)$ for all $n >= n0 \}$

Using the O-notation, we can often describe the running time of an algorithm merely by inspecting the algorithm's overall structure. For example, a double nested loop structure of the following algorithm immediately yields $O(n^2)$ upper bound on the worst case running time.

```
for i=0 to n
    for j=0 to n
        print i,j
        next j
    next i
```

What we mean by saying "the running time is $O(n^2)$" is that the worst case running time (which is a function of n) is $O(n^2)$. Or equivalently, no matter what the particular input of size n is chosen for each value of n, the running time on that set of inputs is $O(n^2)$.

Rules for Using the Big-O Notation

Big-O bounds, because they ignore constants, usually allow for very simple expression for the running time bounds. Below are some properties of big-O that allow bounds to be simplified. The most important property is that big-O gives an upper bound only. If an algorithm is O (N^2), it does not have to take N^2 steps (or a constant multiple of N^2). But it cannot take more than N^2. So any algorithm that is O (N) is also an O (N^2) algorithm. If this seems confusing, think of big-O as being like "<". Any number that is < N is also <N^2.

1. Ignoring constant factors: O (c f (N)) = O (f (N)), where c is a constant; for example O (20 N^3) =O (N^3)

2. Ignoring smaller terms: If $a<b$ then O ($a+b$) = O (b), for example, O (N^2+N) = O (N^2)

3. Upper bound only: If $a<b$ then an O (a) algorithm is also an O (b) algorithm.

   For example, an O (N) algorithm is also an O (N^2) algorithm (but not vice versa).

4. N and log N are bigger than any constant, from an asymptotic view (that means for large enough N). So if k is a constant, an O ($N + k$) algorithm is also O (N), by ignoring smaller terms. Similarly, an O (log $N + k$) algorithm is also O (log N).

5. Another consequence of the last item is that an O (N log $N + N$) algorithm, which is O (N (log $N + 1$)), can be simplified to O (N log N).

Worst and Average Case Behavior of an Algorithm

Worst and average case behaviors of the algorithm are the two measures of performance that are usually considered. These two measures can be applied to both space and time complexity of an algorithm. The worst case complexity for a given problem of size n corresponds to the maximum complexity encountered among all problems of size n. For determination of the worst case complexity of an algorithm, we choose a set of input conditions that force the algorithm to make the least possible progress at each step towards its final goal. In many practical applications it is very important to have a measure of the expected complexity of an algorithm rather than the worst case behavior. The expected complexity gives a measure of the behavior of the algorithm averaged over all possible problems of size n.

As a simple example: Suppose we wish to characterize the behavior of an algorithm that linearly searches an ordered list of elements for some value x.

1 2 3 4 5 … … … …. N

In the worst case, the algorithm examines all n values in the list before terminating.

In the average case, the probability that x will be found at position 1 is $1/n$, at position 2 is $2/n$ and so on.

Therefore, average search cost = $1/n$ (1 + 2+ 3 + ……+ n) = $1/n$ (n/2 (n + 1)) = (n + 1) /2

STRUCTURED PROGRAMMING

Structured programming is one of the several different ways in which a programming language can be constructed. It was originally introduced as a means of getting away from the 'spaghetti' code that was used in the early days and to provide some means by which programmers could more easily follow the code written by other programmers. Structured programming is a procedure oriented method of designing and coding a program.

A structured programming language does not allow just things to happen in any order within the code. There are a limited number of constructs that can be used within the code to define the execution flow. These additional constructs are to be avoided if a true structured program is to be produced.

A structured program may be written out using pseudocode prior to being translated into whatever programming language that the program is to be written in. This pseudocode forms part of the program specification and is readable by anyone who understands structured programming regardless of whether or not they know the specific language in which the program has been written.

Constructs in Structured Programming

Structured programming provides a number of constructs that are used to define the sequence in which the program statements are to be executed.

Consecutive

Statements within a structured program are normally executed in the same sequence as they are listed within source code. If a code fragment consists of three statements following one another then statement one will execute first, statement two second, and statement three last. To change from this straight consecutive execution sequence requires the use of one of the other structured programming constructs which are described below.

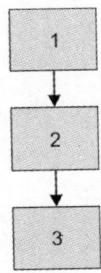

The pseudocode of the above construct is as follows:

Statement-1
statement-2
statement-3

Example

Input a
b = 5 + 2 * a
print b

Block

Statements may be blocked together. A block of statements may be substituted wherever a single statement is allowed. The symbol or keyword used to indicate the start and end of each block differs depending on the programming language used.

The pseudocode of the above construct is as follows:

{
Statement-1
statement-2
statement-3
}

Subroutine

A subroutine is a code segment that has been separated from the preceding and following code. A subroutine usually consists of a series of statements that perform a particular task. The task performed is usually identified by the name given to the subroutine. Once a subroutine has been defined it can then be called from one or more places within the program. This allows a program to perform the same task a number of times without having to repeat the same code. A single call statement replaces (stands in for) all of the statements contained within the subroutine. Parameters can be passed to a subroutine which will supply the data required to perform the task and perhaps to return values for use by the subsequent processing. A subroutine can either be compiled with (internal to) the calling program or separately (external). Graphically it is represented as follows:

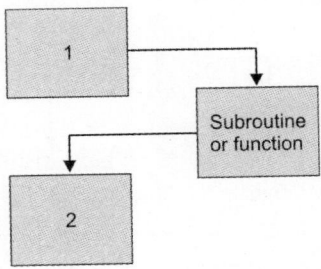

The pseudocode of the above construct is as follows:

Call subroutine-1 (var1, var2)
subroutine-1 (var1, var2)
{
statement-1
statement-2
}

Example

call output (a, b)
output (a, b)
{
print a
print b
}

Branching in Structured Programming

There are two types of branching statements. The if statement is used for conditional execution of a single statement or to select which of two statements is to be executed. The case statement (sometimes referred to as the select statement) allows for selection of one statement out of three or more that should be executed.

If Statement

Statements can be executed conditionally by using an if statement. The if statement specifies a condition which gets tested when the if statement is executed. If the condition is true then the following statement is executed otherwise processing skips that statement.

Optionally a second statement can be attached to the if statement that will be executed if the condition is false.

Graphically an if statement is represented as follows:

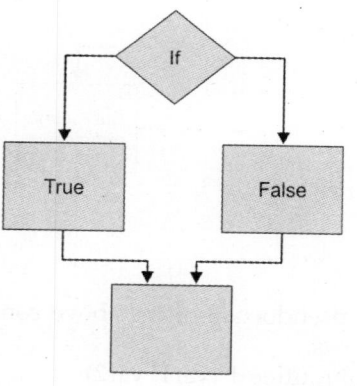

The pseudocode of the above construct is as follows:

if condition
true-statement
else
false-statement

Example

if x > y
print "x is bigger than y"
else
print "x is not bigger than y"

Case Statement

A case or select statement allows for one of a number of statements to be executed depending on the value of a field. There is usually also an additional statement which is executed if none of the specified values is matched. The graphical representation of case statement is as follows:

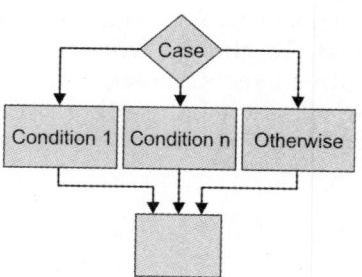

The pseudocode of the above construct is as follows:

case fieldname
  value1: statement-1
  value2: statement-2
  value3: statement-3
  otherwise: other-statement

Example

case size
  1: print "small"
  2: print "medium"
  3: print "large"
  otherwise: print "unknown"

Loops in Structured Programming

Loops allow for the same statement to be executed a number of times in succession. There are three different loop constructs that can be used depending on whether the number of repetitions is known and also (where the number of repetitions is not known and is dependent on a condition) whether the loop is allowed to be bypassed if the termination condition is met before the loop is first executed.

For Loop

A for loop allows a statement to be executed a specified number of times. Graphically a for loop statement is represented as follows:

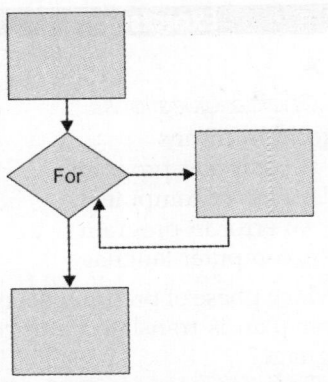

The for loop begins with a loop control variable assigned a specific initial value. This control variable in then incremented (or decremented) by a specified amount each time around the loop until a specified terminating value is reached at which time the statement following the loop is then executed.

The pseudocode of the above construct is as follows:

for (initial-value, final-value, increment)
statement-1

Example

for (a = 3, a > 12, a = a + 2)
 print a
The above code segment will print 3 5 7 9 11.

While Loop

A while loop allows a statement to be executed until a given condition is met. If the condition is met prior to executing the loop then the loop will not be executed. As soon as the condition is met, execution continues with the statement following the loop. Graphically it is represented as follows:

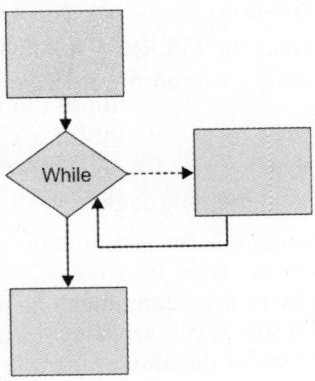

The pseudocode of the above construct is as follows:
while not condition
 statement-1

Example

while not end-of-file
{
 read record
 write record
}

Do Until Loop

An until loop also allows a statement to be executed until a given condition is met but the condition will not be tested until after the loop has been executed once. Once the condition is met the statement following the loop will be executed. Graphically it is represented as follows:

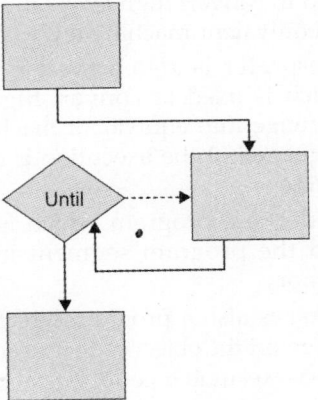

The pseudocode of the above construct is as follows:
do
statement-1
until condition

Advantages of Structured Programming

The following are the advantages of structured programming:

- The foremost advantage is reduced complexity.
- Modularity allows the programmer to tackle problems in a logical fashion.
- Also, the use of logical structures ensures that the flow of control is clear.

- Another advantage is an increase in productiveness.
- Modularity allows multiple programmers to work on a project at the same time.
- Modules can be re-used many times, which saves time and reduces complexity, and increases reliability.
- It enables easier updates or fixing the program by replacing individual modules rather than larger amounts of code.

SUMMARY

- Computer programming (often shortened to programming or coding) is the process of writing, testing, debugging/troubleshooting, and maintaining the source code of computer programs.
- Compiler is a system software which is used to convert high level language into an equivalent machine level language.
- Interpreter is also a system software which is used to convert higher level language into equivalent machine level language but the execution is done line by line
- Loader is a program which is used to load the program segment into main memory
- Linker is also a program which is used to convert the object code into equivalent direct executable code.
- Programming languages can be classified into first-generation language, second generation language, third-generation language, fourth generation language and fifth-generation language.
- The first-generation language is also known as machine language.
- The second-generation language is also known as assembly language
- The third-generation language is also known as higher level language.
- An algorithm is a set of instructions to solve any problem step by step.
- A flow chart is the pictorial representation to solve any problem.

Exercises

Section A

Multiple Choice Questions

1. Algorithm means
   (a) A computer program
   (b) A plan of computer
   (c) An error in program
   (d) A computer language

2. In which phase of program development paper plan is translated into computer language?
   (a) Coding
   (b) Compilation
   (c) Debugging
   (d) Algorithm development

3. Which of the following is the first step of program development life cycle?
   (a) Analysis   (b) Designing
   (c) Coding   (d) Maintenance

4. In which of the following solution of a problem is depicted using graphical symbol?
   (a) Flow chart
   (b) Algorithm
   (c) Top-down development
   (d) None of the above

5. Which of the following should not be there in an algorithm?
   (a) Definite start point
   (b) English sentences
   (c) Definite end point
   (d) None of the above

6. Which symbol is not used in flow chart?
   (a) Ellipse   (b) Rectangle
   (c) Polygon   (d) Arrow lines

7. Which of the following is an example of low level programming language?
   (a) C
   (b) C++
   (c) Java
   (d) Machine language

8. C language is the example of
   (a) Higher level language
   (b) Lower level language
   (c) Middle level language
   (d) None of the above

9. Which of the following software converts high level language in machine language?
   (a) Operating system
   (b) Debugger
   (c) Compiler
   (d) Text editor

10. Using which programming techniques, complex program is divided into smaller problems?
    (a) Structured programming
    (b) Modular programming
    (c) Top-down programming
    (d) None of the above

11. Which of the following is not a good computer programing practice?
    (a) Using line breaks
    (b) Using indentation
    (c) Writing comments in programs
    (d) Using alphabets for variable names

12. Java is the example of
    (a) First generation language
    (b) Second generation language
    (c) Fourth generation language
    (d) Fifth generation language

13. Which of the following is strictly restricted in structured programming?
    (a) Go to statement
    (b) Looping
    (c) Subroutine
    (d) Functions

14. Which of the following does not relate to reusability of code?
    (a) Subroutine   (b) Loop
    (c) Decision   (d) Function

15. Which of the following does not makes use of condition in its execution?
    (a) Function
    (b) Loop
    (c) Case
    (d) Decision

Fill in the Blanks

1. _____ is an example of middle level programming language.
2. _____ is an example of lower level programming language.
3. BASIC is an example of _____.

4. _____ converts high level language program into machine language program, step by step and executes it also.
5. _____ is always written in english like sentences for depicting the solution of a given problem.
6. In flow chart _____ symbol is used for start/end.
7. In flow chart _____ symbol is used for processing.
8. In flowchart _____ symbol is used for decision making.
9. _____ is used to show looping in the flowchart.
10. Efficiency of an algorithm depends upon _____.

Section B

Short Answer Type Questions

1. Define the concept of algorithm.
2. Differentiate between algorithm and flow chart with suitable example.
3. Write all steps of program development life cycle.
4. Name any four constructs suggested by structured programming with suitable example.
5. Differentiate between compiler and Interpreter with suitable example.
6. Differentiate between first generation language and second generation language with suitable example.
7. Define the term structured programming with suitable example.
8. How we measure the efficiency of the algorithm?
9. Write a algorithm to input 2 numbers and print the maximum number between them.
10. Draw a flow chart to input three numbers and print the maximum number between them.

Section C

Long Answer Type Questions

1. Define the term program development life cycle and also define all the stages of

program development life cycle with suitable example.

2. Define the term compiler, interpreter, loader and Linker with suitable example.

3. Define the term programming language and also provide the classification of programming language with suitable example.

4. What do you mean by flow chart and define all the symbols used in flow chart and give one example?

5. Define the term structured programming with suitable example. Also define the various constructs used in structured programming.

6. Differentiate between higher level programming language and lower level programming language with suitable example.

7. Describe top-down and bottom-up approaches of system development.

8. What do you understand by efficiency of algorithm? How can it be evaluated?

9. Draw a flow chart to input 10 numbers and print the maximum number between them.

10. Write the algorithm to input 10 numbers and print the maximum number between them.

11. Write an algorithm and draw a flow chart to input 10 numbers and print the sum of that numbers.

12. Write an algorithm and draw a flow chart to input a number and check whether that number is prime or not.

13. Write an algorithm and draw a flow chart to input a number and check whether that number is palindrome or not.

14. Write an algorithm and draw a flowchart to input a number and check whether that number is Armstrong or not.

15. Write an algorithm and draw a flowchart to input a number and print the reverse of that number.

4 Number System

Number system is used to represent information in quantitative form. Some of the common number systems are binary, octal, decimal and hexadecimal. A number system of base (also called radix) r is a system, which has r distinct symbols for r digits. A string of these symbolic digits represents a number. To determine the value that a number represents, we multiply the number by its place value that is an integer power of r depending on the place it is located and then find the sum of weighted digits.

The number system we are all companion with is the **decimal numeration system**, or **base 10** number system. Computers, on the other hand, use the **binary numeration system**, which is **base 2**. Other number systems, notably **hex** (base 16) and, to a lesser degree, **octal** (base 8) are also used in the study and computer programming of computers, but only because they provide a more compact notation for representing binary values. Internally the computing device *only knows the binary system*.

CLASSIFICATION OF NUMBER SYSTEM

The following are the classification of number system:

Decimal Number System
It is a well-known number system with base 10. The numbers include in the decimal number system are 0, 1, 2, 3, 4, 5, 6, 7, 8, 9.

Binary Number System
It is a well-known number system with base 2. The numbers include in the binary number system are 0,1.

Octal Number System
It is a well-known number system with base 8. The numbers include in the octal number system are 0, 1, 2, 3, 4, 5, 6, 7.

Hexadecimal Number System
It is a well-known number system with base 16. The numbers include in the hexadecimal number system are 0, 1, 2, 3, 4, 5, 6, 7, A, B, C, D, E, F.

DECIMAL NUMBER SYSTEM

Decimal number system is the most commonly used number system, to the base ten. Decimal numbers do not necessarily contain a decimal point; 563, 5.63, and "563 are all decimal numbers. Other systems are primarily used in computing and admit the binary number system, octal number system, and hexadecimal number system.

The decimals 0.3, 0.51, and 0.023 can be expressed as the decimal fractions 3/10, 51/100, and 23/1,000. They are all finishing decimals. These fractions can equally be expressed as the percentages 30%, 51%, and 2.3%.

Decimal to Binary Conversion

Divide the number repeatedly by 2 (answer being a whole number and a remainder, The same as Modula-2 Div and Mod commands). The First remainder is the LSB (Least Significant Bit or Digit, the digit on the right of the number) of the Binary number. The next remainder is the next digit, and so on until the value is 0.

We write the remainders from bottom to top.

Example 1

Convert $(145)_{10}$ into base 2.

- $145 \div 2 =$
  - $- 72$
  - $-$ Rem 1
- $72 \div 2 =$
  - $- 36$
  - $-$ Rem 0
- $36 \div 2 =$
  - $- 18$
  - $-$ Rem 0
- $18 \div 2 =$
  - $- 9$
  - $-$ Rem 0
- $9 \div 2 =$
  - $- 4$
  - $-$ Rem 1
- $4 \div 2 =$
  - $- 2$
  - $-$ Rem 0
- $2 \div 2 =$
  - $- 1$
  - $-$ Rem 0
- $1 \div 2 =$
  - $- 0$
  - $-$ Rem 1

So 145 =10010001

Example 2

Convert 147_{10} to binary

$$147 \div 2 = 73 \dots 1$$
$$73 \div 2 = 36 \dots 1$$
$$36 \div 2 = 18 \dots 0$$
$$18 \div 2 = 9 \dots 0$$
$$9 \div 2 = 4 \dots 1$$
$$4 \div 2 = 2 \dots 0$$
$$2 \div 2 = 1 \dots 0$$
$$1 \div 2 = 0 \dots 1$$

Reading the remainders from bottom to top, we have: $147_{10} = 10010011_2$.

Converting Decimal Fractions to Binary

A beneficial way to coordinate the problem is to use a table, with columns **old**, **bit**, and **new**. We begin with the number we are planning to convert in the top row **old** entry.

Now, we bring across each row, and multiply the **old** entry by 2 to get a new value. Now we accept the integer part of the new value and put it in the **bit** entry, and put the fraction part in the **new** entry. If the **new** entry is non-zero, we simulate it into the **old** entry in the next row down and continue (if it is 0, we are done).

The result is the contents of the **bit** column, with the MSB in the top row.

Example

Convert $(0.375)_{10}$ to $(?)_2$

Old	Bit	New	Notes
.375	0	.75	Multiply .375 by 2, get 0.75
.75	1	.5	Multiply .75 by 2, get 1.5
.5	1	0	Multiply .5 by 2, get 1.0

So the result is .011.

Decimal to Octal Conversion

To convert decimal to octal is somewhat harder. The typical method to change over from decimal to octal is repeated division by 8. While we may also use reoccurred subtraction by the weighted position value, it is more difficult for large decimal numbers.

Repeated Division by 8 Method

For this method, divide the decimal number by 8, and write the remainder on the side as the least significant digit. This process is proceeded by dividing the quotient by 8 and writing the remainder until the quotient is 0. When doing the division, the remainders which will constitute the octal equivalent of the decimal number are written beginning at the least significant digit (right) and each new digit is written to the next more significant digit (the left) of the previous digit.

Example 1

Convert $(44978)_{10}$ to octal.

Solution

Division	Quotient	Remainder	Octal Number
44978 / 8	5622	2	2
5622 / 8	702	6	62
702 / 8	87	6	662
87 / 8	10	7	7662
10 / 8	1	2	27662
1 / 8	0	1	127662

Example 2

Convert $(25)_{10}$ to octal number system.

Solution

The division problem will be setup as $2510 \div 8$

$2510 \div 8 = 3$

$2510 - 24 =$ Remainder 1 -----> 1 (LSD)

$3 \div 8 = 0$

$3 - 0 =$ Remainder 3 -----> 3 (MSD)

Now, write the number from MSD to LSD as $(31)_8$.

Example 3

Convert $(258)_{10}$ to octal.

The division problem will be setup as $25810 \div 8$

$25810 \div 8 = 32$

$18 - 16 =$ Remainder 2 -----> 2 (LSD)

$32 \div 8 = 4$

$32 - 32 =$ Remainder 0 -----> 0

$4 \div 8 = 0$

$4 - 0 =$ Remainder 4 -----> 4 (MSD)

Now, write the number from MSD to LSD. Now the Result is $(402)_8$

Converting Decimal Fractions to Octal

To convert a decimal fraction to octal, you must multiply the fraction number by 8. Then, you need to draw out the number (s) that appear to the left of the decimal point. The first number that is distilled will be the MSD and will follow the decimal point. The last number that is drew out will be the LSD. This process is continue to desired accuracy. This type of problem may result in results comprising a mixed decimal number, which demands the whole number and the fraction to get split in order to achieve its equivalent octal conversion.

Example

Convert $(0.175)_{10}$ to octal.

The multiplication problem will be represented as follows:

.017510

$\times 8$

0.1400 -----> 0 (MSD)

.1400

$\times 8$

1.1200 -----> 1

.1200

$\times 8$

0.9600 -----> 0

.9600

$\times 8$

7.6800 -----> 7

.6800

$\times 8$

5.4400 -----> 5

.4400

$\times 8$

3.5200 -----> 3

.5200

$\times 8$

4.1600 -----> 4

.1600

$\times 8$

1.2800 -----> 1 (LSD)

Writing from MSD to LSD, you should end with $(.01075341)_8$.

Decimal to Hexadecimal Conversion

The following are the steps which we have to follow for converting a decimal number into equivalent hexadecimal number.

1. Divide the given number by 16. Handle the division as an integer division.
2. Now write down the remainder (in hexadecimal).
3. Divide the result once again by 16. Handle the division as an integer division.
4. Repeat step 2 and 3 until the result is 0.
5. The hex value is the digit sequence of the remainders from the top to bottom.

Converting Decimal Fractions to Hexadecimal

To convert a decimal fraction to hexadecimal, you must multiply the fraction by 16. Then, you need to draw out the number(s) that come along to the left of the decimal point. The first number that is distilled will be the MSD and will follow the decimal point. The last number that is distilled will be the LSD. Continue to desired accuracy. This type of problem may answer in results containing a mixed decimal number, which requires the whole number and the fraction to become split in order to achieve its equivalent octal conversion.

Example 1

Convert the number **1128** decimal to hexadecimal.

Notes	Division	Result	Remainder (in Hex)
Start by dividing the number by 16.			
In this case, 1128 divided by 16 is 70.5. So the integer division answer is 70 (throw out anything after the decimal point). The remainder is (70.5 – 70) multiplied with 16; or (0.5 times 16), which is 8.	1128 / 16	70	8
Then, divide the result again by 16 (the number 70 on the DIVISION column comes from the previous RESULT). In this case, 70/16 = 4.375. So the integer division result is 4 (throw out anything after the decimal point) The remainder is (0.375 multiplied with 16, which is 6.	70 / 16	4	6
Repeat. Note here that 4/16=0.25. So the integer partition result is 0. The remainder is (0.25–0) multiplied with 16, which is 4.	4 / 16	0	4
Stop because the result is already 0 (0 divided by 16 will always be 0).			
Well, here is the answer. These numbers come from the REMAINDER column values (read from bottom to top.			468

Example 2
Convert the number **256** decimal to hexadecimal

Division	Result	Remainder (in Hex)
256 / 16	16	0
16 / 16	1	0
1 / 16	0	1
Answer		100

Example 3
Convert the number **921** decimal to hexadecimal

Division	Result	Remainder (in Hex)
921 / 16	57	9
57 / 16	3	9
3 / 16	0	3
Answer		399

Example 4
Convert the number **188** decimal to hexadecimal

Division	Result	Remainder (in Hex)
188 / 16	11	C (12 decimal)
11 / 16	0	B (11 decimal)
Answer		BC

Note that here; the answer would not be 1112, but BC. Remember that you have to write the remainder in hexadecimal not in decimal.

Example 5
Convert the number **590** decimal to hexadecimal

Division	Result	Remainder (Hex)
590 / 16	36	E (14 decimal)
36 / 16	2	4 (4 decimal)
2 / 16	0	2 (2 decimal)
Answer		24E

Example

Convert 0.695_{10} to hexadecimal number system.

Solution

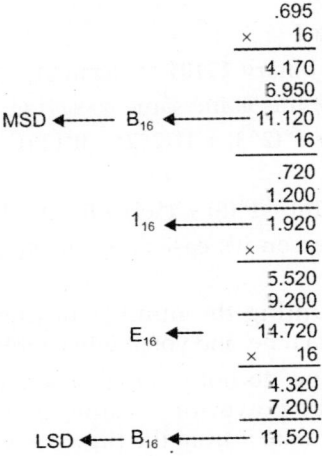

```
                                    .695
                               ×     16
                               ─────────
                                  4.170
                                  6.950
MSD ◄──── B₁₆ ◄──────────── 11.120
                                     16
                               ─────────
                                   .720
                                  1.200
       1₁₆ ◄──────────── 1.920
                               ×     16
                               ─────────
                                  5.520
                                  9.200
       E₁₆ ◄──────── 14.720
                               ×     16
                               ─────────
                                  4.320
                                  7.200
LSD ◄── B₁₆ ◄──────── 11.520
```

So the solution is $B1EB_{16}$

BINARY NUMBER SYSTEM

The basic parts of a computing machine are the central processing unit (CPU), memory, a keyboard or other input device and a screen or other output device. But how does the computer know how to add and subtract, and how can its memory remember the answers it computes? We recognize that the computing machine does not have a real brain inside. It reality, it is made up generally of plastic, metal and silicon. Yet, a computing device acts in many ways as though it does have a real brain.

To find the response, we must take a close look at how we realize numbers. We have got ten digits in our number system: 0, 1, 2, 3, 4, 5, 6, 7, 8, 9. It is interesting that digit also means a finger or toe. A number system grounded on ten is called a decimal system.

Computers do not use the ten digits of the decimal system for calculating and arithmetic. Their CPU and computer memory are made up of millions of tiny switches that can be either ON or OFF. Two digits, 0 and 1, can be used to represent the two states of ON and

OFF. So we can see that computing devices could work with a number system based on two digits. This type of system is called a binary numerating system. How does this type of numeration system work? First, let us look more closely at our own decimal system. As you have acquired in your arithmetic classes, our decimal system is based on place, or position. That is, the place of each digit assures you the value of that digit. For example, the number 17 has a 7 in the one's place and a 1 in the ten's place. The number 138 has a 1 in the hundred's place, a 3 in the ten's place, and an 8 in the one's place. Written in numerals this is $(1×100) + (3×10) + (8×1) = 138$.

The binary system forms in exactly the same way, except that its place value is based on the number two. In the binary system, we have the one's place, the two's place, the four's place, the eight's place, the sixteen's place, and so on. Each place in the number represents two times ($2×$'s) the place to its right.

Here's a comparison of decimal and binary numbers:

Decimal	Binary	Decimal	Binary
0	0	6	110
2	10	7	111
3	11	8	1000
4	100	9	1001
5	101	10	1010

Since the computer is really made up of tiny switches that can be either OFF or ON, you can look at a binary number as a series of light switches. A1 represents a switch that is ON, and a 0 means a switch that is OFF.

As you can see, numbers can get rather longin the binary system. For example, to show the number 10, we need four light changes, or four places. However, the real switches inside a computing machine are tiny and they are able to turn on and off very rapidly. The binary numeration system suits a computer majorly well.

Binary to Decimal Conversion

To convert binary into decimal is very simple and can be done as shown below:

Say we want to convert the 8 bit value **10011101** into a decimal value, we can use a formula like that below:

128	64	32	16	8	4	2	1
1	0	0	1	1	1	0	1

As you can see we have placed the numbers 1, 2, 4, 8, 16, 32, 64, 128 (powers of two) in reverse numerical order and then written the binary value below. To convert you simply take a value from the top row wherever there is a 1 below and add the values together, for instance in our example we would have **128 + 16 + 8 + 4 + 1 = 157**. For a 16 bit value you would employ the decimal values 1, 2, 4, 8, 16, 32, 64, 128, 256, 512, 1024, 2048, 4096, 8192,

16384, 32768 (powers of two) for the conversion.

Because we know binary is base 2 then the above could be written as:

$1*2^7 + 0*2^6 + 0*2^5 + 1*2^4 + 1*2^3 + 1*2^2 + 0*2^1 + 1*2^0 = 157$.

Example 1

Convert binary **11101** to decimal

Basically, this is the same as saying:

$1*(2^4) + 1*(2^3) + 1*(2^2) + 0*(2^1) + 1*(2^0)$

or

$1*(16) + 1*(8) + 1*(4) + 0*(2) + 1*(1)$

The reason it's easier to start backward is because:

- Counting the number of digits takes extra time, and you might count wrongly.
- If you do not remember what a particular power-of-2 value, it is easy to calculate it from the previous value. For instance, if you do not recall what the value of $2*2*2$ is, then just double the value of $2*2$

Notes	Multiplication	Result
start from the last digit, which is 1, multiply that digit with 2^0, note that the power of 0 of any number is always 1	$1*(2^0)$	1
11101 (current digit is in bold)		
process the previous digit, which is 0, multiply that digit with the increasing power of 2	$0*(2^1)$	0
11101 (current digit is in bold)		
process the previous digit, which is 1, note that 2^2 means $2*2$	$1*(2^2)$	4
11101 (current digit is in bold)		
process the previous digit, which is 1, note that 2^3 means $2*2*2$	$1*(2^3)$	8
11101 (current digit is in bold)		
process the previous digit, which is 1, note that 2^4 means $2*2*2*2$	$1*(2^4)$	16
11101 (current digit is in bold)		
here, we stop because there is no more digit to process		
this number comes from the sum of the RESULTS	ANSWER	29

Example 2
Convert binary **1010** to decimal

Multiplication	Result
0* (2^0)	0
1* (2^1)	2
0* (2^2)	0
1* (2^3)	8
Answer	10

Is constructing a table like above required? No, it just depends upon your preference. Some people are visual, and the table might help. Without a table, it is also easy. If you want to be a speed counter, just remember that the value of the multiplier is always the double of the previous one.

1, 2, 4, 8, 16, 32, 64, 128, 256, 512, ...

Power of 2s	Result
2^0	1
2^1 = 2	2
2^2 = 2*2	4
2^3 = 2*2*2	8
2^4 = 2*2*2*2	16

Example 3
Convert binary **1010001** to decimal.
Again, I am starting backward here:
$(1*1) + (0*2) + (0*4) + (0*8) + (1*16) + (0*32) + (1*64) =$
$1 + 0 + 0 + 0 + 16 + 0 + 64 = 81$

Converting Binary Fractions to Decimal
Remember your binary place value. Starting at the decimal point and moving right you have 1/2 place, then 1/4, 1/8, 1/16, and 1/32. If there is a 1 in a particular place, simply add that fraction to the result to get the result.

So 0.11 (binary) is 1/2 + 1/4 or .75 decimal. 0.101 binary is 1/2 + 1/8 = 0.625 decimal.

The thing that makes binary easy is to do the same thing in binary you do in base 10.

Binary to Octal Conversion

Changing binary to octal is also a simple process. Break off the binary digits into groups of three beginning from the binary point and convert each group into its appropriate octal digit. For integers, it may be essential to add a zero as the MSB in order to complete a grouping of three bits. Note that this does not vary the value of the binary number. Similarly, when constituting fractions, it may be necessary to add a trailing zero in the LSB in order to form a complete grouping of three.

Example 1
Converting $(010111)^2$ to octal.
111 = 7 (LSB)
010 = 2 (MSB)
thus, $(010111)^2 = (27)^8$

Example 2
Converting $(0.110101)^2$ to octal.
110 = 6 (MSB)
101 = 5 (LSB)
Thus, $(0.110101)^2 = (0.65)^8$

Binary to Hexadecimal Conversion

Converting binary to hexadecimal is another simple process. Break off the binary digits into groups of four beginning from the binary point and convert each group into its appropriate hexadecimal digit. For integers, it may be essential to add a zero as the MSB in order to complete a grouping of four bits. Note that this addition does not vary the value of the binary number. Likewise, when representing fractions, it may be necessary to add a trailing zero in the LSB in order to form a complete grouping of four.

Example 1
Convert $(1010111)_2$ to hexadecimal

Solution
0111 = 7 (LSB)
0101 = 5 (MSB)
Thus $(1010111)_2 = (57)_{16}$

Example 2
Convert $(0.00111111)_2$ to hexadecimal

Solution
0011 = 3 (MSB)
1111 = F (LSB)
Thus $(0.00111111)_2 = (0.3F)_{16}$

BINARY ARITHMETIC

Binary arithmetic is key to all computers and most other digital systems.

In this context, we will discuss the concept of dinary addition and binary subtraction.

BINARY ADDITION

In particular, addition is the most important binary arithmetic process because it can be used to perform all other arithmetic operations such as subtraction, multiplication and division. Thus, it is important to fully empathize binary addition.

The following table shows the four basic rules for binary addition.

Addition Rules

A + B	Sum	Carry
0 + 0	0	0
0 + 1	1	0
1 + 0	1	0
1 + 1	0	1

Example 1

Decimal	Binary
5	101
+ 3	+ 011
8	1000

Example 2

Decimal	Binary
74	1001010
+ 19	+ 10011
93	1011101

Binary Subtraction

The process of binary subtraction may be viewed as the addition of a negative number. For example 9 −4 may be viewed as 9 + (−4). However to determine the negative representation of a binary number, one must become familiar with 1's and 2's complement.

1's Complement of a Binary Number

The 1's complement of binary number is found by changing all the 1's to 0's and vice versa as illustrated by the examples below:

Number	1's Complement
10001	01110
101001	010110

2's Complement of a Binary Number

The 2's complement of binary number is found by adding 1 to the 1's complement

representation as illustrated by the examples below:

Number	1's Complement	2's Complement
10001	01110	01110
		+ 1
		01111
101001	010110	010110
		+ 1
		010111

Subtracting Using 1's Complement

For subtracting a smaller number from a larger number, the 1's complement method is as follows:

1. Determine the 1's complement of the smaller number
2. Add the 1's complement to the larger number
3. Take out the final carry and add it to the result. This step is named the end-around carry.

Example

11001–10011

Solution

Result from Step 1: 01100
Result from Step 2: 100101
Result from Step 3: 00110'
To verify, note that 25 − 19 = 6

For subtracting a larger number from a smaller number, the 1's complement method is as follows:

1. Determine the 1's complement of the larger number
2. Add the 1's complement to the smaller number
3. If there is no carry. The result has the opposite sign from the answer and is the 1's complement of the answer
4. Change the sign and take the 1's complement of the result to get the final answer.

Example

1001 – 1101

Solution

Result from Step 1: 0010
Result from Step 2: 1011

Result from Step 3: – 0100
To verify, note that 9 – 13 = – 4

Subtracting Using 2's Complement

For subtracting a smaller number from a larger number, the 2's complement method is as follows:

1. Determine the 2's complement of the smaller number
2. Add the 2's complement to the larger number
3. Discard the final carry (there is always one in this case).

Example

11001 – 10011

Solution

Result from Step 1: 01101
Result from Step 2: 100110
Result from Step 3: 00110
Again, to verify, note that 25 – 19 = 6

For subtracting a larger number from a smaller number, the 2's complement method is as follows:

1. Determine the 2's complement of the larger number.
2. Add the 2's complement to the smaller number.
3. There is no carry of the left-most column. The result is in 2's complement form and is negative.
4. Change the sign and take the 2's complement of the result to get the final answer.

Example

1001 – 1101

Solution

Result from Step 1: 0011
Result from Step 2: 1100
Result from Step 3: – 0100
Again to verify, note that 9 – 13 = –4

Binary Multiplication

Multiplication in the binary system works the same way as in the decimal system:

- $1 * 1 = 1$
- $1 * 0 = 0$
- $0 * 1 = 0$

Example

```
   101
 *  11
------
   101
  1010
------
  1111
```

Note that multiplying by two is extremely easy. To multiply by two, just add up a 0 on the end.

Binary Division

Binary division follows the same rules as in decimal division. For the sake of simplicity, throw away the remainder.

Example

```
     111011/11
     10011 r 10
   ------------
11)111011
   – 11
   -----
    101
   – 11
   -----
    101
     11
   -----
     10
```

OCTAL NUMBER SYSTEM

Although this was once a democratic number base, particularly in the Digital Equipment Corporation PDP/8 and other old computing system, it is rarely used today. The octal system is founded on the binary system with a 3-bit boundary. The octal number system:

- uses base 8
- includes only the digits 0 to 7 (any other digit would make the number an invalid octal number).

The weighted values for each position are as follows:

8^5	8^4	8^3	8^2	8^1	8^0
32768	4096	512	64	8	1

Octal to Decimal Conversion

To convert from octal to decimal, multiply the value in each position by its octal weight and add each value.

Example 1

Convert $(127662)_8$ into decimal form.

Solution

1*8^5	2*8^4	7*8^3	6*8^2	6*8^1	2*8^0
1*32768	2*4096	7*512	6*64	6*8	2*1
32768	8192	3584	384	48	2

Answer is 32768 + 8192 + 3584 + 384 + 48 + 2 = 44978

Example 2

Convert $(237.04)_8$ to decimal form.

Solution

Weights	8^2	8^1	8^0	8^{-1}	8^{-2}
Weight Value	64	8	1	0.125	0.015625
Octal Number	2	3	7	0	4
Decimal Value	128	24	7	0	0.0625 Total $(159.0625)_{10}$

Octal to Binary Conversion

The primary application of octal numbers is representing binary numbers, as it is easier to read large numbers in octal form that in binary form. Because each octal digit can be represented by a three-bit binary number (see the following table), it is very easy to convert from octal to binary. Simply exchange each octal digit with the appropriate three-bit binary number as indicated in the examples below.

The following table shows the octal digit and the equivalent binary digit.

Octal digit	Binary digit
0	000
1	001
2	010
3	011
4	100
5	101
6	110
7	111

Example 1

Convert $(13)_8$ into equivalent binary number.

Solution

$(13)_8 = (001011)_2$

Example 2

Convert $(37.12)_8$ into equivalent binary number.

Solution

$(37.12)_8 = (011111.\ 001010)_2$

Example 3

Convert $(127662)_8$ into equivalent binary number.

1	2	7	6	6	2
001	010	111	110	110	010

This gives the binary number 001010 111110110010 or 00 1010 1111 1011 0010 in our more readable format.

Octal to Hexadecimal Conversion

This is intermediate conversion. The following steps are to be followed to implement the octal to hexadecimal conversion.

Step 1: Convert octal to binary or decimal number.

Step 2: Convert decimal/binary to hexadecimal number.

Example

Convert $(53.7)_8$ into equivalent hexadecimal number system.

Solution

The first step to convert the 537 into equivalent binary form which is as follows:

5	3 .	7_8
101	011 .	111_2

Now regroup the binary digits into groups of four and add zeros where needed to complete groups; then convert the binary to hexadecimal.

0010	1011.	1110_2
2	B .	E_{16}

So the solution of: 53.7_8 equal $2B.E_{16}$

HEXADECIMAL NUMBER SYSTEM

Just like the octal number system, the hexadecimal (or base-sixteen) number system provides a convenient way to express binary numbers. The following table shows the weighting for the hexadecimal number system up to 3 decimal places before and 2 places after the *hexadecimal* point. Based on the trend in previous number systems, the methods used to convert hexadecimal to decimal and vice versa should be intuitive

Weights 16^2 16^1 16^0 . 16^{-1} 16^{-2}

The following table lists the equivalent decimal, binary and hexadecimal representations for the decimal numbers ranging from 0 to 15.

Decimal	Binary	Hexadecimal
0	0000	0
1	0001	1
2	0010	2
3	0011	3
4	0100	4
5	0101	5
6	0110	6
7	0111	7
8	1000	8
9	1001	9
10	1010	A
11	1011	B
12	1100	C
13	1101	D
14	1110	E
15	1111	F

Hexadecimal to Decimal Conversion

The following steps are to be followed to convert hexadecimal into decimal number system.

1. Get the last digit of the hexadecimal number; call this digit the **current digit**

2. Make a variable, let's call it **power**. Set the value to 0.
3. Multiply the **current digit** with ($16^\wedge$**power**), store the result.
4. Increment **power** by 1.
5. Set the **current digit** to the previous digit of the hex number.
6. Repeat from step 3 until all digits have been multiplied.
7. Sum the result of step 3 to get the answer number.

Example 1
Convert the number **1128** hexadecimal to Decimal.

Solution
Once discerned, notice that the above process is essentially performing this calculation:
$1 \times (16^\wedge 3) + 1 \times (16^\wedge 2) + 2 \times (16^\wedge 1) + 8 \times (16^\wedge 0)$

Example 2
Convert the number **589** hexadecimal to decimal.

Solution

Multiplication	Result
$9 \times (16^\wedge 0)$	9
$8 \times (16^\wedge 1)$	128
$5 \times (16^\wedge 2)$	1280
Answer	1417

If you want to be a speed counter, it is beneficial to memorize the values of the smaller power of 16s, such as in this table

Multiplication	Result	Notes
$8 \times (16^\wedge 0)$	8	Start from the last digit of the number. In this case, the number is 1128. The last digit of that number is 8. Note that the power of 0 of any number is always 1
$2 \times (16^\wedge 1)$	32	Process the previous, which is 2.

$$0010 \mid 1011. \mid 1110_2$$
$$2 \mid B . \mid E_{16}$$

Multiplication	Result	Notes
		Multiply that number with an increasing power of 16.
$1 \times (16^\wedge 2)$	256	Process the previous digit, which is 1, note that $16^\wedge 2$ means 16×16
$1 \times (16^\wedge 3)$	4096	Process the previous digit, which is 1, note that $16^\wedge 3$ means $16 \times 16 \times 16$ Here, we stop because there is no more digit to process
Answer	4392	This number comes from the sum of the RESULTS (8 + 32 + 256 + 4096) = 4392

Power of 16's	Result
16^0	1
16^1 = 16	16
16^2 = 16 × 16	256
16^3 = 16 × 16 × 16	4096
16^4 = 16 × 16 × 16 × 16	65536

Example 3

Convert the number **1531** hexadecimal to decimal.

Solution

Multiplication	Result
1 × 1	1
3 × 16	48
5 × 256	1280
1 × 4096	4096
Answer	5425

Example 4

Convert the number **8F** hexadecimal to decimal.

Solution

Division	Result
F × 1	15
8 × 16	128
Answer	143

Example 5

Convert the number **A0** hexadecimal to decimal.

Solution

Division	Result
0 × 1	0
A × 16	160
Answer	160

Example 6

Convert the number **12** hexadecimal to decimal.

Solution

Division	Result
2 × 1	2
1 × 16	16
Answer	18

Example 7

Convert the number 35432 hexadecimal to decimal.

Solution

$2 \times (16^0) + 3 \times (16^1) + 4 \times (16^2) + 5 \times (16^3)$
$+ 3 \times (16^4)$
$= 2 + 3 \times 16 + 4*256 + 5*4096 + 3*65536$
$= 2 + 48 + 1024 + 20480 + 196608 = 218162$

Hexadecimal to Binary Conversion

To convert hexadecimal to binary, at each hexadecimal digit we associate the 4 bits binary number using the following table:

0	0000
1	0001
2	0010
3	0011
4	0100
5	0101
6	0110
7	0111
8	1000
9	1001
A	1010
B	1011
C	1100
D	1101
E	1110
F	1111

Example 1

Convert $(F8)_{16}$ into equivalent binary form.

Solution

Write down the binary for F first, then the binary for 8.

F	8
1111	1000

So, the answer is 11111000.

Example 2

Convert hex number 1A to binary.

1	A
0001	1010

So, the answer is 00011010. (Note: Once you got the answer, you can dismiss the zeros at the beginning, so this can be also written as 11010.)

Example Table

The following table represents the conversion between hexadecimal to binary conversion.

Hex	D	C	C	
Bin	1101	1100	1100	
Hex	1	0	0	3
Bin	0001	0000	0000	0011
Hex	F	3	A	2
Bin	1111	0011	1010	0010

Hexadecimal to Octal Conversion

(1) Convert octal (hexadecimal) to binary first.
(2a) Reorganize the binary number in 3 bits a group **starts from the LSB** if Octal is required.
(2b) Reorganize the binary number in 4 bits a group from the LSB if hexadecimal is required.

Example
Convert $5A8_{16}$ to octal.

$5A8_{16}$ = 0101 1010 1000 (binary)
        = 2 6 5 0 (octal)

BINARY CODES

The usual way of expressing a decimal number in terms of a binary number is known as *pure binary coding*. A number of other proficiencies can be used to constitute a decimal number. These are summarized as follows:

8421 BCD Code

In the 8421 binary coded decimal (BCD) representation each decimal digit is converted to its 4-bit pure binary equivalent.

For example: 57_{dec} = **0101 0111$_{bcd}$**

Addition is analogous to decimal addition with normal binary addition taking place from right to left. For example:

6	0110	BCD for 6	42	0100 0010	BCD for 42
+3	0011	BCD for 3	+27	0010 0111	BCD for 27
	1001	BCD for 9		0110 1001	BCD for 69

Where the result of any addition exceeds 9 (1001) then six (0110) must be added to the sum to account for the six invalid BCD codes that are available with a 4-bit number. This is illustrated in the example as follows:

8	1001	BCD for 8
+7	0111	BCD for 7
	1111	exceeds 9 (1001) so
	0110	add six (0110)
	0001 0101	BCD for 15

Note that in the last example the 1 that carried forward from the first group of 4 bits has made a new 4-bit number and so represents the "1" in "15".

In the above examples, the BCD numbers are break at every 4-bit boundary to build reading them easier. This is not essential when writing a BCD number down.

This coding is an instance of a binary coded (each decimal number maps to four bits) weighted (each bit represents a number, 1, 2, 4, etc.) code.

4221 BCD Code

The 4221 BCD code is another binary coded decimal code where each bit is weighted by 4, 2, 2 and 1 respectively. Unlike BCD coding there are no incapacitate representations. The decimal numbers 0 to 9 have the following 4221 equivalents

Decimal	4221	1's complement
0	0000	1111
1	0001	1110
2	0010	1101
3	0011	1100
4	1000	0111
5	0111	1000
6	1100	0011
7	1101	0010
8	1110	0001
9	1111	0000

The 1's complement of a 4221 representation is important in decimal arithmetic. In forming the code remember the following rules

- Below decimal 5 use the right-most bit representing 2 first
- Above decimal 5 use the left-most bit representing 2 first
- Decimal 5 = 2 + 2 + 1 and not 4 + 1.

Gray Code

Gray coding is an important code and is used for its speed, it is also relatively free from errors. In pure binary coding or 8421 BCD then calculating from 7 (0111) to 8 (1000) requires 4 bits to be exchanged at the same time. If this does not encounter then various numbers could be momently generated during the transition so creating spurious numbers which could be read.

Gray coding avoids this since only one bit alters among subsequent numbers. To construct the code there are two mere rules. First begin with all 0s and then continue by changing the least significant bit (LSB) which will bring about a new state.

The first 16 gray coded numbers are showed below.

Decimal	Gray Code
0	0000
1	0001
2	0011
3	0010
4	0110
5	0111
6	0101
7	0100
8	1100
9	1101
10	1111
11	1110
12	1010
13	1011
14	1001
15	1000

ASCII CODE

It is a very well-known fact that computers can manage internally only 0s (zeros) and 1s (ones). This is true, and by intends of sequences of 0s and 1s the computer can express any numerical value as its binary translation, which is a very simple mathematical operation.

However, there is no such evident way to represent letters and other non-numeric characters with 0s and 1s. Hence, in order to do that, computing device use *ASCII tables*, which are tables or lists that comprise all the letters in the roman alphabet plus some additional characters. In these tables each character is ever constituted by the same order number. For example, the ASCII code for the capital letter "A" is always represented by the order number 65, which is easily representable using 0s and 1s in binary: 65 expressed as a binary number is 1000001.

The standard ASCII table defines 128 character codes (from 0 to 127), of which, the first 32 are control codes (non-printable), and the remaining 96 character codes are representable characters.

Because most systems nowadays work with 8 bit bytes, which can represent 256 different values, in addition to the 128 standard ASCII codes there are other 128 that are known as *extended ASCII*, which are platform- and locale-dependent. So there is more than one extended ASCII character set.

EBCDIC

EBCDIC (Extended Binary Coded Decimal Interchange Code) is a binary code for alphabetic and numeric characters that IBM developed for its larger operating systems. It is the code for text files that is used in IBMs OS/390 operating system for its S/390 servers and that thousands of corporations use for

*	0	1	2	3	4	5	6	7	8	9	A	B	C	D	E	F
0	NUL	SOH	STX	ETX	EOT	ENQ	ACK	BEL	BS	TAB	LF	VT	FF	CR	SO	SI
1	DLE	DC1	DC2	DC3	DC4	NAK	SYN	ETB	CAN	EM	SUB	ESC	FS	GS	RS	US
2		!	"	#	$	%	&	'	(	.	*	+	,	-	.	/
3	0	1	2	3	4	5	6	7	8	9	:	;	<	=	>	?
4	@	A	B	C	D	E	F	G	H	I	J	K	L	M	N	O
5	P	Q	R	S	T	U	V	W	X	Y	Z	[	\	]	^	_
6	'	a	b	c	d	e	f	g	h	i	j	k	l	m	n	o
7	p	q	r	s	t	u	v	w	x	y	z	{			}	~	

* This panel is organized to be easily read in hexadecimal: row numbers represent the first digit and the column numbers represent the second one. For example, the "A" character is located at the 4th row and the 1st column, for that it would be represented in hexadecimal as 0 × 41·(65).

their legacy applications and databases. In an EBCDIC file, each alphabetic or numeric character is represented with an 8-bit binary number (a string of eight 0's or 1's). 256 possible characters (letters of the alphabet, numerals, and special characters) are defined.

IBM's PC and workstation operating systems do not use IBM's proprietary EBCDIC. Instead, they use the industry standard code for text, ASCII. Conversion programs allow different operating systems to change a file from one code to another.

SUMMARY

- Number system means how the numbers are represented in computer memory.
- Number system is classified into four categories:
  (a) Decimal number system
  (b) Binary number system
  (c) Octal number system
  (d) Hexadecimal number system
- Decimal number system holds 10 numbers, i.e. from 0 to 9 and its base is 10.
- Binary number system holds only two numbers, i.e. either 0 or 1 and its base is 2.
- Octal number system holds eight numbers, i.e. from 0 to 7 and its base is 8.
- Hexadecimal number system holds 16 numbers/characters i.e. from 0 to 9 and A to F and its base is 16.
- BCD means binary coded decimal and in BCD we maintain the groups of 4 digits.
- Alphanumeric codes are the combinations of digits and characters.
- ASCII means American Standard Code for Information Interchange.
- EBCDIC means Extended Binary Coded Decimal Interchange Code.

Exercises

Section A

Multiple Choice Questions

1. Binary number system has the base
   (a) 2   (b) 10
   (c) 8   (d) None of the above

2. Decimal number system has the base
   (a) 10   (b) 2
   (c) 16   (d) All of the above

3. Octal number system has the base
   (a) 8   (b) 2
   (c) 16   (d) All of the above

4. Hexadecimal number system has the base
   (a) 16   (b) 2
   (c) 10   (d) All of the above

5. Binary numbers can be subtracted with the help of
   (a) 1's complement
   (b) 2's complement
   (c) Both a and b
   (d) None of the above

6. BCD means
   (a) Binary coded decimal
   (b) Binary code decimal
   (c) Binary compart decimal
   (d) None of the above

7. The following is a valid octal number
   (a) 198   (b) 156
   (c) 199   (d) 289

8. The following is a valid hexadecimal number
   (a) X2y
   (b) a2b2
   (c) oi2
   (d) None of the above

9. The following is a valid binary number
   (a) 00100
   (b) 12300
   (c) 23400
   (d) None of the above

10. The following is a valid decimal number
    (a) 234
    (b) de3
    (c) we3
    (d) None of the above

State True/False

1. Binary number system has the base 3.
2. Decimal number system has the base 10.
3. Octal number system has the base 8
4. Hexadecimal number system has the base 15.
5. The total number of digits in binary number system is 2

6. The total number of digits in octal number system is 8.
7. The total number of digits/characters in hexadecimal number system is 16.
8. The total number of digits in decimal number system is 10.
9. AB2 is a valid hexadecimal number.
10. 278 is a valid octal number.

Fill in the blanks

1. The base of binary number system is _____.
2. The base of octal number system is _____.
3. The base of Hexadecimal number system is _____.
4. The base of decimal number system is _____.
5. The binary equivalent of $(3456)_{10}$ is _____.
6. The Octal equivalent of $(1001010)_2$ is _____.
7. The Hexadecimal equivalent of $(1001 01101)_2$ is _____.
8. The Hexadecimal equivalent of $(567)_8$ is _____.
9. The result of $(1001)_2 + (1000)_2$ is _____.
10. The 1's complement of 1001001 is _____.

Section B

Short Answer Type Questions

1. Define the term number system.
2. Define the term decimal number system with suitable example.
3. Define the term binary number system with suitable example.
4. Define the term octal number system with suitable example.
5. Define the term hexadecimal number system with suitable example.
6. Define the process of binary addition with suitable example.
7. Define the process of binary subtraction with suitable example.
8. Define the process of binary multiplication with suitable example.

9. Define the process of binary division with suitable example.
10. Differentiate between binary number system and BCD codes with suitable example.

Section C

Long Answer Type Questions

1. Define the 1's complement and 2's complement process with suitable example.
2. Convert the following binary numbers into equivalent octal numbers:
   (a) 110001100   (b) 100011001
   (c) 100010111   (d) 10111011
3. Convert the following decimal numbers into equivalent octal numbers:
   (a) 1789.456   (b) 1234.678
   (c) 1345.789   (d) 1456.890
4. Convert the following hexadecimal numbers into equivalent octal numbers:
   (a) AB23   (b) BC23
   (c) AE45   (d) DF35
5. Convert the following octal numbers into equivalent decimal numbers:
   (a) 1473   (b) 1345
   (c) 1657   (d) 1275
6. Convert the following octal numbers into equivalent binary numbers:
   (a) 1473   (b) 1345
   (c) 1657   (d) 1275
7. Convert the following octal numbers into equivalent hexadecimal numbers:
   (a) 1473   (b) 1345
   (c) 1657   (d) 1275
8. Convert the following binary numbers into equivalent decimal numbers:
   (a) 11000.0010   (b) 11111.011
   (c) 1110.001   (d) 1100.01
9. Convert the following hexadecimal number system into equivalent decimal number system:
   (a) AB23   (b) AE45
   (c) BE45   (d) DF34
10. Differentiate between 1's complement and 2's complement process with suitable example.

5   Fundamental of C

In the earlier chapter, we introduced you to the concepts of problem-solving, especially as they pertain to computer programming. In this chapter, we present C language — a standardized, industrial-strength programming language known for its power and portability as an implementation vehicle for these problem-solving techniques using computer.

A language is a mode of communication between two people. It is necessary for those two people to understand the language in order to communicate. But even if the two people do not understand the same language, a translator can help to convert one language to the other, understood by the second person. Similar to a translator is the mode of communication between a user and a computer in a computer language. One form of the computer language is understood by the user, while in the other form it is understood by the computer. A translator (or compiler) is needed to convert from user's form to computer's form. Like other languages, a computer language also follows a particular grammar known as the syntax.

PROGRAM AND PROGRAMMING LANGUAGES

We have seen in the previous chapter that a computer has to be fed with a detailed set of instructions and data for solving a problem. Such a procedure which we call an algorithm is a series of steps arranged in a logical sequence. Also we have seen that a flow chart is a pictorial representation of a sequence of instructions given to the computer. It also serves as a document explaining the procedure used to solve a problem. In practice it is necessary to express an algorithm using a programming language.

A procedure expressed in a programming language is known as a computer program.

Computer programming languages are developed with the primary objective of facilitating a large number of people to use computers without the need for them to know in detail the internal structure of the computer. Languages are designed to be machine-independent. Most of the programming languages ideally designed, to execute a program on any computer regardless of who manufactured it or what model it is.

Programming languages can be divided into two categories:

i. **Low-level Languages or Machine Oriented Languages:**
The language whose design is governed by the circuitry and the structure of the machine is known as the **machine language**. This language is difficult to learn and use. It is specific to a given computer and is different for different computers, i.e. these programming languages are **machine-dependent**. These languages have been designed to give a better machine efficiency, i.e. faster program execution. Such languages are also known as low-level languages. Another type of low-level language is the assembly language. We will code the assembly language program in the form of mnemonics. Every machine provides a different set of mnemonics to be used

for that machine only depending upon the processor that the machine is using.

ii. **High-level Languages or Problem Oriented Languages:**

These languages are particularly oriented towards describing the procedures for solving the problem in a concise, precise and unambiguous manner. Every high level language follows a precise set of rules. They are developed to allow application programs to be run on a variety of computers. These programming languages are *machine independent*. Languages falling in this category are FORTRAN, BASIC, and PASCAL, etc. They are easy to learn and programs may be written in these languages with much less effort. However, the computer cannot understand them and they need to be translated into machine language with the help of other programs known as compilers or translators.

C LANGUAGE—INTRODUCTION

C is a programming language formulated at AT & T's Bell Laboratories of USA in 1972. It was projected and written by a man named Dennis Ritchie. In the late 70s, C began to replace the more intimate languages of that time like PL/I, ALGOL, etc. No one forced C. It was not made the 'official' Bell Laboratories programming language. Thus, without any ad C's reputation spread and its pool of users grew. 'Ritchie seems to have been rather surprized that so many programmers favoured C to older languages like FORTRAN or PL/I, or the newer ones like Pascal and APL. But, that is what happened.

The programming language C was primitively developed by Dennis Ritchie of Bell Laboratories and was planned to run on a PDP-11 with a UNIX operating system. Although it was earlier designated to run under UNIX, there has been a great interest in running it under the MS-DOS operating system on the IBM PC and compatibles. It is an first-class language for this surroundings because of the simplicity of expression, the compactness of the code, and the wide range of applicability. Also, due to the simplicity and ease of writing a C compiler, it is usually the first high level language available on any new computer, including microcomputers, minicomputers, and mainframe computer.

C is not the best beginning language because it is somewhat cryptical in nature. It allows the programmer a wide range of procedures from high level down to a very low level, approaching the level of assembly language. There seems to be no limit to the tractability available. One experienced C programmer made the statement, "You can program anything in C", and the statement is well supported by my own feel with the language. Along with the resulting freedom however, you take on a great deal of duty because it is very easy to write a program that destroys itself due to the silly little errors that a good Pascal compiler will flag and call a fatal error. In C, you are a lot on your own as you will soon find. In 1972, C programming language was developed at Bell Laboratories by Dennis Ritchie. C is a simple programming language with a comparatively simple to understand syntax and few keywords. C is useless. C itself has no input/output commands, does not have support for strings as a key data type. There is no predefined useful math functions. C demands the use of libraries as C is useless by itself. This increases the complexity of the C programming language. The use of ANSI libraries and other methods, the issue of standard libraries is settled.

Why Use C?

C has been used successful for every type of programming problem conceivable from operating systems to spreadsheets to expert systems and efficient compilers are available for machines ranging in power from the Apple Macintosh to the Cray supercomputers. The largest measure of C's success seems to be based on purely practical considerations:

- The portability of the compiler
- The standard library concept
- A powerful and varied repertoire of operators
- An elegant syntax

- Ready access to the hardware when needed
- And the ease with which applications can be optimized by hand-coding isolated procedures.

C is often called a "Middle Level" programming language. This is not a reflection on its lack of programming power but more a reflection on its capability to access the system's low-level functions. Most high-level languages (e.g. FORTRAN) supplies everything the programmer might want to do already built into the language. A low-level language (e.g. assembler) supplies nothing other than access to the machines basic instruction set. A middle level language, such as C, probably does not supply all the constructs found in high-languages but it provides you with all the building blocks that you will need to produce the results you want!

HISTORY OF C

C is a programming language developed at AT&T's Bell Laboratory of USA in 1972. It was designed and written by Dennis Ritchie. As compared to other programming languages such as Pascal, C allows a precise control of input and output.

Now let us see its historical development. The late 1960s were a churning era for computing system research at Bell Telephone Laboratories. By 1960, many programming languages came into existence, almost each for a specific purpose.

For example COBOL was being used for commercial or business applications, FORTRAN for Scientific applications and so on. So, people started thinking why could not there be a one general purpose language. Therefore, an International Committee was set up to develop such a language, which came out with the invention of ALGOL60. But this language never became popular because it was too abstract and too general. To improve this, a new language called combined programming language (CPL) was developed at Cambridge University. But this language was very complex in the sense that it had too many features and it was very difficult to learn. Martin Richards at Cambridge University reduced the features of CPL and developed a new language called basic combined programming language (BCPL). But unfortunately it turned out to be much less powerful and too specific. Ken Thompson at AT & T's Bell Labs, developed a language called B at the same time as a further simplification of CPL. But like BCPL this was also too specific. Ritchie inherited the features of B and BCPL and added some features on his own and developed a language called C. C proved to be quite compact and coherent. Ritchie first implemented C on a DEC PDP-11 that used the UNIX operating system. For many years the *de facto* standard for C was the version furnished with the UNIX version 5 operating system. The growing popularity of microcomputers led to the creation of large number of C implementations. At the source code level most of these implementations were highly compatible. However, since no standard existed there were discrepancies. To overcome this situation, ANSI established a committee in 1983 that defined an ANSI standard for the C language.

USES OF C

C was initially used for system development work, in particular the programs that make-up the operating system. Why use C? Mainly because it develops code that runs nearly as fast as code written in assembly language. Some examples of the use of C might be:

- Operating systems
- Language compilers
- Assemblers
- Text editors
- Print spoolers
- Network drivers
- Modern programs
- Data bases
- Language interpreters
- Utilities

In recent years, C has been used as a general-purpose language because of its popularity with programmers. It is not the world's easiest language to learn and you will certainly benefit if you are not learning C as your first programming language! C is trendy (I nearly said sexy)—many well established programmers are switching to C for all sorts of reasons, but mainly because of the portability that writing standard C programs can offer.

PROPERTIES OF C LANGUAGE

The increasing popularity of C is due to its many desirable qualities. It is a robust language whose rich set of built-in functions and operators can be used to compose any complex computer program. The C compiler combines the capabilities of an assembly language with features of a high-level language and therefore it is well suited for writing both system software and business packages.

It has only 32 keywords and its strength lies in its built-in functions. C language is well suited for structured programming, thus requiring the user to think of a problem in terms of function modules or blocks. Another important feature of C is its ability to extend itself. A C program is basically a collection of functions that are supported by the C library. We can endlessly add our own functions to the C program library.

SALIENT FEATURES OF C

C is a general purpose, structured programming language. Among the two types of programming languages discussed earlier, C lies in between these two categories. That's why it is often called a *middle level language.* It means that it combines the elements of high level languages with the functionality of assembly language. It provides relatively good programming efficiency (as compared to machine oriented language) and relatively good machine efficiency as compared to high level languages). As a middle level language, C allows the manipulation of bits, bytes and addresses—the basic elements with which the computer executes the inbuilt and memory management functions. C code is very portable, that it allows the same C program to be run on machines with different hardware configurations. The flexibility of C allows it to be used for systems programming as well as for application programming.

C is commonly called a structured language because of structural similarities to ALGOL and Pascal. The recognizing feature of a structured programming language is compartmentalization of code and data. Structured language is one that divides the entire program into modules using top-down approach where each module executes one job or task. It is easy for debugging, testing, and maintenance if a language is a structured one. C supports several control structures such as **while, do-while** and **for** and various data structures such as **strucs, files, arrays,** etc. as would be seen in the later units. The basic unit of a C program is a **function**—C's standalone subroutine.

The structural component of C makes the programming and maintenance easier.

GETTING STARTED WITH C

Communicating with a computer involves speaking the language the computer understands, which immediately rules out english as the language of communication with computer. However, there is a close analogy between learning english language and learning C language. The classical method of acquiring english is to first learn the alphabets used in the language, then learn to fuse these alphabets to form words, which in turn are combined to form sentences and sentences are combined to form paragraphs. Learning C is like and easier. Instead of straight-away acquiring how to write programs, we must first recognize what alphabets, numbers and special symbols are used in C, then how using them constants, variables and keywords are constructed, and finally how are these combined to form an instruction. A group of statements would be aggregated later on to form a program. This is illustrated in the following figure.

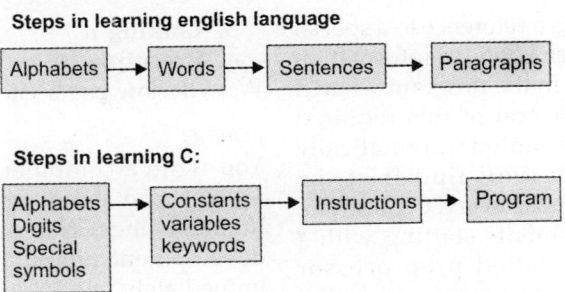

Steps in learning english language

Alphabets → Words → Sentences → Paragraphs

Steps in learning C:

Alphabets Digits Special symbols → Constants variables keywords → Instructions → Program

STRUCTURE OF A 'C' PROGRAM

As we have already seen, to solve a problem there are three main things to be considered. Firstly, what should be the output? Secondly, what should be the inputs that will be required to produce this output and thirdly, the steps of instructions which use these inputs to produce the required output. As stated earlier, every programming language follows a set of rules; therefore, a program written in C also follows predefined rules known as syntax. C is a case sensitive programming language. All C programs comprise one or more functions. One function that must be present in every C program is **main ( )**. This is the first function called up when the program execution begins. Basically, **main ( )** outlines what a program does. Although **main** is not given in the keyword list, it cannot be used for naming a variable. The structure of a C program is illustrated in the following where functions func1 ( ) through funcn ( ) represent user defined functions.

```
Preprocessor directives
Global data declarations
main ( ) /* main function*/
{
Declaration part;
Program statements;
}
/*User defined functions*/
func1 ( )
{
............
}
func2 ( )
{
............
}
.
.
.
funcn ( )
{
............
}
```

Structure of a C Program.

A SIMPLE C PROGRAM

From the above sections, you have become familiar with, a programming language and structure of a C program. It is now time to write a simple C program.

This program will illustrate how to print out the message "This is my First Program".

Example

Write a program to print a message on the screen.

```
/*Program to print a message*/
#include <stdio.h> /* header file*/
void main ( ) /* main function*/
{
printf ("This is my First Program\n") ; /* output statement*/
}
```

The explanation of the above program is as follows:

Every C program contains a function called **main ( )**. This is the starting point of the computer program. This is the point from where the execution begins. It will usually call other functions to help perform its job, some that we write and others from the standard libraries provided.

#include <stdio.h> is a reference to a special file called stdio.h which contains information that must be included in the program when it is compiled. The inclusion of this required information will be handled automatically by the compiler. You will find it at the commencement of almost every C program. Basically, all the statements starting with # in a C program are called preprocessor directives. These will be considered in the later units. Just remember, that this statement allows you to use some predefined functions such as, *printf ( )*, in this case.

main ( ) declares the start of the function, while the two curly brackets { } shows the start and finish of the function. Curly brackets in C are used to group statements together as a function, or in the body of a loop. Such a grouping is acknowledged as a compound statement or a block. Every statement within a function ends with a terminator semicolon (;).

printf ("This is my First Program\n"); prints the words on the screen. The text to be printed is put in double quotes. The\n at the end of the text tells the program to print a new line as part of the output. That means now if we give a second printf statement, it will be printed in the next line.

Comments may appear anywhere within a program, as long as they are placed within the delimiters /* and */. Such comments are helpful in identifying the program's principal features or in explaining the underlying logic of various program features. While useful for teaching, such a simple computer program has few practical uses. Let us think something rather more practical. Let us look into the example given below, the complete program development life cycle.

RUNNING C PROGRAMS

If we want to run the C programs then we have to use the concept of The Edit-Compile-Link-Execute Process.

The Edit-Compile-Link-Execute Process

Developing a program in a compiled language such as C requires at least four steps:

1. **Editing** (or writing) the program
2. **Compiling** it
3. **Linking** it
4. **Executing** it

We will now cover each step separately.

Editing

You write a computer program with words and symbols that are understandable to human beings. This is the *editing* part of the development phase. You type the program immediately into a window on the screen and save the resulting text as a separate file. This is often concerned to as the *source file* (you can read it with the **TYPE** command in **DOS** or the **cat** command in **UNIX**). The tradition is that the text of a C program is stored in a file with the extension .c for C programming language.

Compiling

You cannot directly execute the source file. To run on any computer system, the source file must be translated into binary numbers understandable to the computer's Central processing unit (for example, the 80*87 microprocessor). This process develops an intermediate object file with the extension .obj, the .obj represents Object.

Linking

The first question that comes to most peoples minds is *why is linking necessary?* The main cause is that many compiled languages come with library routines which can be added to your program. Theses functions are written by the manufacturer of the compiler to perform a variety of tasks, from input/output to complicated mathematical functions. In the case of C the standard input and output routines are contained in a library (stdio.h) so even the most basic program will require a library function. After linking the file extension is .exe which is denoted as executable files.

Executing

Thus the text editor produces .c source files, which go to the compiler, which produces .obj object files, which go to the linker, which produces .exe executable file. You can then run .exe files as you can other applications, simply by typing their names at the **DOS** prompt or run utilizing windows menu.

Graphically it is represented as follows:

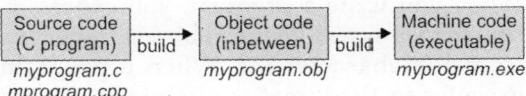

myprogram.c
mprogram.cpp
myprogram.obj
myprogram.exe

Language Using Microsoft Windows Environment

If we want to use Microsoft Windows/DOS environment then we have to follow the following steps for execute any C program:

Edit Stage

The direct executable file of C language is TC.exe then firstly we have to find tc.exe file and then double click on it then automatically the editor of Turbo C environment is open and we have to write the program on this environment.

Compile and Link

For the Turbo C environment, firstly we have to save the program with filename.c and then compile it with the help of ALT+F9 and if any errors are occurred then that particular errors are reported to the user and then we link the program with the help of CTRL+F9 then automatically the direct executable file is created.

Execute

When the source file is compiled and then the .obj file is created then we have to link it then automatically the direct executable file is created and the program is automatically executed.

Errors

First error highlighted. Use **Next Error** from **Search** menu for advance errors if applicable.

If you get an error message, or you find that the program does not work when you finally run it (at least not in the way you anticipated) you will have to go back to the source file, the .c file, to make alters and go through the whole development process again!

C Language Using UNIX Environment

on all UNIX systems further help on the C compiler can be obtained from the on-line manual. Type

Man CC

on your local UNIX system for more information.

Please note that UNIX is a case sensitive operating system and files named firstprog.c and FIRSTPROG.c are handled as two separate files on these system. By default the UNIX system compiles and links a program in one step, as follows:

cc firstprog.c

This command creates an executable file called a.out that overwrites any existing file called a.out. Executable files on UNIX are run by typing their name. In this case the program is executed as follows:

a.out

To change the name of the executable file type:

cc-o firstprog firstprog.c

This produces an executable file called firstprog which is run as follows:

Firstprog

GENERAL INSTRUCTIONS FOR WRITING A PROGRAM IN TURBO C ENVIRONMENT

The following are the general instructions for writing a program in TURBO C environment.

1. Firstly we have to write a program using TURBO C environment and then save it as first.c where first is the primary name of the file and .c is the extension of that file.
2. Then we have to compile it with the help of ALT+F9 then automatically the object file is created, i.e. First.obj that is written in machine language.
3. Then we have to LINK it with the help of CTRL+F9 then automatically the direct executable file is created i.e. First. exe and the program is automatically executed.

ADD COMMENTS TO A PROGRAM

A **comment** is a note to yourself (or others) that you put into your source code. All comments are dismissed by the compiler. They exist solely for your benefit. Comments are used mainly to document the intending and purpose of your source code, so that you can remember later how it functions and how to use it. You can also apply a comment to temporarily remove a line of code. Simply surround the line(s) with the comment symbols.

In C, the start of a comment is signalled by the /* character pair. A comment is ended by */. For example, this is a syntactically correct C comment:

/ This is a comment. */*

Comments can extend over several lines and can go anywhere except in the middle of any C keyword, function name or variable name. In C you cannot have one comment within another comment. That is comments may not be nested. Let us now look at our first program one last time but this time with comments:

main ( ) /* main function heading */
{
printf ("\n Hello, World! \n") ; /* Display message on */
}   **/* the screen */**

This program is not large enough to warrant comment statements but the principle is still the same.

DIAGRAMMATIC REPRESENTATION OF PROGRAM EXECUTION PROCESS

The following figure shows the diagrammatic representation of the program execution process.

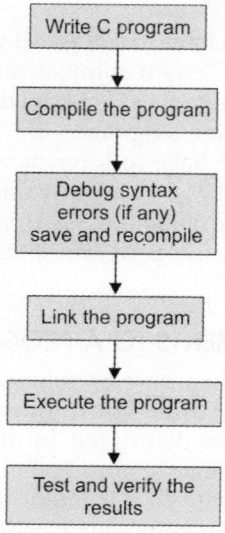

Program execution process

C CHARACTER SET

When you write a program, you express C source files as text lines containing characters from the character set. When a program executes in the target environment, it uses characters from the character set. These character sets are associated, but need not have the same encoding or all the same members.

Every character set comprises a distinct code value for each character in the **basic C character set**. A character set can also contain additional characters with other code values. The C language character set has alphabets, numbers, and special characters as shown below:

1. Alphabets including both lowercase and uppercase alphabets—a-z and A-Z.
2. Numbers 0–9
3. Special characters include:

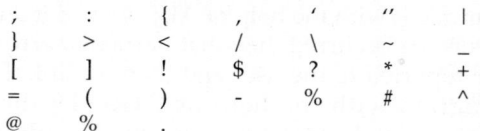

IDENTIFIERS AND KEYWORDS

Identifiers are the names given to various program elements such as constants, variables, function names and arrays, etc. Every element in the program has its own distinct name but one cannot select any name unless it conforms to valid name in C language. Let us study first the rules to define names or identifiers.

Rules for Forming Identifiers

Identifiers are defined according to the following rules:

1. It consists of letters and digits.
2. First character must be an alphabet character or underscore.
3. Both upper and lower cases are allowed. Same text of different case is not equivalent

For Example: **TEXT** is not same as **text**.

4. Except the special character underscore (_), no other special symbols can be used.

For example, some valid identifiers are as follows:

X
XY123
_XI
Temp1
tax_rate1

For example, some invalid identifiers are shown below:

123 First character to be alphabet.

"X." Not allowed.

order-no Hyphen allowed.

error flag Blankspace allowed.

Keywords

Keywords are reserved words which have standard, predefined meaning in C. They cannot be used as program-defined identifiers. The following are the lists of C keywords:

char	white	do	typedef	auto
int	if	else	switch	case
printf	double	struct	break	static
long	enum	register	extern	return
union	const	float	short	unsigned
continue	for	signed	void	default
goto	sizeof	volatile		

Generally, all keywords are in lower case although uppercase of same names can be used as identifiers.

SUMMARY

- C language is a middle level language that means we can do the higher level programming as well as lower level programming in C language.
- C language is developed by Dennis Ritchie in AT & T Bell Laboratories.
- We can compile the program in C language with the help of Alt+F9 key in Windows Environment.
- We can link the program in C language with the help of Ctrl + F9 key in Windows Environment.
- #include is a preprocessor directive which tells the compiler that we want to include the header files in our program.
- main ( ) is a predefined function which tells the compiler that from here main ( ) program starts.
- We can add the comments in C language with the help of // or /* and */

- C language is a case sensitive language that means all the predefined keywords must be written in small letters.
- ; is used to terminate the statement.
- Some examples of the use of C might be:
  - Operating systems
  - Language compilers
  - Assemblers
  - Text editors
  - Print spoolers
  - Network drivers
  - Modern programs
  - Data bases
  - Language interpreters
  - Utilities.

Exercises

Section A

Multiple Choice Questions

1. C is often called
   (a) Middle level language
   (b) High level language
   (c) Lower level language
   (d) Third generation language

2. Some of the examples of the use of C might be
   (a) Print spoolers
   (b) Network drivers
   (c) Both a and b
   (d) Website designing

3. In C, we can
   (a) Use and add functions in library
   (b) Not use and add functions in library
   (c) Use but cannot functions from library
   (d) Not use but can add functions in library

4. The sequence in which C programs are run is
   (a) Editing| Compiling| Executing| Linking
   (b) Compiling|Editing|Linking|Executing
   (c) Editing|Compiling|Linking|Executing
   (d) Compiling|Linking|Executing|Editing

5. Executable files are created by
   (a) Linker   (b) Compiler
   (d) Text editor   (d) All of the above

6. _____ is not a keyword
   (a) Union   (b) Continue
   (c) Start   (d) Default
7. C's character set does not includes
   (a) + – / * ^
   (b) Space, . ; $
   (c) Navigation keys
   (d) # & ! {} [] |
8. Pre-processor directives
   (a) Are not free layout
   (b) They begin with #
   (c) Should be in first column
   (d) All of the above
9. Main function in C program
   (a) Must be present in all C programs
   (b) Calls other functions in the C program
   (c) Both a and b
   (d) None of the above
10. Command printf ("Hello World/n");
    would
    (a) Give syntax error
    (b) Print hello world/n
    (c) Print hello world and new line
    (d) Depends on compiler

State True/False

1. C was originally developed to be used on UNIX.
2. Compiler compiles source code to .obj object.
3. There are 36 keywords in C language.
4. C is a high-level language.
5. C language is well suited for structured programming.
6. Linking is not required in case of simple programs of C.
7. We compile the C program with Alt+F9.
8. The first function that is called while executing is main.
9. The library that contains printf is stdio.h.
10. Every thing written as comments is ignored by compilers.

Section B

Short Answer Type Questions

1. Define the history of C programming language.
2. How we impose the comment in C languages define it?
3. Why we use C language define with proper example?
4. Define the uses of C language.
5. Define the character sets used in C language.
6. Define the various stages of executing a C program.
7. Define the various keywords used in C language.
8. Define the various rules for creating a Identifier.
9. Define the various rules for creating an integer type variable.
10. Define the various rules for constructing a character type variable.

Section C

Long Answer Type Questions

1. C is a middle level language explains it with proper explanation.
2. Define the basic structure of C programming language.
3. Define the EDIT COMPILE LINK and EXECUTE process in C language.
4. C is a structured programming language? Explain it with proper explanation.
5. Define the various properties of C programming language.
6. Define the term program and programming language explain it with proper explanation.
7. Define the diagrammatic representation of program execution process.
8. Define the term comment and how we impose the comment in C language?
9. Define the Edit-Compile-Link process in Windows environment.
10. Define the Edit-Compile-Link process in UNIX environment.

6 Data Types in C

All the data types defined by C are made up of units of memory called *bytes*. On most architectures a byte is built up of eight *bits*, each bit stores a one or a zero. These eight bits with two states contribute 256 combinations (2^8). So an integer which takes up two bytes can store a number between 0 and 65535 (0 and 2^{16}. Usually however, integer variables use the first *bit* to store whether the number is positive or negative so their value will be between −32768 and +32767.

As we mentioned, there are eight basic data types determined in the C language. Five types for storing integers of changing sizes and three types for storing *floating point* values (values with a decimal point). C does not allow for a basic data type for text. Text is built up of individual characters and characters are constituted by numbers. In the last example, we applied one of the integer types: **int**. This is the most usually used type in the C language.

The majority of data applied in computer programs is made up of the integer types, we shall talk about the floating point types a little later. In order of size, beginning with the smallest, the integer types are **char**, **short**, **int**, **long** and **long long**. The smaller types have the advantage of adopting less memory, the larger types obtain a performance penalty. Variables of type **int** store the largest potential integer which does not incur this performance penalty. For this reason, **int** variables can be different depending what type of computing machine you are using.

The **char** data type is commonly one byte, it is so called because they are usually used to store single characters. The size of the other types is depending on the hardware of your computer. Most computers are "32-bit", this refers to the size of data that they are designed for processing. On "32-bit" machines the **int** data type takes up 4 bytes (2^{32}). The **short** is usually smaller, the **long** can be larger or the same size as an **int** and finally the **long** is for handling very large numbers.

The type of variable you use broadly does not have a big impact on the speed or memory usage of your application. Unless you have a special need you can just use **int** variables. We will try to point out the few cases where it can be important in this chapter. A decade ago, most computing machines had 16-bit processors; this limited the size of **int** variables to 2 bytes. At the time, **short** variables were commonly also 2 bytes and **long** would be 4 bytes. Nowadays, with 32-bit machines, the default type (**int**) is commonly large enough to satisfy what used to require a variable of type **long**. The **long** type was introduced more recently to handle very large numeric values.

Some computers are better at dealing actually big numbers so the size of the data types will be bigger on these machines. To find out the size of each data type on your computing machine compile and run this piece of code. It uses one new language concept **sizeof ( )**. This tells you how many bytes a data type accepts.

DATA TYPES AND STORAGE

To store data inside the computer we need to first identify the type of data elements we need in our program. There are several different types

119

of data, which may be represented differently within the computer memory.

The data type specifies two things:

1. Permissible range of values that it can store.
2. Memory requirement to store a data type.

C Language provides four basic data types viz. int, char, float, double and long double. Using these, we can store data in simple ways as single elements or we can group them together and use different ways (to be discussed later) to store them as per requirement.

The four basic data types are described in the following table:

Data type	Type of data	Memory	Range
Int	Integer	2 bytes	–32, 768 to 32, 767
Char	Character	1 byte	–128 to 128
Float	Floating point number	4 bytes	3.4e – 38 to 3.4e + 38
Double	Floating point number with higher precision	8 bytes	1.7e – 308 to 1.7e + 308

Memory requirements or size of data associated with a data type indicates the range of numbers that can be stored in the data item of that type.

DATA TYPE QUALIFIERS

Short, long, signed, unsigned are called the data type qualifiers and can be used with any data type. A **short int** requires less space than **int** and **long int** may require more space than **int**. If **int** and **short int** takes 2 bytes, then **long int** takes 4 bytes.

Unsigned bits use all bits for magnitude; therefore, this type of number can be larger.

For example, **signed int** ranges from –32768 to +32767 and **unsigned int** ranges from 0 to 65,535. Similarly, **char** data type of data is used to store a character. It needs 1 byte. **Signed char** values range from –128 to 127 and **unsigned char** value range from 0 to 255.

The above discussions can be summarized as follows:

Data type	Size (bytes)	Range
Short int or int	2	–32768 to 32,767
Long int	4	–2147483648 to 2147483647

Signed int	2	–32768 to 32767
Unsigned int	2	0 to 65535
Signed char	1	–128 to 127
Unsigned char	1	0 to 255

The following table represents the data type name and the memory allocation of that data type and the range of that data type on 16 bit machine.

Type	Size (bits)	Range
Char or signed char	8	–128 to 127
Unsigned char	8	0 to 255
Int or signed int	16	–32768 to 32767
Unsigned int	16	0 to 65535
Short int or signed Short int	8	–128 to 127
Unsigned short int	8	0 to 255
Long int or signed Long int	32	–2147483648 to 2147483647
Unsigned long int	32	0 to 4294967295
Float	32	3.4 e – 38 to 3.4 e+38
Double	64	1.7e – 308 to 1.7e + 308
Long double	80	3.4 e – 4932 to 3.4e + 4932

The following table gives the complete details of the data types used in C language.

VARIABLES

Variable is an identifier whose value changes from time to time during execution. It is a named data storage location in your computer's memory. By using a variable's name in your program, you are, in effect, referring to the data stored there. A variable represents a single data item, i.e. a numeric quantity or a character constant or a string constant. Note that a value must be assigned to the variables at some point of time in the program which is termed as assignment statement. The variable can then be accessed later in the program. If the variable is accessed before it is assigned a value, it may give garbage value. The data type of a variable does not change whereas the value assigned to can change.

Attributes of the Variables

All variables have three essential attributes:

1. The name
2. The value
3. The memory, where the value is stored.

Data type	Reserve words	Bits	Minimum	Maximum	printf and scanf Format Specifier
Signed character	**Char**	8	−128	127	%hd, %d, %c
Unsigned character	**Unsigned char**	8	0	255	%u
Signed short integer	**Short**, short int, Signed short, Signed short int	16	−32,768	32767	%hd, %d
Unsigned short Integer	**Unsigned short**, Unsigned short int	16	0	65535	%u
Signed integer	**Int**, signed, Signed int	32	−2,147,483,648	2,147,483,647	%d
Unsigned integer	Unsigned, **Unsigned int**	32	0	4,294,967,295	%u
Signed long integer	**Long**, long int, Signed long, Signed long int	32/64	−9,223,372,036, 854,775,808	9,223,372,036, 854,775,807	%ld
Unsigned long integer	**Unsigned long**, Unsigned long int	32/64	0	18,446,74 4,073,709, 551,615	%lu
Floating-point	**Float**	32	1.175494351e−38	3.402823 466e+38	%f, %e
Double floating-pt	**Double**	64	2.225073858507 2014e−308	1.797693 1348623158 e+308	%lf, %le (%f, %e for printf)
Long double Flt-pt	**Long double**	64/128	3.36210314311 20935063e−4932	1.189731495 357231765 e+4932	%Lf, %Le

For example, in the following C program *a*, *b*, *c*, *d* are the variables but variable *e* is not declared and is used before declaration. After compiling the source code and look what is the output of the following code?

```
void main( )
{
int a, b, c;
char d;
a = 30;
b = 50;
c = a + b;
d = 'a';
e = d;
..........
..........
}
```

After compiling the code, this will generate the message that variable *e* not defined.

Declaring Variables

Before any data can be stored in the memory, we must assign a name to these locations of memory. For this we make declarations. Declaration associates a group of identifiers with a specific data type. All of them need to be declared before they appear in program statements, else accessing the variables results in junk values or a diagnostic error. The syntax for declaring variables is as follows:
<data- type> <variable-name>;

Example

If we define int a;
The meaning of this line is that a is a variable of integer type and automatically two bytes memory is allocated for this variable.

Integer Number Variables

The first type of variable we need to know about is of class type **int**—short for integer. An **int** variable can store a value in the range

–32768 to +32767. You can think of it as a largest positive or negative whole number: no fractional part is allowed. To declare an **int** you use the instruction:

int *variable name*;

For example:

int a;

declares that you want to create an **int** variable called **a**.

To assign a value to our integer variable we would use the following C statement:

a=10;

The C programming language uses the "=" character for *assignment*. A statement of the form **a=10;** should be interpreted as *take the numerical value 10 and store it in a memory location associated with the integer variable a*. The "=" character should not be seen as an equality otherwise writing statements of the form:

a=a+10;

will get mathematicians blowing fuses! This statement should be interpreted as *take the current value stored in a memory location associated with the integer variable a; add the numerical value 10 to it and then replace this value in the memory location associated with a*.

Decimal Number Variables

As described above, an integer variable has no fractional part. Integer variables tend to be used for counting, whereas *real* numbers are used in arithmetic. C uses one of two keywords to declare a **variable** that is to be associated with a decimal **number: float** and **double**. They are each offer a different level of precision as outlined below.

Float

A float, or floating point, number has about seven digits of precision and a range of about 1.E–36 to 1.E+36. A float takes four bytes to store.

Double

A double, or double precision, number has about 13 digits of precision and a range of about 1.E–303 to 1.E+303. A double takes eight bytes to store.

Example

If we want to define a variable total of float type then we can define it as follows:

float total;

if we want to define a variable sum of double type then we can define it as follows:

double sum;

To assign a numerical value to our floating point and double precision variables we would use the following C statement:

total=0.0;

sum=12.50;

Character Variables

C only has a concept of numbers and characters. It very often comes as a surprise to some programmers who learnt a beginner's language such as BASIC that C has no understanding of *strings* but a string is only an *array* of characters and C does have a concept of arrays which we shall be meeting later in this course.

To declare a variable of type character we use the keyword **char**. — A single character stored in one byte.

Example

char c;

To assign, or store, a character value in a **char** data type is easy — a character variable is just a symbol enclosed by single quotes. For example, if **c** is a **char** variable you can store the letter **A** in it using the following C statement:

c='A'

Notice that you can only store a single character in a **char** variable. Later we will be discussing using character strings, which has a very real potential for confusion because a string constant is written between double quotes. But for the moment remember that a **char** variable is **'A'** and not **"A"**.

INITIALIZING VARIABLES

When variables are declared initial, values can be assigned to them in two ways:

Within a Type Declaration

The value is assigned at the time of declaration

Example

int a = 100;
float b = 0.4 e –5;
char c = 'a12';

Using Assignment Statement
The values are assigned just after the declarations are made.

Example
int a,b,c;
a = 1000;
b = 0.4 *e* –5;
c = '*a*';

CONSTANTS
A constant is an identifier whose value cannot be changed throughout the execution of a program whereas the variable value keeps on changing.

C Constants Classification
C constants can be divided into two major categories:
(a) Primary constants
(b) Secondary constants
These constants are further classified in the following diagram.

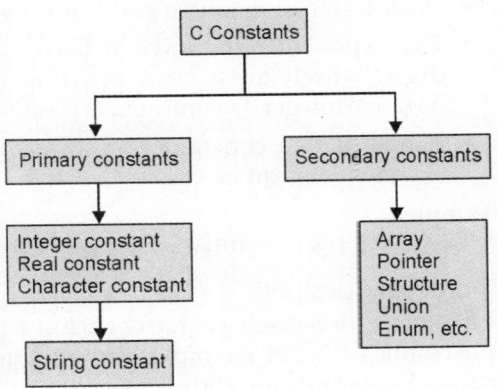

Now we will discuss only primary constants, namely, integer, real, character and string constants.

Integer Constants
Rules for Constructing Integer Constants
The following are the rules that we have to follow to form an integer constant.
- An integer constant must have at least one digit.
- It must not accept a decimal point.
- It can be either positive or negative.

- If no sign antecedes an integer constant it is assumed to be positive.
- No commas or blanks are allowed for within an integer constant.
- The permissible range for integer constants is –32768 to 32767.

Truly talking the range of an integer constant depends upon the compiler. For a 16-bit compiler like Turbo C or Turbo C++ the range is –32768 to 32767. For a 32-bit compiler the range would be even bigger.

Classification of Integer Constants
Further, these constant can be classified according to the base of the numbers which are as follows:

1. Decimal integer constants
These consist of digits 0 through 9 and first digit should not be 0.

Example
1   44   3277 are valid decimal integer constants
The following are the invalid decimal constants
12, 45	, not allowed
36.0	illegal char
1 010	blankspace not allowed
10 – 10	illegal char –
0900	the first digit should not be a zero

2. Octal integer constants
These consist of digits 0 through 7. The first digit must be zero in order to identify the constant as an octal number.

Example
Valid octal integer constants are as follows:
0   01   0743   0777
Invalid octal integer constants are as follows:
743	does not begin with 0
0438	illegal character 8
0777.77	illegal char.

3. Hexadecimal integer constants
These constants begin *with 0x or OX* and are followed by combination of digits taken from hexadecimal digits 0 to 9, a to f or A to F.

The following are the valid hexadecimal constants:
OX0   OX1   OXF77   Oxabcd.

The following are the invalid hexadecimal constants:

OBEF	x is not included
Ox.4bff	illegal char (.)
OXGBC	illegal char G

Maximum values these constants can have are as follows:

Integer constants	Maximum value
Decimal integer	32767
Octal integer	77777
Hexadecimal integer	7FFF

4. Unsigned interger constants

Exceed the ordinary integer by magnitude of 2, they are not negative. A character U or u is prefixed to number to make it unsigned.

5. Long integer constants

These are used to exceed the magnitude of ordinary integers and are appended by L.

Example

50000U	decimal unsigned
1234567889L	decimal long
0123456L	octal long
0777777U	octal unsigned

Floating Point Constants or Real Constants

Real constants are also named floating point constants. The real constants could be written in two forms — Fractional form and Exponential form.

Rules for Constructing Floating Point Constants

The following rules must be observed while constructing real constants expressed in fractional form:

- A real constant must have at least one digit.
- It must accept a decimal point
- It could be either positive or negative
- Default sign is positive
- No commas or blanks are allowed within a real constant

Example of valid floating point constants

+325.38   426.56   −32.78   −48.589

Examples of invalid floating point numbers are as follows:

1	decimal or exponent required
1,00.0	comma not allowed
2E+10.2	exponent is written after integer quantity
3E 10	no blank space.

The exponential form of representation of real constants is usually used if the value of the constant is either too small or too large.

In exponential form of representation, the real constant is represented in two parts. The part appearing before 'e' is called mantissa, whereas the part following 'e' is called exponent.

Rules for constructing floating point constants when it is treated as an exponent form

The following rules must be observed while constructing real constants expressed in exponential form:

- The mantissa part and the exponential part should be separated by a letter e.
- The mantissa part may have a positive or negative sign.
- Default sign of mantissa part is positive.
- The exponent must have at least one digit, which must be a positive or negative integer. Default sign is positive.
- Range of real constants expressed in exponential form is −3.4e38 to 3.4e38.

Examples

+3.2e−5   4.1e8   −0.2e+3   3..2e−5

Character Constants

This constant is a single character enclosed in apostrophes ' '. For example, some of the character constants are shown below:

'B', 'X', '4', '$'

'\0' is a null character having value zero.

Character constants have integer values associated depending on the character set adopted for the computer. ASCII character set is in use which uses 7-bit code with $2^7 = 128$ different characters. The digits 0-9 are having ASCII value of 48-56 and 'A' have ASCII value from 65 and 'a' having value 97 are sequentially ordered.

For example,

'A' has 65, blank has 32

Rules for Constructing Character Constants

The following rules must be observed while constructing character constants:

- A character constant is a single alphabet, a single digit or a single special symbol enclosed within single inverted commas. Both the inverted commas should point to the left. For example, 'A' is a valid character constant whereas 'A' is not.
- The maximum length of a character constant can be 1 character. Ex.: 'C'

Escape Sequence

There are some non-printable characters that can be printed by preceding them with '\' backslash character. Within character constants and string literals, you can write a variety of **escape sequences**. Each escape sequence determines the code value for a single character.

You can use escape sequences to represent character codes which are as follows:

- You cannot otherwise write (such as \n)
- That can be difficult to read properly (such as \t)
- That might change value in different target character sets (such as \a)
- That must not change in value among different target environments (such as \0)

The following is the list of the escape sequences:

Escape sequence	Character
\n	New line
\r	Carriage return
\t	Tab
\v	Vertical tab
\b	Backspace
\f	Form feed (page feed)
\a	Alert (beep)
\'	Single quote (')
\"	Double quote (")
\?	Question mark (?)
\\	Backslash (\)

String Constants

String constants consist of collection of characters enclosed within double quotes.

Example

"Red light" "Blue Sea hundred" "hello world".

Symbolic Constants

Symbolic constant is a name that substitutes for a sequence of characters or a numeric constant, a character constant or a string constant. When program is compiled each occurrence of a symbolic constant is replaced by its corresponding character sequence.

The syntax is as follows:

#define name text

Where **name** implies symbolic name in caps, **Text** implies value or the text.

Example

#define Q 100
#define MAX1 100
#define TRUE 1
#define FALSE 0
#define SIZE 10

The # character is used for preprocessor commands. A **preprocessor** is a system program, which comes into action prior to Compiler, and it replaces the replacement text by the actual text.

Advantages of using symbolic constants

The advantages of using symbolic constants are as follows:

- They can be used to assign names to values
- Replacement of value has to be done at one place and wherever the name appears in the text it gets the value by execution of the preprocessor. This saves time. If the symbolic constant appears 20 times in the program; it needs to be changed at one place only.

USER DEFINED TYPE DECLARATION

In C language a user can define an identifier that represents an existing data type. The user defined data type identifier can later be used to declare variables.

The general syntax of defining a user defined type is as follows:

typedef type identifier;

where type represents existing data type and 'identifier' refers to the name given to the data type.

Example

typedef int salary;
typedef float average;

Here salary symbolizes int and average symbolizes float. They can be later used to declare variables as follows:

salary dept1, dept2;

Average section1, section2;

Therefore, dept1 and dept2 are indirectly declared as integer datatype and section1 and section2 are indirectly float data type.

ENUMERATED TYPE DECLARATION

An enumeration consists of a set of named integer constants. An enumeration type declaration gives the name of the (optional) enumeration tag and defines the set of named integer identifiers (called the "enumeration set," "enumerator constants," "enumerators," or "members"). A variable with enumeration type stores one of the values of the enumeration set defined by that type.

Variables of **enum** type can be used in indexing expressions and as operands of all arithmetic and relational operators. Enumerations provide an alternative to the **#define** preprocessor directive with the advantages that the values can be generated for you and obey normal scoping rules.

In ANSI C, the expressions that define the value of an enumerator constant always have **int** type; thus, the storage associated with an enumeration variable is the storage required for a single **int** value. An enumeration constant or a value of enumerated type can be used anywhere the C language permits an integer expression.

The general syntax for defining an enumerated data type is as follows:

enum identifier {value1, value2 .... value n};

The identifier is a user defined enumerated data type which can be used to declare variables that have one of the values enclosed within the braces. After the definition we can declare variables to be of this 'new' type as below. enum identifier V1, V2, V3, ... Vn

The enumerated variables V1, V2, ... Vn can have only one of the values value1, value2 ..... value n

Example

enum DAY /* Defines an enumeration type */

{

saturday,

sunday = 0,

monday,

tuesday,

wednesday,

thursday,

friday

} workday;

The value 0 is associated with Saturday by default. The identifier Sunday is explicitly set to 0. The remaining identifiers are given the values 1 through 5 by default.

STORAGE CLASSES

The variables can be characterized by their **data type** and by their **storage class**. One way to classify a variable is according to its data type and the other can be through its storage class. **Data type** refers to the type of value represented by a variable whereas **storage** class refers to the **permanence** of a variable and its scope within the program, i.e. portion of the program over which variable is recognized.

The storage class associated with a variable can sometimes be established by the location of the variable declaration within the program or by prefixing keywords to variables declarations.

Moreover, a variable's storage class tells us:

- Where the variable would be stored.
- What will be the initial value of the variable, if initial value is not specifically assigned (i.e. the default initial value).
- What is the scope of the variable, i.e. in which functions the value of the variable would be available.
- What is the life of the variable, i.e. how long would the variable exist.

C programming language consists four storage classes which are as follows:

(a) Automatic storage class

(b) Register storage class

(c) Static storage class

(d) External storage class

Now we will discuss these storage classes one by one.

Automatic Storage Class

The following are the key points for the automatic storage class

- They are declared at the start of a program's block such as in the curly braces ({ }). Memory is allocated automatically upon entry to a block and freed automatically upon exit from the block.
- The scope of automatic variables is local to the block in which they are declared, including any blocks nested within that block. For these reasons, they are also called **local variables**.
- No block outside the defining block may have direct access to automatic variables (by variable name) but, they may be accessed indirectly by other blocks and/ or functions using pointers.
- Automatic variables may be specified upon declaration to be of storage class auto. However, it is not required to use the keyword auto because **by default**, storage class within a block is auto.
- Automatic variables declared with initializes are initialized every time the block in which they are declared is entered or accessed.

The features of a variable defined to have an automatic storage class are as follows:

Storage	Memory
Default initial value	An unpredictable value, which is often called a garbage value
Scope	Local to the block in which the variable is defined
Life	Till the control remains within the block in which the variable is defined

The following program shows how an automatic storage class variable is declared, and the fact that if the variable is not initialized it contains a garbage value.

```
void main( )
{
auto int i, j;
printf ( "\n%d %d", i, j ) ;
}
```

The output of the above program could be...
1223 1224

Where, 1223 and 1224 are garbage values of **i** and **j**. When you run this program you may get different values, since garbage values are unpredictable. So always make it a point that you initialize the automatic variables properly, otherwise you are likely to get unexpected results. Note that the keyword for this storage class is **auto**, and not automatic.

Scope and life of an automatic variable is illustrated in the following program.

```
void main( )
{
        auto int i = 2;
        {
        {
        printf ("\n%d ", i );
        }
                printf ( "%d ", i ) ;
        }
printf ( "%d", i ) ;
}
}
```

The output of the above program is:
2 2 2
This is because, all **printf ( )** statements occur within the outermost block (a block is all statements enclosed within a pair of braces) in which **i** has been defined. It means the scope of **i** is local to the block in which it is defined. The moment the control comes out of the block in which the variable is defined, the variable and its value is irretrievably lost.

To understand this go through the following program segment

```
void main ( )
{
    auto int i = 4;
{
    auto int i = 5 ;
{
    auto int i = 6;
    printf ( "\n%d ", i );
}
        printf ( "%d ", i ) ;
}
        printf ( "%d", i ) ;
}
```

The output of the above program would be:
6 5 4

The explanation of the above code is as follows:

Note that the compiler treats the three **i**'s as totally different variables, since they are defined in different blocks. Once the control comes out of the innermost block the variable **i** with value 6 is lost, and hence the **i** in the second **printf( )** refers to **i** with value 5. Similarly, when the control comes out of the next innermost block, the third **printf( )** refers to the **i** with value 4.

Register Storage Class

The following are the key points for the register storage class:

- Automatic variables are allocated storage in the main memory of the computer; however, for most computers, accessing data in memory is considerably slower than processing directly in the CPU.
- Registers are memory located within the CPU itself where data can be stored and accessed quickly. Normally, the **compiler determines** what data is to be stored in the registers of the CPU at what times.
- However, the C language provides the storage class register so that the programmer can suggest to the compiler that particular automatic variables should be allocated to CPU registers, if possible and it is not an obligation for the CPU to do this.
- Thus, register variables provide a certain control over efficiency of program execution.
- Variables which are used repeatedly or whose access times are critical may be declared to be of storage class register.
- Variables can be declared as a register as follows:

register int var;

The features of a variable defined to be of **register** storage class are as follows:

Storage	CPU registers
Default initial value	Garbage value
Scope	Local to the block in which the variable is defined
Life	Till the control remains within the block in which the variable is defined

A value stored in a CPU register can always be accessed faster than the one that is stored in memory. Therefore, if a variable is used at many places in a program it is better to declare its storage class as **register**. A good example of frequently used variables is loop counters. We can name their storage class as **register**.

Example

```
void main( )
{
register int i ;
for ( i = 1 ; i <= 100 ; i++ )
printf ("\n%d", i);
}
```

Here, even though we have declared the storage class of **i** as **register**, we cannot say for sure that the value of **i** would be stored in a CPU register. Why? Because the number of CPU registers are limited, and they may be busy doing some other task. What happens in such an event... the variable works as if its storage class is **auto**. Not every type of variable can be stored in a CPU register.

For example, if the microprocessor has 16-bit registers then they cannot hold a **float** value or a **double** value, which require 4 and 8 bytes respectively. However, if you use the **register** storage class for a **float** or a **double** variable you will not get any error messages. All that would happen is the compiler would treat the variables to be of **auto** storage class.

External Storage Class

The following are the key points for the external storage class:

- All variables we have seen so far have had limited scope (the block in which they are declared) and limited lifetimes (as for automatic variables).
- However, in some applications it may be useful to have data which is accessible from within any block and/or which remains in existence for the entire execution of the program. Such variables are called **global variables**, and the C language provides storage classes which can meet these requirements; namely, the **external** (extern) and **static** (static) classes.

- Declaration for external variable is as follows:

  . extern int var;

- External variables may be declared outside any function block in a source code file the same way any other variable is declared; by specifying its type and name (extern keyword may be omitted).
- Typically if declared and defined at the beginning of a source file, the extern keyword can be omitted. If the program is in several source files, and a variable is defined in let say **file1.c** and used in **file2.c** and **file3.c** then the extern keyword must be used in **file2.c** and **file3.c**.
- But, usual practice is to collect extern declarations of variables and functions in a separate header file (**.h** file) then included by using #include directive.
- Memory for such variables is allocated when the program begins execution, and remains allocated until the program terminates. For most C implementations, every byte of memory allocated for an external variable is **initialized to zero**.
- The scope of external variables is **global**, i.e. the entire source code in the file following the declarations. All functions following the declaration may access the external variable by using its name. However, if a local variable having the same name is declared within a function, references to the name will access the local variable cell.
- External variables may be initialized in declarations just as automatic variables; however, the initializers must be **constant expressions**. The initialization is done only once at compile time, i.e. when memory is allocated for the variables.
- In general, it is a good programming practice to avoid using external variables as they destroy the concept of a function as a 'black box' or independent module.
- The black box concept is essential to the development of a modular program with modules. With an external variable, any function in the program can access and

alter the variable, thus making debugging more difficult as well. This is not to say that external variables should never be used.

- There may be occasions when the use of an external variable significantly simplifies the implementation of an algorithm. Suffice it to say that external variables should be used rarely and with caution.

The features of a variable whose storage class has been defined as external are as follows:

Storage	Memory
Default initial value	Zero
Scope	Global
Life	As long as the program's execution does not come to an end

External variables differ from those we have already discussed in that their scope is global, not local. External variables are declared outside all functions, yet are available to all functions that care to use them.

The following example illustrates this fact

```
int i=0 ;
void main( )
{
        printf ("\ni = %d", i ) ;
        incre( ) ;
        incre( ) ;
        decre( ) ;
        decre( ) ;
}
void incre( )
{
i = i + 1 ;
printf ( "\non incrementing i = %d", i ) ;
}
void decre( )
{
i = i - 1 ;
printf ( "\non decrementing i = %d", i ) ;
}
```

The output of the above code is as follows

```
i = 0
on incrementing i = 1
on incrementing i = 2
```

on decrementing i = 1
on decrementing i = 0
The explanation of the above output is as follows:

The value of i is available to the functions **incre( )** and **decre( )** since i has been declared outside all functions.

Let us take a look of the following program segment

```
int x =120 ;
void main( )
{
extern int y;
printf ( "\n%d %d", x, y ) ;
}
int y = 130 ;
```

Here, x and y both are global variables. Since both of them have been defined outside all the functions both enjoy external storage class.

Note the difference between the following:

```
extern int y ;
int y = 130 ;
```

Here the first statement is a declaration, whereas the second is the definition. When we declare a variable no space is reserved for it, whereas, when we define it space gets reserved for it in memory. We had to declare y since it is being used in **printf ( )** before it's definition is encountered. There was no need to declare x since its definition is done before its usage. Also remember that a variable can be declared several times but can be defined only once.

Now let us take a look of the following program segment and what is the output of the following code segment

```
int x = 100;
void main( )
{
    int x = 200 ;
    printf ( "\n%d", x ) ;
    display( ) ;
}
void display( )
{
printf ( "\n%d", x ) ;
}
```

The output of the above code is as follows:
200
100

The explanation of the above code segment is as follows:

Here x is defined at two places, once outside **main ( )** and once inside it. When the control reaches the **printf ( )** in **main ( )** which x gets printed? Whenever such a conflict arises, it's the local variable that gets preference over the global variable. Hence the **printf ( )** outputs 200. When **display ( )** is called and control reaches the **printf ( )** there is no such conflict. Hence this time the value of the global x, i.e. 100 gets printed.

A static variable can also be declared outside all the functions. For all practical purposes it will be treated as an extern variable. However, the scope of this variable is limited to the same file in which it is declared. This means that the variable would not be available to any function that is defined in a file other than the file in which the variable is defined.

Static Storage Class

The following are the key points for the static storage class

- As we have seen, external variables have global scope across the entire program (provided extern declarations are used in files other than where the variable is defined), and have a lifetime over the entire program run.

- Similarly, static storage class provides a lifetime over the entire program, however; it provides a way to limit the scope of such variables, and static storage class is declared with the keyword static as the class specifier when the variable is defined.

- These variables are **automatically initialized to zero** upon memory allocation just as external variables are. Static storage class can be specified for automatic as well as external variables such as:

    . static extern varx;

- Static automatic variables continue to exist even after the block in which they are defined terminates. Thus, the value of a static variable in a function is retained between repeated function calls to the same function.

- The scope of static automatic variables is identical to that of automatic variables, i.e. it is local to the block in which it is defined; however, the storage allocated becomes permanent for the duration of the program.
- Static variables may be initialized in their declarations; however, the initializers must be **constant expressions**, and initialization is done only once at compile time when memory is allocated for the static variable.

The features of a variable defined to have a **static** storage class are as follows:

Storage	Memory
Default initial value	Zero
Scope	Local to the block in which the variable is defined
Life	Value of the variable persists between different function calls

Compare the two programs and their output given in the following figure to understand the difference between the **automatic** and **static** storage classes.

void main ( )
{
increment( ) ;
increment( ) ;
increment( ) ;
}
void increment( )
{
auto int i = 1 ;
printf ("%d\n", i) ;
i = i + 10 ;
}
	void main ( )
	{
	increment( ) ;
	increment( ) ;
	increment( ) ;
	}
	increment( )
	{
	static int i = 1 ;
	printf ("%d\n", i) ;
	i = i + 10 ;
	}

The output of the above program segment is as follows	The output of the above program segment is as follows
1 1 1	1 11 21

The explanation of the above program segment is as follows:

The above programs consist of two functions **main ( )** and **increment ( )**. The function **increment ( )** gets called from **main ( )** thrice.

Each time it increments the value of **i** and prints it. The only difference in the two programs is that one uses an **auto** storage class for variable **i**, whereas the other uses **static** storage class.

Like **auto** variables, **static** variables are also local to the block in which they are declared. The difference between them is that **static** variables do not disappear when the function is no longer active. Their values persist. If the control comes back to the same function again the **static** variables have the same values they had last time around.

In the above example, when variable **i** is **auto**, each time **increment ( )** is called it is reinitialized to one. When the function terminates, **i** vanishes and its new value of 10 is lost. The result: no matter how many times we call **increment ( )**, **i** is initialized to 1 every time.

On the other hand, if **i** is **static**, it is initialized to 1 only once. It is never initialized again. During the first call to **increment ( )**, **i** is incremented to 10. Because **i** is static, this value persists. The next time **increment ( )** is called, **i** is not re-initialized to 1; on the contrary its old value 10 is still available. This current value of **i** (i.e. 10) gets printed and then **i = i + 10** adds 1 to **i** to get a value of 11. When **increment ( )** is called the third time, the current value of **i** (i.e. 21) gets printed and once again **i** is incremented. In short, if the storage class is **static** then the statement **static int i = 1** is executed only once, irrespective of how many times the same function is called.

When to Use Which Storage Class

Dennis Ritchie has made available to the C programmer a number of storage classes with varying features, believing that the programmer is in a best position to decide which one of these storage classes is to be used when. We can make a few ground rules for usage of different storage classes in different programming situations with a view to:

(a) Economize the memory space consumed by the variables

(b) Improve the speed of execution of the program

The rules are as follows:

- Use static storage class only if you want the value of a variable to persist between different function calls.
- Use register storage class for only those variables that are being used very often in a program. Reason is, there are very few CPU registers at our disposal and many of them might be busy doing something else. Make careful utilization of the scarce resources. A typical application of register storage class is loop counters, which get used a number of times in a program.
- Use extern storage class for only those variables that are being used by almost all the functions in the program. This would avoid unnecessary passing of these variables as arguments when making a function call. Declaring all the variables as extern would amount to a lot of wastage of memory space because these variables would remain active throughout the life of the program.
- If you do not have any of the express needs mentioned above, then use the auto storage class. In fact, most of the times we end up using the auto variables, because often it so happens that once we have used the variables in a function we do not mind loosing them.

SUMMARY

- Data types are the predefined keywords which are used to define the type of data.
- Int, float, char, double and long are the basic data types used in C language.
- 2 bytes memory is allocated for integer type variable
- 4 bytes memory is allocated for float type variable.
- 1 byte per character is allocated for character type variable.
- 8 bytes memory is allocated for long type variable
- 10 bytes memory is allocated for double type variable.
- Int is the data type which holds the numeric value and the range of integer type variable is –32768 to +32767.
- Float is the data type which holds the decimal value.
- Char is the data type which holds the character type value.
- Auto, extern, static and register are the storage classes used in C language.
- Automatic variables may be specified upon declaration to be of storage class auto. However, it is not required to use the keyword auto because **by default**, storage class within a block is auto.
- Variables which are used repeatedly or whose access times are critical may be declared to be of storage class register.
- Static storage class provides a lifetime over the entire program, however; it provides a way to limit the scope of such variables, and static storage class is declared with the keyword static as the class specifier when the variable is defined.
- Extern storage class is used to define the global variable.

Exercises

Section A

Multiple Choice Questions

1. All C processors accepts the following data types
   - (a) Void
   - (b) Floating point
   - (c) Both a and b
   - (d) None of the above

2. Int has a range of
   - (a) –32768 to 32768
   - (b) –128 to 127
   - (c) 0 to 65535
   - (d) 0 to 255

3. Vertical tab is represented by
   - (a) \t
   - (b) \v
   - (c) Both a and b
   - (d) \a

4. In C programs the storage classes are
   - (a) Auto
   - (b) External
   - (c) Register
   - (d) All of the above

5. _____ determines which data is to be stored in registers.
   - (a) CPU
   - (b) Compilers
   - (c) Memory
   - (d) None of these

6. Which is not a data type
   - (a) Signed long integer
   - (b) Signed floating point

(c) Double floating point

(d) Long double floating point

7. Which of the following is false regarding extern variable
   (a) Its scope is global
   (b) Extern variables are stored in separate .h files
   (c) If declared and defined in beginning extern can be omitted.
   (d) None of the above

8. A Boolean variable can take
   (a) True or false value
   (b) Strings
   (c) 0 or 1
   (d) All of the above

9. Size of long double is
   (a) 64   (b) 32
   (c) Both a and b   (d) 80

10. Default type of floating point literals is
    (a) Double   (b) Float
    (c) Long
    (d) None of the above

State True/False

1. Char data type is only of one byte.
2. Double number has about 13 digits of precision and takes 8 bytes to store.
3. Range of unsigned int is 0 to 65535 and takes 8 bytes to store.
4. The #define is a preprocessor directive used to define constant.
5. Static variables are automatically initialized to null.
6. Specifier %f means converts the next value to the signed decimal integer.
7. Scope of all the external variable is global.
8. Printf and Scanf takes a constant number of parameters.
9. %x is the unsigned integer format specifier.
10. Unsigned int takes 32 bits to store.

Section B

Short Answer Type Questions

1. Define the various data type used in C language and also defines the range of the data types.
2. Define the various storage classes used in C language with proper example.

3. Differentiate between auto and static storage class with proper explanation.
4. Differentiate between static and extern storage class with proper explanation.
5. Differentiate between register and static storage class with proper explanation.
6. Define the automatic storage class in C language and when we use the concept of automatic storage class.
7. Define the static storage class in C language and when we use the concept of static storage class.
8. Define the extern storage class in C language and when we use the concept of extern storage class.
9. Define the various rules when to use which storage class.
10. Define the ranges of various data types used in C programming language.

Section C

Long Answer Type Questions

1. When we use the long data type in place of int data type? Explain.
2. Define the standard input function used in C language.
3. Define the standard output function used in C language.
4. Write a program to input the temperature in Fahrenheit and convert it into equivalent Celsius.
5. Write a program to print the Hello World on the screen.
6. Define the term data type and also define the classification of data type.
7. Define the concept of enumerated data type and how we define the enumerated data type in C language.
8. Define the concept of user defined data type used in C language with proper explanation.
9. Differentiate between predefined data type and user defined data type with proper explanation.
10. Define the general syntax of typedef keyword with proper explanation.

7

Operators and Expression

We have already learnt variables, constants, data types and how to declare them in C programming. The next step is to use those variables in expressions. For writing an expression we need operators along with variables.

An *expression* is a sequence of operators and operands that does one or a combination of the following:

- Specifies the computation of a value
- Designates an object or function
- Generates side effects.

An *operator* performs an operation (evaluation) on one or more operands. An *operand* is a sub expression on which an operator acts. This unit focuses on different types of operators available in C including the syntax and use of each operator and how they are used in C.

ASSIGNMENT STATEMENT

We know that variables are basically memory locations and they can hold certain values. But, how to assign values to the variables? C provides an assignment operator for this purpose. The function of this operator is to assign the values or values in variables on the right hand side of an expression to variables on the left hand side.

The syntax of the assignment expression is as follows:

Variable = constant / variable/ expression;

The data type of the variable on left hand side should match the data type of constant/ variable/expression on right hand side with a few exceptions where automatic type conversions are possible. Some examples of assignment statements are as follows:

b = c ; /* b is assigned the value of c */
b = 10 ; /* b is assigned the value 10*/
b = c+5; /* b is assigned the value of expr c+5 */

The expression on the right hand side of the assignment statement can be:

- An arithmetic expression
- A relational expression
- A logical expression
- A mixed expression.

The abovementioned expressions are different in terms of the type of operators.

Connecting the variables and constants on the right hand side of the variable.

Arithmetic operators, relational operators and logical operators are discussed in the following sections.

Example

```
int a;
float b,c, avg, t;
avg = (b+c) / 2; /*arithmetic expression */
a = b && c; /*logical expression*/
a = (b+c) && (b<c); /* mixed expression*/
```

ARITHMETIC OPERATORS

The basic arithmetic operators in C are the same as in most other computer languages, and correspond to our usual mathematical/ algebraic symbolism.

The following arithmetic operators are present in C language.

Operator	Meaning
+	Addition
−	Subtraction
*	Multiplication
/	Division
%	Modular division (for taking the remainder)

Some of the examples of algebraic expressions and their C notation are given below.

Expression	C notation
(a^3+bc)	$((a*a*a) + (b*c))$
(a^2+cd)	$((a*a) + (c*d))$

The arithmetic operators are all binary operators i.e. all the operators have two operands. The integer division yields the integer result. For example, the expression 10/3 evaluates to 3 and the expression 15/4 evaluates to 3. C provides the modulus operator, %, which yields the remainder after integer division. The modulus operator is an integer operator that can be used only with integer operands. The expression $x\%y$ yields the remainder after x is divided by y. Therefore, 10%3 yields 1 and 15%4 yields 3. An attempt to divide by zero is undefined on computer system and generally results in a run-time error. Normally, arithmetic expressions in C are written in straight-line form. Thus 'a divided by b' is written as a/b.

The operands in arithmetic expressions can be of integer, float, double type. In order to effectively develop C programs, it will be necessary for you to understand the rules that are used for implicit conversation of floating point and integer values in C language.

The rules are as follows:

- An arithmetic operator between an integer and integer always yields an integer result.
- Operator between float and float yields a float result.
- Operator between integer and float yields a float result.

If the data type is double instead of float, then we get a result of double data type.

Example

Operation	Result
7/3	2
7.0/3	2.33
7/3.0	2.33
7.0/3.0	2.33

Parentheses can be used in C expression in the same manner as algebraic expression.

Example

b * (c + d).

It may so happen that the type of the expression and the type of the variable on the left hand side of the assignment operator may not be same. In such a case the value for the expression is promoted or demoted depending on the type of the variable on left hand.

side of = (assignment operator).

Example

Consider the following assignment statements:
int i;
float b;
i = 5.6;
b = 30;
In the first assignment in the first assignment statement, float (5.6) is demoted to int. Hence *i* gets the value 5. In the second statement int (30) is promoted to float, *b* gets 30.0.

If we have a Complex expression like:
float a, b, c;
int s;
s = a * b / 6.0 * c;

Where some operands are integers and some are float, then int will be promoted or demoted depending on left hand side operator. In this case, demotion will take place since s is an integer.

Rules of Arithmetic Precedence

The rules of arithmetic precedence are as follows:

1. Parentheses are at the "highest level of precedence". In case of nested parenthesis, the innermost parentheses are evaluated first.

Example

If we have the following arithmetic expression
(((4+5)*6)/7)
The order of evaluation is given below
(((4+5) * 6) /7)

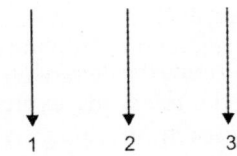

1   2   3

2. Multiplication, division and modulus operators are evaluated next. If an expression contains several multiplication, division and modulus operators, evaluation proceeds from left to right. These three are at the same level of precedence.

Example

If we have the following arithmetic expression

6*5+7*7

The order of evaluation is given below.

6*5+7*7

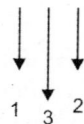

1 3 2

Firstly 6*5 is evaluated then 7*7 is evaluated and then the result of 6*5 and 7*7 is added.

3. Addition, subtraction are evaluated last. If an expression contains several addition and subtraction operators, evaluation proceeds from left to right. Or the associatively is from left to right.

Example

If we have the following arithmetic expression

9/5–7+5/2

The order of evaluation is as follows:

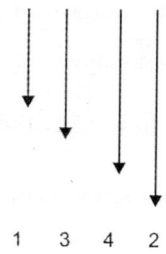

1 3 4 2

9/5–7+5/2

Example: Determine the hierarchy of operations and evaluate the following expression:

k = 3 * 3 / 4 + 5 /5 + 8 – 2 + 5 / 8

Solution

k = 9 / 4 + 5 / 5 + 8 – 2 + 5 / 8	operation: *
k = 2 + 5 /5 + 8 – 2 + 5 / 8	operation: /
k = 2 +1+ 8 - 2 + 5 / 8	operation: /
k = 2 + 1 + 8 – 2 + 0	operation: /
k = 3 + 8 – 2 + 0	operation: +
k = 11 – 2 + 0	operation: +
k = 9 + 0	operation: –
k =9	operation: +

RELATIONAL OPERATORS

In order to evaluate a comparison between two expressions we can use the relational and equality operators. The outcome of a relational operation is a Boolean value that can only be true or false, according to its Boolean result.

We may want to equate two expressions, for example, to know if they are equal or if one is greater than the other. Here is a list of the relational and equality operators that can be used in C.

The following are the relational operators:

==	Equal to
!=	Not equal to
>	Greater than
<	Less than
>=	Greater than or equal to
<=	Less than or equal to

The following table represents the relational operator, condition and the meaning of that operator.

Relational operator	Condition	Meaning
==	x==y	x is equal to y
!=	x!=y	x is not equal to y
<	x<y	x is less than y
<=	x<=y	x is less than or equal to y
>	x>y	x is greater than y
>=	x>=y	x is greater than or equal to y

Examples

(7 == 5) // measures to false.
(5 > 4) // measures to true.
(3 != 2) // measures to true.
(6 >= 6) // measures to true.
(5 < 5) // measures to false.

Of course, instead of using only numeric constants, we can use any valid expression, including variables. Suppose that $a = 2$, $b = 3$ and $c = 6$.

(a == 5) // measures to false since a is not equal to 5
($a*b$ >= c) // measures to true since ($2*3$ >= 6) is true
($b+4$ > $a*c$) // measures to false since ($3+4$ > $2*6$) is false
((b=2) == a) // measures to true

Note

Be careful! The operator = (one equal sign) is not the like as the operator == (two equal signs), the first one is an assignment operator (assigns the value at its right to the variable at its left) and the other one (==) is the equality operator that equates whether both expressions in the two sides of it are equal to each other. Thus, in the last expression ((b=2) == a), we first allotted the value 2 to b and then we compared it to a, that also stores the value 2, so the result of the operation is true.

LOGICAL OPERATORS

Logical operators in C, as with other computer languages, are used to evaluate expressions which may be true or false. Expressions which involve logical operations are evaluated and found to be one of two values: **true** or **false**. So far we have studied simple conditions. If we want to test multiple conditions in the process of making a decision, we have to perform simple tests in separate IF statements (will be introduced in detail in the next chapter). C provides logical operators that may be used to form more complex conditions by combining simple conditions.

In C language the following logical operators exists.

Operator	Meaning
&&	Logical AND
||	Logical OR
!	Logical NOT

NOT Operator !

The Operator ! is the C operator to perform the Boolean operation NOT, it has only one operand, located at its right, and the only thing that it does is to inverse the value of it, producing

false if its operand is true and true if its operand is false. Basically, it gives the opposite Boolean value of evaluating its operand.

In C language NOT operator is represented by ! Sign.

The truth table of NOT operator is as follows:

a	$!a$
True	False
False	True

The above table shows that NOT operator is used to converse the result.

Example

!(5 == 5) // measures to false because the expression at its right (5 == 5) is true.

!(6 <= 4) // measures to true because (6 <= 4) would be false.

!true // measures to false

!false // measures to true.

AND Operator &&

The logical operators && and || are used when evaluating two expressions to obtain a single relational result. The operator && corresponds with Boolean logical operation AND. This operation results true if both its two operands are true and false otherwise.

The Truth table of AND operator is as follows:

a	b	$a \&\& b$
True	True	True
True	False	False
False	True	False
False	False	False

The above table shows that if we impose the AND operator between two conditions then the result of the final condition is true when both the conditions are true.

OR Operator

The operator || corresponds with Boolean logical operation OR. This operation results true if either one of its two operands is true, thus being false only when both operands are false themselves. Here are the possible results of a || b:

The truth table of OR operator is as follows

| | OPERATOR

a	b	a		b
True	True	True
True	False	True
False	True	True
False	False	False

The above table shows that if we impose the AND operator between two conditions then the result of the final condition is true when either one condition is true or both the conditions are true.

Example

((5 = 5) && (3 > 6)) // evaluates to false (true && false).
((5 = 5) | | (3 > 6)) // evaluates to true (true | | false).

The following table shows the operator precedence and associativity of logical operators in C language.

Operator	Associativity
!	Right to left
&&	Left to right
			Left to right

CONDITIONAL OPERATOR

C language provides a conditional operator (?:) which is intimately related to the **If/else** structure. The conditional operator is Cs only ternary operator—it takes three operands. The operands together with the conditional operator build a conditional expression. The first operand is a condition, the second operand represents the value of the entire conditional expression if the condition is true and the third operand is the value for the entire conditional expression if the condition is false.

The general syntax of conditional operator is as follows:

Condition? result1 : result2

If condition is true the expression will return result1, if it is not it will return result2.

Example 1

x = (y<25)? 9: 10;

This means, if (y<25), then x = 9 else x = 10;

Example 2

7==5? 4: 3 // returns 3, since 7 is not equal to 5.

Example 3

(a>b)? printf ("a is greater than b \n"): printf ("b is greater than a \n");

If a is greater than b, then first printf statement is executed else second printf statement is executed.

Example 4

7==5+2? 4 : 3 // returns 4, since 7 is equal to 5+2.

Example 5

5>3? a : b // returns the value of a, because 5 is greater than 3.

Programming Problem

The following program represents the concept of conditional operator.

Program	Output
// conditional operator	7
#include<stdio.h>	
#include<conio.h>	
void main ()	

```
{
int a,b,c;
clrscr();
a=2;
b=7;
c = (a>b)? a : b;
printf("\n The value of c is %d",c);
getch();
}
```

In this example a was 2 and b was 7, so the expression being evaluated (a>b) was not true, thus the first value specified after the question mark was discarded in favor of the second value (the one after the colon) which was b, with a value of 7.

BITWISE OPERATORS

When you learn to program in a high-level language like C (although C is fairly low-level, as high-level languages go), the idea is to avoid worrying too much about the hardware. You want the power to present mathematical abstractions, such as sets, etc. and have high-level language characteristics like threads, higher-order functions, exceptions.

High-level languages, mostly, try to make you as unaware of the hardware as possible. Clearly, this is not entirely true, because efficiency is still a major thoughtfulness for some programming languages

C, in particular, was produced to make it easier to write operating systems. Rather than write UNIX in assembly, which is slow procedure (because assembly code is tedious to write), and not very portable, the goal was to have a language that allowed for good control-flow, some abstractions (structures, function calls), and could be efficiently compiled and run quickly.

Writing operating systems wants the manipulation of data at addresses, and this wants manipulating individual bits or groups of bits.

That is where two sets of operators are useful: *bitwise* operators and *bitshift* operators.

You can find these operators in C, C++, and Java (and presumably C#, since it is basically Java). Bitwise operators permit you to read and manipulate bits in variables of certain types.

Even though such characteristics are available in C, they are not often taught in an introductory level computer programming course. That is because intro level courses prefer to emphasize abstraction. With many departments using Java, there is a trend to increase what is abstract, and not get into the representation.

For example, some languages have support for stacks, queues, hash tables, and so forth. These "canned" data structures are meant to provide you, the computer programmer, with objects that execute certain tasks, while relieving you of the tedium and detail of understanding how the data is represented.

However, if you mean to do some work in systems programming, or other forms of low-level coding (operating systems, device drivers, socket programming, network programming), acknowledging how to access and manipulate bits is important.

Bitwise AND

This makes more sense if we apply this to a specific operator. In C/C++/Java, the & operator is bitwise AND. The following is a chart that defines &, defining AND on individual bits.

x_i	y_i	x_i & y_i
0	0	0
0	1	0
1	0	0
1	1	1

Example

Variable	b_3	b_2	b_1	b_0
x	1	1	0	0
y	1	0	1	0
$z = x$ & y	1	0	0	0

Bitwise OR

The | operator is bitwise OR (it is a single vertical bar). The following is a chart that defines |, defining OR on individual bits.

x_i	y_i	x_i	y_i
0	0	0
0	1	1
1	0	1
1	1	1

Example

Variable	b_3	b_2	b_1	b_0
x	1	1	0	0
y	1	0	1	0
$z = x$	y	1	1	1	0

Bitwise XOR

The ^ operator is bitwise XOR. The usual bitwise OR operator is *inclusive* OR. XOR is true only if just one of the two bits is true. The XOR operation is quite interesting, but we defer talking about the interesting things you can do with XOR until the next set of notes.

The following is a chart that defines ^, defining XOR on individual bits.

x_i	y_i	x_i ^ y_i
0	0	0
0	1	1
1	0	1
1	1	0

Example

Variable	b_3	b_2	b_1	b_0
x	1	1	0	0
y	1	0	1	0
$z = x \wedge y$	0	1	1	0

Bitwise NOT

There is only one unary bitwise operator, and that is bitwise NOT. Bitwise NOT pitches all of the bits.

There is not that much to say about it, other than it is not the same operation as unary minus.

The following is a chart that defines ~, defining NOT on an individual bit.

x_i	$\sim x_i$
0	1
1	0

Example

Variable	b_3	b_2	b_1	b_0
x	1	1	0	0
$z = \sim x$	0	0	1	1

Bitwise Shift Operators

In the bitwise shift operators we have two Operators (i) <<(Bitwise Left-Shift) and (ii) >> (Bitwise Right-Shift)

(a) << - Bitwise Left-Shift

Bitwise Left-Shift is useful when we want to multiply an integer (not floating point numbers) by a power of 2.

The operator, like many others, takes 2 operands like this:

$a << b$

This expression returns the value of a multiplied by 2 to the power of b.

Why is it named a left-shift?

Answer: Take the binary internal representation of a, and add b number of zeros to the right, consequently "shifting" all the bits b places to the left.

Example: 4 << 2.

4 is 100 in binary number system. Adding 2 zeros to the end gives 10000, which is 16, i.e. $4*2^2 = 4*4 = 16$.

What is 4 << 3? Simply add 3 zeros to get 100000, which is $4*2^3 = 4*8 = 32$.

Notice that shifting once to the left multiplies the total by 2. Multiple shifts of 1 to the left, ensues in multiplying the number by 2 over and over again. In other words, reproducing by a power of 2.

Examples

$5 << 3 = 5*2^3 = 5*8 = 40$
$8 << 4 = 8*2^4 = 8*16 = 128$
$1 << 2 = 1*2^2 = 1*4 = 4$

Bitwise shifts are said to be more efficient than the normal arithmetic operations, but I am still yet to write a computer game :)

Bitwise shifting of negative numbers requires knowledge of the binary representation of negative numbers.

Examples

8 << 6 returns 512
1 << 5 returns 32
5 << 4 returns 80
4 << 3 returns 32
4 << 8 returns 1024
3 << 7 returns 384
6 << 5 returns 192
0 << 3 returns 0
2 << 8 returns 512
4 << 8 returns 1024

(b) >> – Bitwise Right-Shift

Bitwise Right-Shift does the opposite, and takes away bits on the right.

Suppose we had:

$a >> b$

This expression returns the value of a divided by 2 to the power of b.

Example: 8 >> 2.

8 is 1000 in binary. Performing a right shift of 2 involves knocking the last 2 bits off: 10~~00~~, which leaves us with 10, i.e. 2.

8 >> 2 is the as doing $8 / 2^2 = 8 / 4 = 2$.

But what happens if we had a left operand that is not a power of 2? Let us try 9 >> 2.

9 is 10~~01~~. Now take off the last 2 bits, allowing for us with 10, which is 2. But this does make sense, since 9 / 4 = 2.25, which rounds down to 2.

Another example: 29 >> 3.

29 is 11101 in binary. Take of the last 3 bits to leave 11, which is 3 in decimal number system. Check: 29 / 8 is 3.625, which rounds down to 3, so we are okay :)

Examples

114 >> 4 returns 7
36 >> 1 returns 18
96 >> 4 returns 6
55 >> 0 returns 55
151 >> 3 returns 18
140 >> 1 returns 70
181 >> 0 returns 181
66 >> 0 returns 66
12 >> 1 returns 6
130 >> 0 returns 130

Uses of Bitwise Operations

Occasionally, you may want to implement a large number of Boolean variables, without using a lot of space.

A 32-bit int can be used to store 32 Boolean variables. Normally, the minimum size for one Boolean variable quantity is one byte. All types in C must have sizes that are products of bytes. However, only one bit is essential to represent a Boolean value.

You can also use bits to present elements of a (small) set. If a bit is 1, then element i is in the set, otherwise it is not.

You can use bitwise AND to implement set intersection, bitwise OR to implement set union.

Facts about Bitwise Operators

Consider the expression x + y. Do either x or y get changed? The answer is no.

Most built-in binary manipulators do not modify the values of the arguments. This uses to logical operators too. They do not modify their arguments.

There are operators that do assignment such as +=, −=, *=, and so on. They utilize to logical operators too. For example, |=, &=, ^=. Nearly all binary operators have a version with = after it.

Example

The following example shows the use of bitwise operators in C language.

```
#include <stdio.h>
/* a demonstration of C bitwise operators */
void main()
{
int d1 = 4,/* 101 */
d2 = 6,/* 110*/
d3;
printf("\nd1=%d",d1);
printf("\nd2=%d",d2);
d3 = d1 & d2; /* 0101 & 0110 = 0100 (=4) */
printf("\n Bitwise AND d1 & d2 = %d",d3);
d3 = d1 | d2;/* 0101 | 0110 = 0110 (=6) */
printf("\n Bitwise OR d1 | d2 = %d",d3);
d3 = d1 ^ d2;/* 0101 & 0110 = 0010 (=2) */
printf("\n Bitwise XOR d1 ^ d2 = %d",d3);
d3 = ~d1; /* ones complement of 0000 0101 is
1111 1010 (-5) */
printf("\n Ones complement of d1 = %d",d3);
d3 = d1<<2;/* 0000 0101 left shift by 2 bits is
0001 0000 */
printf("\n Left shift by 2 bits d1 << 2 =
%d",d3);
d3 = d1>>2;/* 0000 0101 right shift by 2 bits is
0000 0001 */
printf("\n Right shift by 2 bits d1 >> 2 =
%d",d3);
getch();
}
```

The output of the above program is as follows:
d1=4
d2=6
Bitwise AND d1 & d2 = 4
Bitwise OR d1 | d2 = 6
Bitwise XOR d1 ^ d2 = 2
Ones complement of d1 = -5
Left shift by 2 bits d1 << 2 = 16
Right shift by 2 bits d1 >> 2 = 1

C SHORTHAND OPERATORS

C has a special shorthand that simplifies coding of certain type of assignment statements.

Example

a = a+2; can be written as: a += 2;

The operator +=tells the compiler that a is assigned the value of a + 2;

This shorthand works for all binary operators in C.

The general syntax is as follows:
variable operator = variable / constant / expression

Operators	Examples	Meaning
+=	a+=2	a=a+2
−=	a−=2	a=a−2
*=	a*=2	a = a*2
/=	a/=2	a=a/2
%=	a%=2	a=a%2
&&=	a&&=c	a=a&&c
||=	a||=c	a=a||c
>>=	a>>=c	a=a>>c
<<=	a<<=c	a=a<<c
&=	a&=c	a=a&c
|=	a|=c	a=a|c

Example

The following table represents the some basic C Shorthand Operators.

Expression	Is equivalent to
Value += increase;	Value = value + increase;
a −= 5;	a = a − 5;
a /= b;	a = a / b;
Price *= units + 1;	Price = price * (units + 1);

Programming Problem

The following program represents the concept of += Operator.

Program	Output
// compound assignment operators	5
#include <stdio.h>	
void main ()	
{	
int a, b=3;	
a = b;	
a+=2; // equivalent to a=a+2	
printf("\n The value of a is %d",a);	
}	

OTHER OPERATORS

Now we will discuss some other operators used in C Programming Language.

(a) Comma operator (,)

The comma operator (,) is used to separate two or more expressions that are included where only one expression is expected. When the set of formulas has to be measured for a value, only the rightmost expression is considered.

Example

a = (b=3, b+2);

Would first assign the value 3 to b, and then assign b+2 to variable a. So, at the end, variable a would hold the value 5 while variable b would contain value 3.

Generally, comma operator (,) is used in the for loop (will be introduced in the next Chapter)

for (i = 0,j = n;i<j; i++,j−)
{
printf ("A");
}

In this example **for** is the looping construct (discussed in the next chapter). In this loop, i = 0 and j = n are separated by comma (,) and i++ and j−are separated by comma (,).

(b) sizeof () operator

This operator accepts one parameter, which can be either a type or a variable itself and returns the size in bytes of that type or object:
a = sizeof(char);
.

In the above declaration a holds 1 byte because a character variable holds one byte memory.

a=sizeof(int);

In the above declaration a holds 2 bytes because an integer variable holds 2 bytes memory.

The value returned by sizeof is a constant, so it is always determined before program execution.

(c) Increase and decrease (++, −)

The increment operator increases the variable by one and decrement operator decreases the variable by one. These operators can be written in two forms i.e. before a variable or after a variable. If an **increment/decrement** operator is written before a variable, it is referred to as **preincrement/predecrement** operators and if it is written after a variable, it is referred to as **postincrement/postdecrement** operator.

Example

a++ or ++a is equivalent to a = a+1 and a− or −a is equivalent to a = a −1

The importance of **pre** and **post** operator occurs while they are used in the expressions. **Preincrementing (predecrementing)** a variable causes the variable to be incremented (decremented) by 1, then the new value of the variable is used in the expression in which it appears. **Postincrementing (postdecrementing)** the variable causes the current value of the variable is used in the expression in which it appears, then the variable value is incremented (decremented) by 1.

The explanation is given below

(1) If we write b=a++

The meaning of this line is that the initial value of *a* is initialized in *b* and then the value of *a* is incremented.

Example

int a=10;
int b;
b=a++;

In this case the value of b is 10 and the value of a is 11.

(2) If we write b=++a

The meaning of this line is that firstly the value of a is incremented and then the incremented value is stored in b.

Example

int a=10;
int b;
b=++a;

In this case the value of b is 11 and the value of a is 11.

(3) If we write b=a—

The meaning of this line is that firstly the initial value of *a* is initialized in *b* and then the value of a is decremented.

Example

int a=10;
int b;
b=a—;

In this case the value of *b* is 10 and the value of *a* is 9.

(4) If we write b=—a

The meaning of this line is that firstly the value of a is decremented and then the decremented value is stored in *b*.

Example

int a=10;
int b;
b=—a;

In this case the value of b is 9 and the value of a is 9.

The precedence of these operators is right to left.

Example 1

int a = 2, b=3;
int c;
c = ++a – b- -;
printf ("a=%d, b=%d,c=%d\n",a,b,c);
OUTPUT
a = 3, b = 2, c = 0.

The explanation is as follows:

Since the precedence of the operators is right to left, first b is evaluated, since it is a postdecrement operator, current value of b will be used in the expression, i.e. 3 and then b will be decremented by 1. Then, a preincrement operator is used with a, so first a is incremented to 3. Therefore, the value of the expression is evaluated to 0.

Example 2

int a = 1, b = 2, c = 3;
int k;
k = (a++)*(++b) + ++a - —c;
printf("a=%d,b=%d, c=%d, k=%d",a,b,c,k);

The output of the above code is as follows:
a = 3, b = 3, c = 2, k = 6
The evaluation is explained as follows:

K	= (a++) * (++b) + ++a - —c	
	= (a++) * (3) + 2 - 2	Step1
	= (2) * (3) + 2 - 2	Step2
	= 6	Final result

(d) Type Cast Operator

We know that when constants and variables of different types are mixed in an expression, they are converted to the same type. That is automatic type conversion takes place.

The following type conversion rules are followed implicitly:

1. All chars and **short ints** are converted to **ints**. All floats are converted to doubles.

2. In case of binary operators.

if one of the two operands is a **long double**, the other operand is converted to **long double**,

else if one operand is **double**, the other is converted to **double**,

else if one operand is **long**, the other is converted to **long**,

else if one operand is **unsigned**, the other is converted to **unsigned**,

C converts all operands "up" to the type of largest operand (largest in terms of memory requirement, e.g. **float** requires 4 bytes of storage and **int** requires 2 bytes of storage so if one operand is **int** and the other is **float**, **int** is converted to **float**).

All the above mentioned conversions are automatic conversions or implicit conversions.

Now we also convert one data type to another data type that is known as explicit conversion.

The general syntax is as follows:

(type) expression

Where *type* is the standard C data type.

For example, if you want to make sure that the expression a/5 would evaluate to type **float** you would write it as

(float) a/5

cast is an unary operator and has the same precedence as any other unary operator.

OPERATORS PRECEDENCE AND ASSOCIATIVITY

When writing complex expressions with several operands, we may have some doubts about which operand is evaluated first and which later. For example, in this expression:

$$a = 5 + 7 \% 2$$

we may doubt if it really means:

$$a = 5 + (7 \% 2) \text{ // with a result of 6, or } a = (5 + 7) \% 2 \text{ // with a result of 0}$$

The correct answer is the first of the two expressions, with a result of 6. There is an established order with the priority of each operator, and not only the arithmetic ones (those whose preference come from mathematics) but for all the operators which can appear in C. From greatest to lowest priority, the priority order is as follows:

Level	Operator	Description	Associativity
1	::	Scope	Left-to-right
2	() [] . -> ++ — dynamic_cast static_cast reinterpret_cast const_cast typeid	Postfix	Left-to-right
3	++ — ~ ! sizeof new delete	Unary (prefix)	Right-to-left
	* &	Indirection and reference	
	(pointers)		
	+ –	Unary sign operator	
4	(type)	Type casting	Right-to-left
5	.* ->*	Pointer-to-member	Left-to-right
6	* / %	Multiplicative	Left-to-right
7	+ –	Additive	Left-to-right
8	<< >>	Shift	Left-to-right
9	< > <= >=	Relational	Left-to-right
10	== !=	Equality	Left-to-right
11	&	Bitwise AND	Left-to-right
12	^	Bitwise XOR	Left-to-right
13	|	Bitwise OR	Left-to-right
14	&&	Logical AND	Left-to-right
15	||	Logical OR	Left-to-right
16	?:	Conditional	Right-to-left
17	= *= /= %= += –= >>= <<= &= ^= |=	Assignment	Right-to-left
18	,	Comma	Left-to-right

Grouping defines the precedence order in which operators are evaluated in the case that there are several operators of the same level in an expression.

All these precedence levels for operators can be manipulated or become more legible by removing possible ambiguities using parentheses signs (and), as in this example:

a = 5 + 7 % 2;

might be written either as:

a = 5 + (7 % 2);

or

a = (5 + 7) % 2;

Depending on the operation that we want to perform.

So if you want to write complicated expressions and you are not entirely sure of the precedence levels, always admit parentheses. It will also become a code lighter to read.

Programming Problems

Problem 1

Write a program to input two numbers and print the addition, subtraction, multiplication and division of both the numbers.

Solution

```
#include<stdio.h>
#include<conio.h>
void main()
{
int a,b,c;
clrscr();
printf("\n Enter the values of a and b");
scanf("%d %d",&a,&b);
c=a+b;
printf("\n The addition is %d",c);
c=a-b;
printf("\n The subustraction is %d",c);
c=a*b;
printf("\n The multiplication is %d",c);
c=a/b;
printf("\n The division is %d",c);
getch();
}
```

Problem 2

Write a program to input the radius of a circle and print the area of that circle.

Solution

```
#include<stdio.h>
#include<conio.h>
void main()
{
float radius,area;
clrscr();
printf("\n Enter the radius");
scanf("%f",&radius);
area=3.14*radius*radius;
printf("\n The area of circle is %f",area);
getch();
}
```

SUMMARY

- Operator is a symbol that represents a specific action.
- Arithmetic operators are used to perform the arithmetic operations between operands.
- Relational operators are used to return either true or false based on some particular condition.
- Logical operators are used to perform some logical operations.
- Assignment operator is used to assign the value of the variable.
- If we impose the and operator between two conditions then the result of the condition is true when the result of both the conditions are true.
- If we impose the or operator between two conditions then the result of the condition is true when either one condition is true or both the conditions are true.
- Not operator is used to converse the result of the given condition.
- ++ and − operator are used to increase and decrease the values.
- Comma operator is used to separate the value of the variables with in loop.
- Bitwise operator is used to perform the low level programming in C language.

Exercises

Section A

Multiple Choice Questions

1. Arithmetic operators does not include
   (a) +   (b) −   (c) *   (d) ^
2. Compound statement includes
   (a) +=   (b) −=
   (c) >>=   (d) All of the above
3. !(5==6) evaluates to
   (a) 1   (b) 5   (c) True   (d) False
4. 1<<5 returnes
   (a) True   (b) False   (c) 32   (d) 1
5. a−= b is equivalent to
   (a) b=a−b   (b) b= b−a
   (c) a= a−b   (d) a= b−a
6. 4 << 3 represents
   (a) 4 is much greater than 3
   (b) 4* 2^3   (c) 3* 2^ 4
   (d) None of the above
7. Priority of the operators in decreasing fashion is
   (a) :: > sizeof() > type > &&
   (b) sizeof() >:: >type> &&
   (c) type>sizeof()>&&>::
   (d) && >::> sizeof()>type
8. a= 5+7%2 will give a result of
   (a) 0   (b) 6
   (c) Both a and b   (d) None of the above
9. >= represents
   (a) Less than equal to
   (b) Greater than equal to
   (b) May be greater may be equal
   (d) None of the above
10. !(true) will evaluate to
    (a) True   (b) False
    (c) Both a and b   (d) Wrong evaluation

State True/False

1. The assignment operators assigns right value to a left value.
2. ^ is the arithmetic operator supported by C language.
3. && is the bitwise operator supported by C.
4. sizeof() operator can accept either type or variable as its parameter.
5. The ^ represents bitwise NOR.
6. /= represents a compound statement.
7. =< represents less than or equal to.

8. << is used when we have to multiply by a power of 2.
9. a= 11%3 assigns 1 to a.
10. ++ operator is equivalent to +=1.

Section B

Short Answer Type Questions

1. Define the various operators used in C language.
2. Define the various bitwise operators used in C language.
3. Define the concept of unary operators used in C language?
4. Define the operator precedence and associability in C language.
5. Define the working of ++ and − operator with proper example.
6. Define the working of bitwise AND operator with suitable example.
7. Define the working of bitwise OR operator with suitable example.
8. Define the working of bitwise Left-Shift operator with suitable example.
9. Define the working of bitwise Right-Shift operator with suitable example.
10. Define the working of bitwise XOR operator with suitable example.

Section C

Long Answer Type Questions

1. Define the working of AND operator with suitable example.
2. Define the working of OR operator with suitable example.
3. Define the working of NOT operator with suitable example.
4. Define the working of bitwise AND operator with suitable example?
5. Define the working of bitwise OR operator with suitable example?
6. Define the working of bitwise NOT operator with suitable example.
7. Define the working of bitwise XOR operator with suitable example.
8. Write a program to implement the working of OR operator.
9. Write a program to implement the working of NOT operator.
10. Write a program to implement the working of AND operator.

8

Input and Output Functions

One of the essential operations performed in a C language programs is to provide input values to the program and output the data produced by the program to a standard output device. We can allot values to variable through assignment statements such as x = 5 a = 0 ; and so on. Another method acting is to use the input then scanf which can be used to read data from a keyboard. For outputting outcomes we have used extensive the function printf which sends out results out to a terminal. There exists several functions in 'C' language that can carry out input/output operations. These functions are conjointly known as standard input/output library. Each program that applies standard input/output function must contain the statement. # include < stdio.h > at the beginning.

TYPES OF INPUT/OUTPUT

In C language input/output (I/O) functions can be classified into two major categories.

Console I/O Function

With the help of Console I/O function we take the input from keyboard and write output to VDU (visual display unit).

FILE I/O Function

With the help of FILE I/O function we take the input from the files and write output to VDU (visual display unit).

In this chapter we only discuss the Console I/O Function and FILE I/O function will be discussed in FILE HANDLING chapter.

CONSOLE I/O FUNCTION

The screen and keyboard together are called a console. Console I/O functions can be further separated into two categories—formatted and unformatted console I/O functions. The basic difference between them is that the formatted functions allow the input read from the keyboard or the output exposed on the VDU to be formatted as per our necessities. For example, if values of average marks and percentage marks are to be displayed on the screen, then the details like where this output would appear on the screen, how many spaces would be present between the two values, the number of places after the decimal points, etc. can be assured using formatted functions.

Console I/O function can be classified in the following figure:

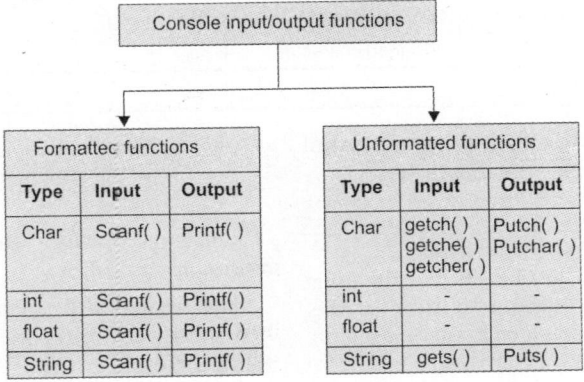

Console input/output functions		

Formatted functions

Type	Input	Output
Char	Scanf( )	Printf( )
int	Scanf( )	Printf( )
float	Scanf( )	Printf( )
String	Scanf( )	Printf( )

Unformatted functions

Type	Input	Output
Char	getch( ) getche( ) getcher( )	Putch( ) Putchar( )
int	-	-
float	-	-
String	gets( )	Puts( )

Now we will discuss these functions one by one.

FORMATTED CONSOLE I/O FUNCTIONS

From the above figure we know that printf () and scanf () are the functions which exists in the category of Formatted Console I/O functions.

These functions permit us to supply the input in a fixed format and let us obtain the output in the specified form.

Now we will discuss these predefined functions one by one.

The printf() Function

The standard library function *printf* is expended for formatted output. It builds the user input a string and an optional list of variables or strings to output. The variables and strings are output granting to the specifications in the printf () function.

The following are the important work that is done by printf() function.

- Printf shows information on screen.
- Printf gives the number of characters printed.

- Printf shows the text you put inside the double quotes.
- Printf needs the backslash character, an escape sequence, to display some special characters.
- Printf can show variables by using the % conversion character.
- Printf formats a string argument followed by any extra arguments.

The general syntax of printf () function is as follows:

printf ("format string", list of variables);

Where the format string can contain:

- Characters that are simply printed as they are
- Conversion specifications that begin with a % sign
- Escape sequences that begin with a \ sign

The following table represents the format specifiers for the basic data types in C language.

Data type		Format specifier
Integer	short signed	%d or %l
	short unsigned	%u
	long signed	%1d
	long unsigned	%1u
	unsigned hexadecimal	%x
	unsigned octal	%o
Real	float	%f
	double	%1f
Character	signed character	%c
	unsigned character	%c
String		%s

We can also provide following optional specifiers in the format specifications:

Specifier	Description
dd	Digits specifying field width
.	Decimal point breaking field width from precision (precision stands for the number of places after the decimal point)
dd	Digits specifying precision

– Minus sign for left excusing the output in the specified field width

The explanations of these optional specifiers are as follows:

The field width specifier assures **printf ( )** how many columns on screen should be used while printing a value.

Example

%5d says, "Print the variable as a decimal integer in a field of 5 columns". If the value to be printed encounters not to fill up the entire field, the value is right justified and is padded with blanks on the left.

If we include the minus sign in format specifier (as in **%-5d**), this means left justification is desired and the value will be padded with blanks on the right.

Programming Example

Specifying the field width can be useful in creating tables of numeric values, as the following program demonstrates.

```
#include<stdio.h>
#include<conio.h>
void main( )
{
clrscr();
printf ("\n%f %f %f", 10.0, 15.5, 135.9 ) ;
printf ( "\n%f %f %f", 310.0, 1220.9, 3102.3 ) ;
getch();
}
```

And here is the output...

```
10.000000 15.500000 135.900000
310.000000 1220.900000 3102.300000
```

Even though the numbers have been printed, the numbers have not been lined up properly and hence are hard to read. A better way would be something like this...

```
void main( )
{
clrscr();
printf ("\n%10.1f %10.1f %10.1f", 10.0, 15.5,
135.9 ) ;
printf ( "\n%10.1f %10.1f %10.1f", 310.0,
1220.9, 3102.3 );
getch();
}
```

This results into a much better output...

```
0123456789012345678901234567890 1
10.0        15.5        135.9
310.0       1220.9      3102.3
```

The format specifiers could be used even while displaying a string of characters. The following program would clarify this point:

```
void main( )
{
```

```
char f[ ] = "Manish" ;
char s[ ] = "Varshney" ;
char f1[ ] = "Vineet" ;
char s1[ ] = "Agrawal" ;
printf ( "\n%20s%20s", f, s ) ;
printf ( "\n%20s%20s", f1, s1 ) ;
getch();
}
```

And here's the output...

```
0123456789012345678901234567890123456789012345678901234567890
            Manish            Varshney
            Vineet            Agrawal
```

The explanation of the above output is as follows:

The format specifier **%20s** reserves 20 columns for printing a string and then prints the string in these 20 columns with right justification. This helps lining up names of different lengths properly. Obviously, the format **%-20s** would have left justified the string.

Use of Escape Sequence in printf () statement

We saw earlier how the new line character, **\n**, when inserted in a **printf ( )**'s format string, takes the cursor to the beginning of the next line. The new line character is an 'escape sequence', so called because the backslash symbol (\) is considered as an 'escape' character—it causes an escape from the normal interpretation of a string, so that the next character is recognized as one having a special meaning.

The following example shows usage of **\n** and a new escape sequence \t, called 'tab'. A **\t** moves the cursor to the next tab stop. A 80-column screen usually has 10 tab stops. In other words, the screen is divided into 10 zones of 8 columns each. Printing a tab takes the cursor to the beginning of next printing zone.

Example

```
#include<stdio.h>
#include<conio.h>
void main ( )
{
clrscr();
printf ("You\t are\t a\t good\t boy " );
getch();
}
```

The output of the above code is as follows
You   are   a   good   boy
The following table represents the complete list of Escape Sequence.

\a	Produces a beep or flash; the cursor position is not changed.
\b	Moves the cursor to the last column of the previous line.
\f	Moves the cursor to start of next page.
\n	New line
\r	Carriage Return (Moves the cursor to the first column of the current line.)
\t	Horizontal Tab
\v	Vertical Tab
\\	Prints single \
\	for quotation
%%	Prints %

The Standard Input Function scanf ( )

scanf ( ) function allows us to enter data from keyboard that will be formatted in a certain way. The conversion specifier argument tells scanf how to convert the incoming data.

The general syntax of **scanf( )** statement is as follows:
scanf ("format string", list of addresses of variables);

The following are the important points related to scanf ( ) function

- scanf starts with a string argument and may contain additional arguments.
- Additional arguments must be pointers.
- scanf returns the number of successful inputs.

Example

scanf ("%d %f %c", &a, &b, &c) ;
Note that we are sending addresses of variables (addresses are obtained by using '**&**' the 'address of' operator) to **scanf ( )** function. This is necessary because the values received from keyboard must be dropped into variables corresponding to these addresses. The values that are supplied through the keyboard must be separated by either blank(s), tab(s), or newline(s). Do not include these escape sequences in the format string.

All the format specifications that we learnt in **printf ( )** function are applicable to **scanf ( )** function as well.

Programming Problems

Problem 1

Write a program to input two numbers and print the addition of both the numbers.

Solution

```c
#include <stdio.h>
#include<conio.h>
void main()
{
int a,b,c;
clrscr();
printf("\n Enter the value of a ");
scanf("%d",&a);
printf("\n Enter the value of b");
scanf("%d",&b);
c=a+b;
printf("The answer is %d \n",c);
}
```

The explanation of the above program is as follows:

The first instruction declares three integer variables: **a**, **b** and **c**. The first two **printf** statements simply display message on the screen asking the user for the values. The **scanf** functions then read in the values from the keyboard into **a** and **b**. These are added together and the result in **c** is displayed on the screen with a suitable message. Notice the way that you can include a message in the **printf** statement along with the value.

Type the program in, compile it and link it and the result should be your first interactive program. Try changing it so that it works out something a little more adventurous. Try changing the messages as well. All you have to remember is that you cannot store values or work out results greater than the range of an integer variable or with a fractional part.

Problem 2

Write a program to input two numbers and print the addition, multiplication, subtraction and division of both the numbers.

Solution

```c
#include<stdio.h>
#include<conio.h>
void main()
{
```

```
int a,b,c;
printf("\n enter the values of a and b");
scanf("%d %d",&a,&b);
c=a+b;
printf("\n the addition is %d"c);
c=a-b;
printf("\n the subtraction is %d",c);
c=a*b;
printf("\n the multiplication is %d"c);
c=a/b;
printf("\n the division is %d",c);
getch();
}
```

UNFORMATTED CONSOLE I/O FUNCTIONS

The basic difference between formatted and unformatted I/O functions is that the formatted functions allow the input read from the keyboard or the output displayed on the VDU to be formatted as per our requirements. For example, if values of average marks and percentage marks are to be exposed on the screen, then the details like where this output would appear on the screen, how many spaces would be present among the two values, the number of places after the decimal points, etc., can be controlled using formatted functions.

In the unformatted console I/O functions we cannot provide the formatting according to our requirement.

In this category, we will discuss the following unformatted I/O functions.

- getchar()
- putchar()
- getch()
- putch()
- getche()
- gets()
- puts()

Getchar() Function

The getchar () function is used to read a single character from the keyboard.

The general syntax of getchar() function is as follows:

int getchar (void);

Example

/* getchar example: typewriter */

```
#include <stdio.h>
#include<conio.h>
void main ()
{
char c;
puts ("Enter text. Include a dot ('.') in a
sentence to exit:");
do
{
c=getchar();
putchar (c);
} while (c != '.');
getch();
}
```

The output of the above program is as follows:

A simple typewriter. Every sentence is echoed once ENTER has been pressed until a dot (.) is included in the text.

Putchar() Function

putchar is a function in C programming language that writes a single character to the standard output stream, stdout. Its prototype is as follows:

int putchar (int character)

The character to be printed is fed into the function as an argument, and if the writing is successful, the argument character is returned. Otherwise, end-of-file is returned.

The putchar function is specified in the C standard library header file **stdio.h**.

Example

```
/* putchar example: printing alphabet */
#include <stdio.h>
#include<conio.h>
void main ()
{
char c;
clrscr();
for (c = 'A' ; c <= 'Z' ; c++)
{
putchar (c);
}
getch();
}
```

Getch() Function

The getch() function is used to get the next available keystroke from the console. The getch() function obtains the next available keystroke from the console. Nothing is echoed on the screen (the function getche() will echo the keystroke, if possible). When no keystroke is available, the function waits until a key is depressed. This function is available in conio.h header file.

The general syntax of getch() function is as follows:

```
int getch( void );
```

Example

WAP TO ENTER CHARACTER BY USING GETCH ( ) FUNCTION

```
#include<stdio.h>
#include<conio.h>
void main ()
{
char ch;
clrscr ();
printf ("Enter any character: ");
ch=getch();
printf ("You have pressed %c",ch);
getch ();
}
```

The output of the above program is as follows:

Enter any character: M
You have pressed M

Putch() Function

putch () function is used to write one character onto the screen. This function is predefined in conio.h header file.

The general syntax of putch() function is as follows:

```
void putch (character)
```

Example

The following example demonstrates the concept of putch() function.

```
#include<stdio.h>
#include<conio.h>
void main()
{
char ch;
clrscr();
printf("\n Enter a character");
scanf("%c",&ch);
putch(ch);
getch();
```

}

The output of the above program is as follows:
Enter a Character C
C

Getche() Function

getche() function reads each keyboard entry and echoes it to the screen without the ENTER key being pressed.

Example

```
#include <stdio.h>
#include <conio.h>
void main(void)
{
char ch;
do {
ch = getche();
printf("%c", ch);
} while(ch != 'q');
getch();
}
```

The output of the above code is as follows:

This program accepts the character from the keyboard and print that characters until we press q.

Difference between getch() and getche() function

getche () is almost same as getch(). Apart from holding the screen or waiting for a character to be entered to exit the output screen, it also prints the input given by the user.

Gets() Function

Function gets can be used to get a single string from the keyboard. Unlike scanf gets will get spaces too. It takes one parameter, a string. Here is the format.

```
gets (string);
```

it will continue to store characters in string until the user hits enter. You should probably use this function for reading single strings rather than scanf.

To use gets you need the stdio.h header file.

Example

Write a program to input a string and print the string with the help of gets () function.

Solution

```
#include<stdio.h>
```

```
#include<conio.h>
void main ()
{
char name [25];
clrscr ();
printf("\nEnter the string");
gets(name);
printf("\n The string is %s",name);
getch();
}
```
The output of the above program is as follows:
Enter the String Manish Varshney
The String is Manish Varshney

Puts() Function

Function puts will display a single string of text on its own line. The string can be a variable or a string constant (text in closed in parenthesis). This function has no special formatting capabilities like printf does. Here is the format.

```
    puts ("this will appear on it's own line");
    //or
    puts (string );
```
Either way will result the string being on its own line.
To use puts you need the stdic.h header file.

Example

Write a program input a string and print the string with the help of puts () function.

Solution

```
#include<stdio.h>
#include<conio.h>
void main ()
{
char name[25];
clrscr();
puts("\n Enter the String");
gets(name);
puts(name);
getch();
}
```

SUMMARY

- Input/Output functions are useful when we want to take the input from the keyboard or we want to print the data on to the monitor.
- Input/Output functions can also be classified into Console Input/Output and File Input/Output.
- Printf() is the predefined function which is used to print the contents to the monitor.
- Scanf() is the predefined function which is used to take the input from the keyboard.
- Printf() and scanf() are the formatted Input/Output functions and these functions are predefined in stdio.h header file.
- In the unformatted Input/Output functions we have getch(), putch(), getche(), getchar(), putchar() functions.

Exercises

Section A

Multiple Choice Questions

1. We ask user to enter data/statistics to be assigned to particular variables, this process is called
   (a) Input
   (b) Output
   (c) Processing
   (d) None of the above

2. Scanf can take data as
   (a) Integer
   (b) Characters
   (c) Float
   (d) All of the above

3. Printf() is a
   (a) Sser defined function
   (b) Statement
   (c) Predefined function
   (d) All of the above

4. Each source of data is treated as
   (a) String
   (b) Pointers
   (c) Stream of characters
   (d) Stream of bytes

5. Any number of characters can be read or written from a movable point, known as
   (a) Pointer
   (b) Address
   (c) File point indicator
   (d) File position indicator

6. This is not a type of buffer
   (a) Un buffered (b) Array buffered
   (c) Fully buffered (d) Line buffered
7. The buffer should be flushed by using
   _____ explicitly
   (a) Flush (b) Fflush
   (c) Both a and b
   (d) None of the above
8. Input function also includes
   (a) Getchar() (b) Gets()
   (c) Scanf() (d) All of the above
9. For output we can use
   (a) Putchar() (b) Printf()
   (c) Both a and b
   (d) None of the above
10. Char a[25]; indicates the compiler to
    (a) Save 26 locations randomly for a
    (b) Save 26 locations consequently for a
    (c) Save 25 locations randomly for a
    (d) Save 25 locations randomly for a

State True/False
1. In C all the I/O function is performed by library functions.
2. Library package for I/O in C is stdio.h.
3. Old C programs should be modified in a UNIX environment.
4. The standard input function used to take input is scanf().
5. The standard output function used to provide output is printf().
6. & symbol prefixes ach variable in case of printf().
7. Each program must have input and output.
8. We takes values from users and assign to variables is known as output.
9. We takes input from users and perform calculation to get the result is called outputting.
10. Inputting and outputting are two consecutive phases.

Section B

Short Answer Type Questions
1. Define the various input output functions in short.
2. Define the working of printf() function in short.
3. Differentiate between printf() and putchar() function in short.
4. Define the working of scanf() function in short.
5. Differentiate between getch() and getche() function in short.
6. Define the working of getch() function with example.
7. Define the working of getche() function with example.
8. Define the working of putch() function with example.
9. Define the working of putchar() function with example.
10. Define the working of getchar() function with example.

Section C

Long Answer Type Questions
1. Define the various functions used for input output in C language.
2. Define the classification of input output functions in C language.
3. Differentiate between getch() and getche() function with proper example.
4. Differentiate between getch() and getchar() function with proper example.
5. Write a program to input a character and print the character with the help of getchar() and putchar() function.
6. Write a program to input a number and print the number with the help of printf() and scanf() function with proper explanation.
7. Define the working of gets() function with the help of proper explanation.
8. Define the working of puts() function with the help of proper explanation.
9. Write a program to input a string and print the string with the help of proper explanation.
10. Write a program to enter character by using getch() function.

Conditional Statements and Looping

9

A program consists of a number of statements to be executed by the computer. Not many of the programs execute all their statements in sequential order from beginning to end as they appear within the program. A C program may require that a logical test be carried out at some particular point within the program. One of the several possible actions will be carried out, depending on the outcome of the logical test. This is predicted as **Branching**. In the **Selection** process, a set of statements will be selected for execution, among the several sets available. Suppose, if there is a need of a group of statements to be executed repeatedly until some logical condition is satisfied, then **looping** is required in the program. These can be carried out using various control statements.

These **Control statements** determine the "flow of control" in a program and enable us to specify the order in which the various instructions in a program are to be executed by the computer. Normally, high-level procedural programming languages require three basic control statements:

- Sequence instruction
- Selection/decision instruction
- Repetition or loop instruction

Sequence instruction means executing one instruction after another, in the order in which they occur in the source file. This is usually built into the language as a default action, as it is with C. If an instruction is not a control statement, then the next instruction to be executed will simply be the next one in sequence.

Selection means executing different sections of code depending on a specific condition or the value of a variable. This allows a program to take different courses of action depending on different conditions. C provides the following selection structures.

- if
- if else
- nested if
- if else if else
- switch case

Repetition/Looping means executing the same section of code more than once. A section of code may either be executed a fixed number of times, or while some condition is true. C provides three looping statements:

- while
- do...while
- for
- Nested Loops

This chapter introduces you the decision and loop control statements that are available in C programming language along with some of the example programs.

DECISION CONTROL STATEMENTS

In a C program, a decision causes a one-time jump to a different part of the program, depending on the value of an expression. Decisions in C can be made in several ways. The most significant is with the **if...else** statement, which prefers between two alternatives. This statement can be applied without the **else**, as a simple **if** statement. Another decision control statement, **switch**, creates branches for multiple alternative sections of code, depending on the value of a single variable.

The if Statement

It is used to execute an *instruction* or sequence/ *block of instructions* only if a *condition* is fulfilled. In *if* statements, expression is evaluated first and then, depending on whether the value of the expression (relation or condition) is *"true"* or *"false"*, it transfers the control to a particular statement or a group of statements.

Different forms of implementation *if-* statement are as follows:

- Simple *if* statement
- *If-else* statement
- *Nested if-else* statement
- *Else if* statement

Simple if statement

It is used to execute an instruction or block of instructions only if a given condition is true.

The syntax of simple if statement is as follows if (condition)

statement;

Where condition is the expression that is to be evaluated. If this condition is **true**, statement is performed. If it is **false**, statement is ignored (not executed) and the program continues on the next instruction after the conditional statement.

The flow chart of the simple if statement is as follows:

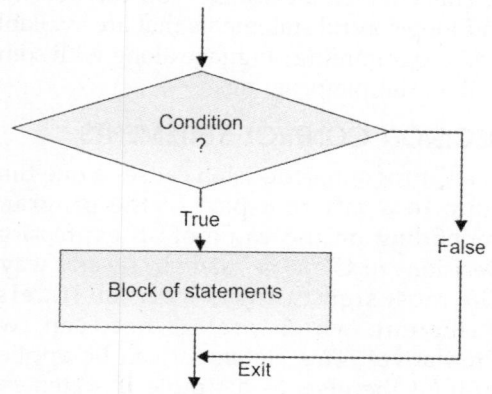

If we want more than one statement to be executed, then we can specify a block of statements within the curly brackets { }.

The syntax is as follows:
if (condition)
{
block of statements;
}

Programming Problem

Problem 1

Write a program to calculate the net salary of an employee. If the gross salary is less than 15000 then the net salary is equal to the gross salary otherwise net salary is equal to the gross salary minus 15% of gross salary.

Solution

```
#include <stdio.h>
#include<conio.h>
void main ( )
{
float gross_salary, net_salary;
clrscr();
printf("Enter gross salary of an employee\n");
scanf("%f ",&gross_salary );
if (gross_salary <15000)
net_salary= gross_salary;
if (gross_salary >= 10000)
net_salary = gross_salary- 0.15*gross_salary;
printf("\nNet salary is Rs.%.2f\n", net_salary);
getch();
}
```

The output of the above code is as follows:
Enter gross salary of an employee
12000
Net salary is Rs.12000.00
Enter gross salary of any employee
16000
Net salary is Rs. 13600.00

If ... Else Statement

If...else statement is used when a different sequence of instructions is to be executed depending on the logical value (True/False) of the condition evaluated.

The general syntax of if else statement is as follows:

if <condition>
{
<Statements1>;

```
}
else
{
<Statements2>;
}
```

The execution of the above if else statement is as follows:

If the result of the above condition is true then the statement of if condition is executed and if the result of the above condition is false then pointer is automatically goes to the else part and the statements of the else is executed. The { } is imposed when we want to execute more than one statements according to if and else statement.

The flow chart of the above statement is as follows:

Programming Problems

Problem 1

Write a program to input two numbers and then print the maximum number between them.

Solution

```
#include<stdio.h>
#include<conio.h>
void main()
{
int a,b;
clrscr();
printf("\n enter the values of a and b");
scanf("%d %d",&a,&b);
if(a>b)
```

```
printf ("\n the maximum number is %d", a);
else
printf("\n the maximum number is %d", b);
getch();
}
```

The output of the above code is as follows:
Enter the values of a and b
23 56
The maximum number is 56.

Problem 2

Write a program to input two numbers and then print the minimum number between them.

Solution

```
#include<stdio.h>
#include<conio.h>
void main()
{
int a,b;
clrscr();
printf("\n Enter the values of a and b");
scanf("%d %d",&a,&b);
if(a<b)
printf("\n the minimum number is %d", a);
else
printf("\n the minimum number is %d", b);
getch();
}
```

The output of the above code is as follows:
Enter the values of a and b
23 56
The minimum number is 23.

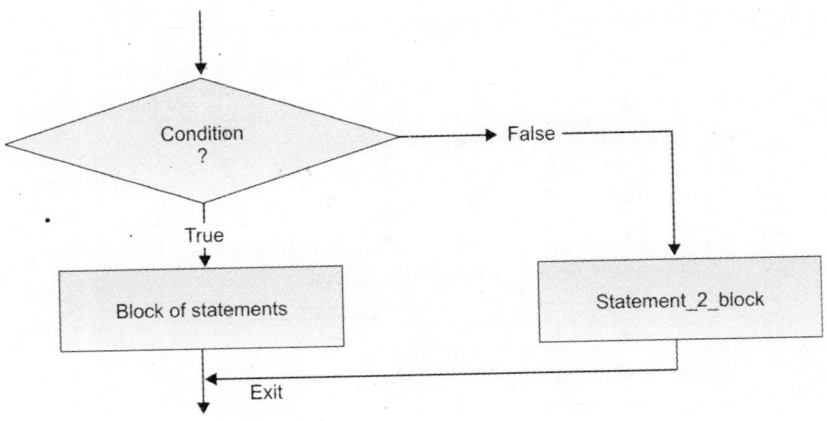

Problem 3

Write a program to print whether the given number is even or odd.

Solution

```c
#include <stdio.h>
#include<conio.h>
void main ( )
{
int x;
clrscr();
printf("Enter a number:\n");
scanf("%d",&x);
if (x % 2 == 0)
printf("\n given number is even\n");
else
printf("\n given number is odd\n");
getch();
}
```

The output of the above code is as follows:

Enter a number:

8

Given number is even

Enter a number

11

Given number is odd

Nested if Else

In *nested if... else statement,* an entire *if...else* construct is written within either the body of the *if* statement or the body of an *else* statement.

The syntax is as follows:

```c
if (condition_1)
{
if (condition_2)
{
Statements_1_Block;
}
else
{
Statements_2_Block;
}
}
else
{
Statements_3_Block;
}
Statements_4_Block;
```

The execution of the above structure is as follows:

Here, condition_1 is evaluated. If it is **false** then Statements_3_Block is executed and is followed by the execution of State- ments_4_Block, otherwise if condition_1 is **true,** then condition_2 is evaluated. Statements_1_Block is executed when condition_2 is **true** other- wise Statements_2_ Block is executed and then the control is transferred to Statements_4_Block.

The flow chart of the Nested if structure is as follows:

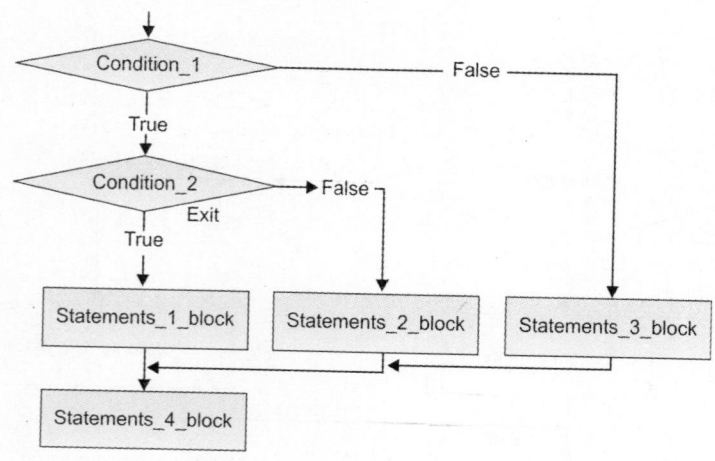

Programming Problems

Problem 1

Write a program to input three numbers and print the maximum number with the help of Nested if.

Solution

```
#include <stdio.h>
#include<conio.h>
void main()
{
int one,two,three;
clrscr();
printf("\n enter three numbers");
scanf("%d%d%d", &one, &two, &three);
if ( one>two)
{
if ( one>three)
printf("\n the maximum number is %d",one);
else
printf("\n the maximum number is %d", three);
}
else
{
if ( two>three)
printf("\n the maximum number is %d",two);
else
printf("\n the maximum number is %d", three);
}
getch();
}
```

The output of the above code is as follows:
Enter three numbers
12
34
55
The maximum number is 55

Problem 2

Write a program to input three numbers and print the minimum number with the help of Nested if.

Solution

```
#include <stdio.h>
#include<conio.h>
void main()
{
int one,two,three;
clrscr();
printf("\n enter three numbers");
scanf("%d%d%d",&one,&two,&three);
if ( one<two)
{
if ( one<three)
printf("\n the minimum number is %d",one);
else
printf("\n the minimum number is %d", three);
}
else
{
if ( two<three)
printf("\n the minimum number is %d", two);
else
printf("\n the minimum number is %d", three);
}
getch();
}
```

The output of the above code is as follows:
Enter three numbers
12
34
55
The minimum number is 12

If else if else Statement

To show a multi-way decision based on several conditions, we use the *else if* statement. This works by cascading of several comparisons. As soon as one of the conditions is true, the statement or block of statements following them is executed and no further comparisons are performed.

The general syntax of if else if else is as follows:

```
if (condition_1)
{
Statements_1_Block;
}
else if (condition_2)
{
Statements_2_Block;
}
```

------

else if (condition_n)

{

Statements_n_Block;

}

else

Statements_x;

The execution of the above code is as follows:

Here, the *conditions* are evaluated in order from top to bottom. As soon as any condition evaluates to *true*, then the statement associated with the given condition is executed and control is transferred to *Statements_x* skipping the rest of the conditions follo-wing it. But if all conditions evaluate *false*, then the statement following final *else* is executed followed by the execution of *Statements_x*.

The flow chart of the if else if else statement is as follows:

Programming Problems

Problem 1

Write a program to input three numbers and then print the maximum number between them with the help of if else if else.

Solution

```
#include <stdio.h>
#include<conio.h>
void main()
{
int a,b,c;
clrscr();
printf("\n enter the values of a b and c");
scanf("%d %d %d",&a,&b,&c);
if(a>b && a>c)
printf("\n the maximum no is %d",a);
else if(b>a && b>c)
printf("\n the maximum number is %d",b);
else
printf("\n the maximum number is %d",c);
getch();
}
```

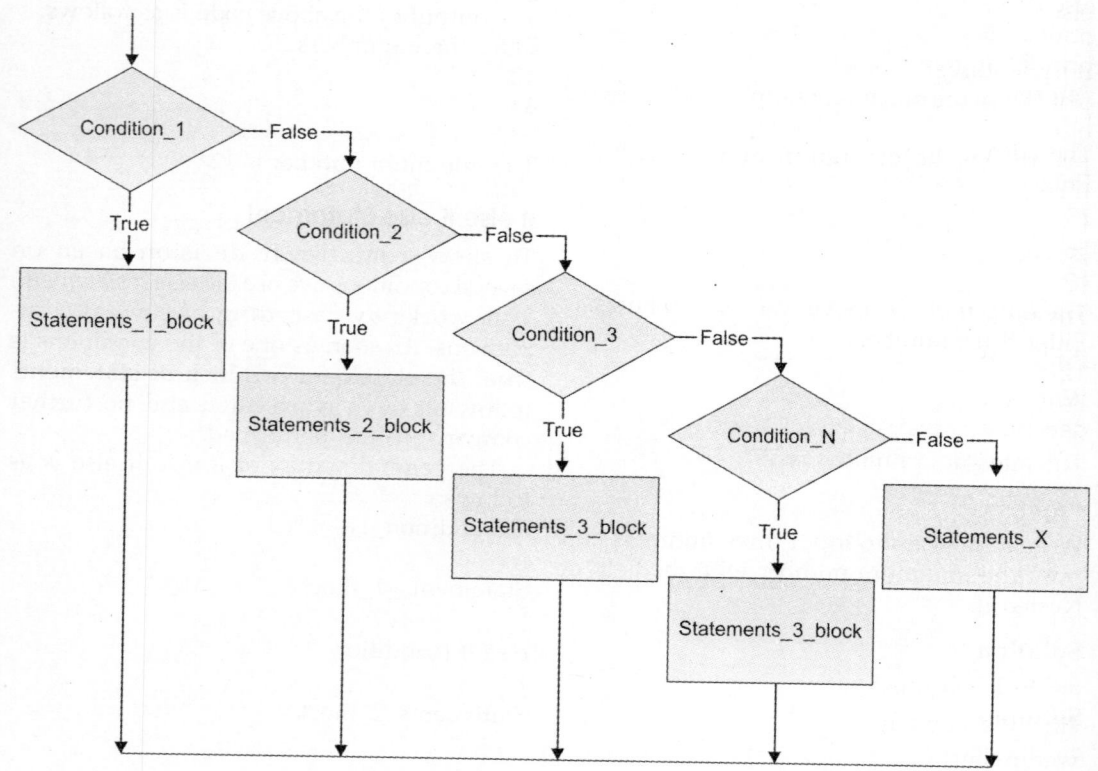

The output of the above code is as follows:
Enter the values of a, b and c
23
56
45
The maximum number is 56.

Problem 2

Write a program to input three numbers and then print the minimum number between them with the help of if else if else.

Solution

```
#include <stdio.h>
#include<conio.h>
void main()
{
int a,b,c;
clrscr();
printf("\n Enter the values of a b and c");
scanf("%d %d %d",&a,&b,&c);
if(a<b && a<c)
printf("\n the minimum number is %d", a);
else if(b<a && b<c)
printf("\n the minimum number is %d", b); else
printf("\n the minimum number is %d", c);
getch();
}
```

The output of the above code is as follows
Enter the values of a, b and c
23
56
45
The minimum number is 23

Problem 3

Write a program to award grades to students depending upon the criteria mentioned below
- Marks less than or equal to 50 are given "D" grade
- Marks above 50 but below 60 are given "C" grade
- Marks between 60 and 75 are given "B" grade
- Marks greater than 75 are given "A" grade.

Solution

```
#include <stdio.h>
#include<conio.h>
void main()
{
int result;
printf ("Enter the total marks of a student:\n");
scanf("%d",&result);
if (result <= 50)
printf ("Grade D\n");
else if (result <= 60)
printf("Grade C\n");
else if (result <= 75)
printf("Grade B\n");
else
printf("Grade A\n");
}
```

The output of the above code is as follows:
Enter the total marks of a student:
80
Grade A

The Switch Statement

While **if** is good for choosing between two alternatives, it quickly becomes cumbersome when several alternatives are needed. Cs solution to this problem is the **switch** statement. The **switch** statement is Cs multiple selection statement. It is used to select one of several alternative paths in program execution and works like this: A variable is successively tested against a list of integer or character constants. When a match is found, the statement sequence associated with the match is executed.

When we want the check the various cases based on a given condition then we have to use the concept of switch case statement. There are times when you will find yourself writing a vast if block that consists of many else if statements.

The switch statement can help simplify things a bit. It permits you to test the value returned by a single expression and then execute the relevant bit of code.

You can have as many cases as you want, including a default case which is measured if all the cases fail.

The general syntax of switch case is as follows:

switch (expression)

```
{
case 1: <statements>;
break;
case 2: <statements>;
break;
case 3: <statements>;
break;
-------------------------------
-------------------------------
-------------------------------
-------------------------------
case n: <statement>;
break;
default: <statement>;
}
```

The execution of the above switch case is as follows:

If the expression satisfies the case 1 then the statements of case 1 is executed and the switch is automatically terminated, if the expression satisfies the case 2 then the statements of case 2 is automatically executed and then the switch is automatically terminated, this process is undergone for (n) case. and if the expression does not satisfy any cases then the pointer is automatically goes to default and the statements of default is automatically executed.

Limitations of Switch Case Statement

Switch statement is a powerful statement used to handle many alternatives and provides good presentation for C program. But there are some restrictions with switch statement which are given as follows:

Logical operators cannot be used with switch statement. For instance
```
case k>=20:
```
is not allowed

Switch case variables can have only int and char data type. So float or no data type is allowed.

For instance in the switch syntax given below:
```
switch(ch)
{
case1:statement-1;
break;
case2:statement-2;
```

```
break;
}
```

In this ch can be integer or char and cannot be float or any other data type.

Use of Break with Switch

break statement is used to terminate the switch statement and to terminate any loop according to the user's requirement.

Here, we will discuss the use of break with switch statement.

Example

The following example shows the use of break with switch statement

If we write the following code:
```
switch(a)
{
case 1: printf("\n You Pressed 1");
case 2: printf("\n You Pressed 2");
case 3: printf("\n You Pressed 3");
}
```

In the above example, if we input 1 to a then the output will be You Pressed 1 You Pressed 2 You Pressed 3 because here we are not using the break statement.

But if we write the following code:
```
switch(a)
{
case 1: printf("\n You Pressed 1");
    break;
case 2: printf("\n You Pressed 2");
    break;
case 3: printf("\n You Pressed 3");
    break;
}
```

In the above example, if we input 1 to a then the output will be you pressed 1 and switch is automatically terminated and if we input 2 to a then the output will be you pressed 2 and switch is automatically terminated and if we input 3 to a then the output will be you pressed 3 and switch is automatically terminated i.e break is used to distinguish the cases in the switch statement.

Use of Default with Switch

If we impose the default statement with the switch statement then the statements of default is executed when none of the conditions are satisfied.

Programming Problems

Problem 1

The program will branch off depending on what is returned by the expression in the parentheses. However, all is not what it seems. Examine the output of this example:

```
#include <stdio.h>
#include<conio.h>
void main()
{
int a;
printf("Pick a number from 1 to 4:\n");
scanf("%d", &a);
switch (a) {
case 1:
printf("You chose number 1\n");
case 2:
printf("You chose number 2\n");
case 3:
printf("You chose number 3\n");
case 4:
printf("You chose number 4\n");
default:
printf("That's not 1,2,3 or 4!\n");
}
}
```

The output of the above code is as follows:

Pick a number from 1 to 4:2
You choose number 2
You choose number 3
You choose number 4
That is not 1,2,3 or 4!

You will notice that the program will select the correct case but will also run through all the cases below it (including the default) until the switch block's closing bracket is reached.

To prevent this from happening, we shall need to insert another statement into our cases i.e break statement that we will discuss later on.

Problem 2

Write a program to input a character and check whether the given character is vowel or not with the help of switch case statement.

Solution

```
#include<stdio.h>
#include<conio.h>
```

```
void main()
{
char c;
clrscr();
printf("\n Enter the character");
scanf("%c",&c);
switch(c)
{
case 'a':
case 'A':    printf("\n The given character is
             Vowel");
             break;
case 'e':
case 'E':    printf("\n The given character is
             Vowel");
             break;
case 'i':
case 'I':    printf("\n The given character is
             Vowel");
             break;
case 'o':
case 'O':    printf("\n The given character is
             Vowel");
             break;
case 'u':
case 'U':    printf("\n The given character is
             Vowel");
             break;
default:     printf("\n The given character is
             consonant");

}
getch();
}
```

Problem 3

The following are simple examples, written in the various languages, that use switch statements to print one of several possible lines, depending on the value of an integer entered by the user. The lack of **break** keywords to cause fall through of program executing from one block to the next is used extensively. For example, if n = 5, the third case statement will develop a match to the control variable. Since there are no statements following this line and no **break** keyword, performance continues through the 'case 7:' line and to the next line, which produces output. The **break** line after this does the switch statement to conclude. If the user types in more than one digit, the **default** block is accomplished, producing an error message.

```
switch(n) {
case 0:
printf("You typed zero.\n");
break;
case 1:
case 9:
printf("n is a perfect square\n");
break;
case 3:
case 5:
case 7:
printf("n is a prime number\n");
break;
case 2:
printf("n is a prime number\n");
case 4:
case 6:
case 8:
printf("n is an even number\n");
break;
default:
printf("Only single-digit numbers are allowed\n");
break;
}
```

Problem 4

Write a program that performs the following, depending upon the choice selected by the user.

  i. Calculate the square of number if choice is 1
  ii. Calculate the cube of number if choice is 2 and 4
  iii. Calculate the cube of the given number if choice is 3
  iv. Otherwise print the number as it is

Solution

```
#include<stdio.h>
#include<conio.h>
#include<math.h>
void main()
{
int n,choice;
printf("\n Enter any number:\n ");
scanf("%d",&n);
printf("Choice is as follows:\n\n");
printf("1. To find square of the number\n");
printf("2. To find square-root of the number\n");
printf("3. To find cube of a number\n");
printf("4. To find the square-root of the number\n\n");
printf("Enter your choice:\n");
scanf("%d",&choice);
switch (choice)
{
case 1 : printf("The square of the number is %d\n",n*n);
break;
case 2 :
case 4 : printf("The square-root of the given number is %f",sqrt(n));
break;
case 3: printf(" The cube of the given number is %d",n*n*n);
default : printf("The number you had given is %d",n);
break;
}
getch();
}
```

The output of the above code segment is as follows:

```
Enter any number:
4
Choice is as follows:
    1. To find square of the number
    2. To find square-root of the number\n");
    3. To find cube of a number
    4. To find the square-root of the number
Enter your choice:
2
The square-root of the given number is 2
```

LOOP CONTROL STATEMENTS

In looping, a sequence of statements are executed until some condition is satisfied which is placed for termination of the loop. A program loop comprises two segments; one is the body of the loop and the other known as the control statement. The control is tested forever for execution of the body of the loop. Depending upon the position of the control statement in the loop, a control may be classified as the entry-controlled loop or as the exit-controlled loop. In the entrycontrolled loop, first the conditions are tried out and if satisfied then only body of loop is executed. In the exit-controlled, the test is made at the end of the body, so the body is executed unconditionally first time.

A looping process, in general, would involve the following four steps:

1. Setting and initialization of a counter.
2. Executing of the statements in the loop.
3. Test for a defined condition for execution of the loop.
4. Incrementing the counter.

The C language supplies for three loop constructs for performing loop operations. They are:

1. The while statement.
2. The do statement.
3. The for statement.

Nested Loops

- The while loop keeps repeating an action until an associated condition returns **false**. This is useful where the programmer does not acknowledge in advance how many times the loop will be traversed.
- The do while loop is similar, but the condition is checked after the loop body is executed. This ensures that the loop body is run at least once.
- The for loop is often used, usually where the loop will be traversed a fixed number of times.

The While Loop

When in a program a single statement or a certain group of statements are to be executed repeatedly depending upon certain test condition, then *while statement* is used.

The general syntax of while loop is as follows:

while (test condition)
{
body of the loop
}

The while is an entry-controlled loop statement. The test condition is measured and only if the condition is true the body is executed. After execution of instrument of the body, the test-condition is once again evaluated and if it is true, the body is executed once again. This process of repeated execution of instrument of the body continues until the test-condition finally becomes false and the control is transferred out of the loop. On exit, the program extends with the statement immediately after the body of the loop. If the body contains only one statement it is not necessary to put the braces, but placing them is a good programming practice.

The flow chart of while loop is as follows:

Example

```
........
........
x = 10; ------- Initialization
while (x < 16) ------ Test condition
{
printf("%d",x); ------ body of the loop
x = x+1;
}
........
```

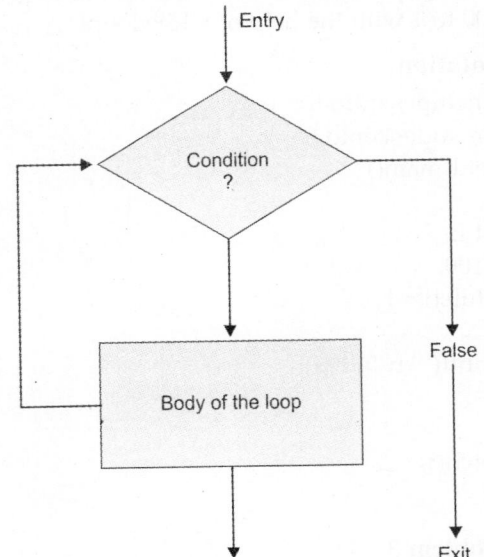

The above example shows a simple position where the test condition is measured first and since the test is satisfied the body is executed. This will proceed exactly six times printing out 10 to 15. Finally when x becomes 16 then test fails & so the loop finishes. The program proceeds with the statements after the loop.

Programming Problems

Problem 1

Write a program to print only the numbers from 1 to 100 with the help of while loop.

Solution

```
#include<stdio.h>
#include<conio.h>
void main()
{
int i;
i=1;
while(i<=100)
{
printf("\n %d",i);
i++;
}
getch();
}
```

Problem 2

Write a program to print the numbers from 100 to 1 with the help of while loop.

Solution

```
#include<stdio.h>
#include<conio.h>
void main()
{
int i;
i=100;
while(i>=1)
{
printf("\n %d",i);
i—;
}
getch();
}
```

Problem 3

Write a program to print only the even numbers from 1 to 100 with the help of while loop.

Solution

```
#include<stdio.h>
#include<conio.h>
void main()
{
int i;
i=1;
while(i<=100)
{
if(i%2==0)
printf("\n %d",i);
i++;
}
getch();
}
```

Problem 4

Write a program to print only the odd numbers from 1 to 100 with the help of while loop.

Solution

```
#include<stdio.h>
#include<conio.h>
void main()
{
int i;
i=1;
while(i<=100)
{
if(i%2==1)
printf("\n %d",i);
i++;
}
getch();
}
```

Problem 5

Program for generating 'N' natural numbers using while loop:

Solution

```
# include < stdio.h >// include the stdio.h file
void main() // Start of your program
{
int n, i=0; //Declare and initialize the variables
printf("Enter the upper limit number"); // Message to the user
scanf("%d", &n); //read and store the number
while(I < = n) // While statement with condition
{ // Body of the loop
printf("\t%d",I); // print the value of i
I++; increment I to the next natural number.
```

```
}
}
```

In the above program the looping concept is used to give n natural numbers. Here n and I are declared as integer variables and I is initialized to value zero. A message is given to the user to enter the natural number until where he wants to give the numbers. The entered number is read and put in by the scanf statement. The while loop then checks whether the value of I is less than n, i.e. the user entered number if it is true then the control enters the loop body and prints the value of I using the printf instruction and increments the value of I to the next natural number this process repeats till the value of I becomes equal to or greater than the number given by the user.

Problem 6

Write a program to input a number and print number of digits of that number with the help of while loop.

Solution

```
#include<stdio.h>
#include<conio.h>
void main()
{
int n,digit=0;
clrscr();
printf("\n enter the number");
scanf("%d",&n);
while(n>0)
{
n=n/10;
digit++;
}
printf("\n the number of digits are %d", digit);
getch();
}
```

Problem 7

Write a program to input a number and print sum of digits of that number with the help of while loop.

Solution

```
#include<stdio.h>
#include<conio.h>
```

```
void main()
{
int n,sum=0,r;
clrscr();
printf("\n enter the number");
scanf("%d",&n);
while(n>0)
{
r=n%10;
n=n/10;
sum=sum+r;
}
printf("\n the sum of digits is %d",sum);
getch();
}
```

Problem 8

Write a program to calculate the factorial of a given input natural number.

Solution

```
#include <stdio.h>
#include <stdio.h>
void main( )
{
int x;
long int fact = 1;
printf("Enter any number to find factorial:\n"); /*read the number*/
scanf("%d",&x);
while (x > 0)
{
fact = fact * x; /* factorial calculation*/
x=x-1;
}
printf("Factorial is %ld",fact);
getch();
}
```

The output of the above code segment is as follows:

Enter any number to find factorial:

5

Factorial is 120

Here, *condition* in *while* loop is evaluated and body of loop is repeated until *condition* evaluates to *false*, i.e. when x goes zero. Then the control is jumped to first statement following *while* loop and print the value of factorial.

The *do...while* Loop

The while statement discussed is an entry-controlled loop structure, in which the loop will not be executed if the test condition comes out to be false. But in some positions it might be necessary to execute the loop, even if the test-condition fails. So these kinds of loops are called the exit-controlled loop structure.

The do loop also performs a block of code as long as a condition is satisfied. The difference between a "do ...while" loop and a "while {} " loop is that the while loop examines its condition before performance of the contents of the loop starts; the "do" loop examines its condition after it has been done at least once. As noted above, if the test condition is false as the while loop is inserted the block of code is never done. Since, the condition is tested at the bottom of a do loop, its block of code is forever performed at least once.

Some people do not like these loops because it is always did at least once. When I ask them "so what?", they usually reply that the loop performs even if the data is wrong. Basically because the loop is always performed, it will execute no matter what value or type of data is supposed to be required.

The general syntax of do while loop is as follows:

```
do
{
statement(s);
} while(test condition);
```

In *do...while* loop, the body of loop is executed at least once before the *condition* is evaluated. Then the loop repeats body as long as *condition* is *true*. However, in *while* loop, the statement does not execute the body of the loop even once, if *condition* is *false*. That is why *do...while* loop is also called *exit-control loop*.

The flow chart of do...while loop is as follows:

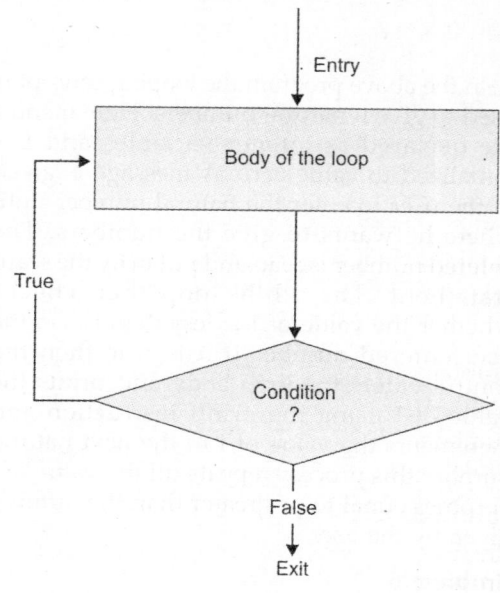

Programming Problems

Problem 1

Write a program to print only the numbers from 1 to 100 with the help of do...while loop.

Solution

```
#include<stdio.h>
#include<conio.h>
void main()
{
int i;
i=1;
do
{
    printf("\n%d",i);
    i++;
}while(i<=100);
getch();
}
```

Problem 2

Write a program to print the numbers from 100 to 1 with the help of do...while loop.

Solution

```
#include<stdio.h>
#include<conio.h>
```

```
void main()
{
int i;
i=100;
do
{
    printf("\n%d",i);
    i—;
}while(i>=1);
getch();
}
```

Problem 3

Write a program to print only the even numbers from 1 to 100 with the help of do...while loop.

Solution

```
#include<stdio.h>
#include<conio.h>
void main()
{
int i;
i=1;
do
{
    if(i%2==0)
    printf("\n%d",i);
    i++;
}while(i<=100);
getch();
}
```

Problem 4

Write a program to print only the odd numbers from 1 to 100 with the help of do...while loop.

Solution

```
#include<stdio.h>
#include<conio.h>
void main()
{
int i;
i=1;
do
{
    if(i%2==1)
    printf("\n%d",i);
    i++;
```

```
}while(i<=100);
getch();
}
```

Problem 5

Program for generating 'N' natural numbers using do...while loop.

Solution

```
# include < stdio.h >
void main()
{
int n, i=0; //Declare and initialize the variables
printf("Enter the upper limit number"); // Message to the user
scanf("%d", &n); //read and store the number
do
{
i=i+1;
printf("\n%d",i);
}while(i<=n);
getch();
}
```

Problem 6

Write a program to input a number and print number of digits of that number with the help of do...while loop.

Solution

```
#include<stdio.h>
#include<conio.h>
void main()
{
int n,digit=0;
clrscr();
printf("\n Enter the number");
scanf("%d",&n);
do
{
n=n/10;
digit++;
}while(n>0);
printf("\n the number of digits are %d",digit);
getch();
}
```

Problem 7

Write a program to input a number and print sum of digits of that number with the help of do...while loop.

Solution

```
#include<stdio.h>
#include<conio.h>
void main()
{
int n,sum=0,r;
clrscr();
printf("\n enter the number");
scanf("%d",&n);
do
{
r=n%10;
n=n/10;
sum=sum+r;
}while(n>0);
printf("\n the sum of digits is %d",sum);
getch();
}
```

Problem 8

Write a program to print first ten even natural numbers.

Solution

```
#include <stdio.h>
#include<conio.h>
void main()
{
int i=0;
int j=2;
clrscr();
do
{
printf("%d",j);
j =j+2;
i=i+1;
} while (i<10);
getch();
}
```

The output of the above code segment is as follows:

2 4 6 8 10 12 14 16 18 20

The for Loop

The for loop is an entry-controlled loop that provides a more concise loop control structure. The general form of the for loop is:

for (initialization; test-condition; increment)
```
{
body of the loop
}
```

The execution of the for statement is as follows:

1. Initialization of the control variables is done first, using assignment statements such as **i = 1** and **count = 0**. The variables i and count are acknowledged as the loop control variables.

2. The value of the control variable is examined using the test-condition. The test-condition is a comparative expression, such as **i<10** that determines when the loop will exit. If the condition is true, the body of the loop is performed; otherwise the loop is terminated and the execution continues with the statement that immediately follows the loop.

3. When the body of the loop is executed, the control is transferred back to the for statement after evaluating the last statement in the loop. Now, the control variable is incremented using an assignment statement such as **i = i + 1** and the new value of the control variable is again tested to see whether it satisfies the loop condition. If the condition is satisfied, the body of the loop is again executed. This process continues till the value of the control variable fails to satisfy the test-condition.

The flow chart of for loop is as follows:

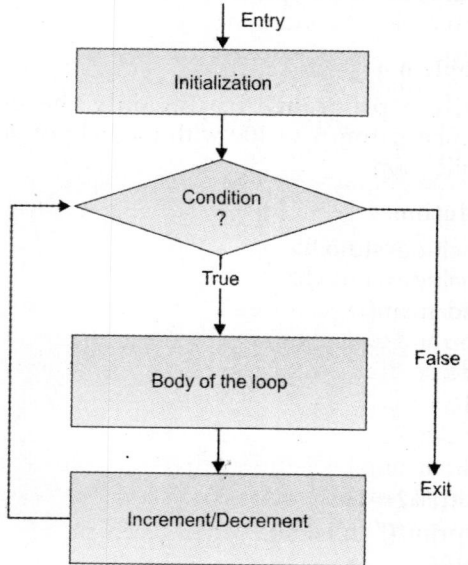

Example

```
for (i = 0; i<10; i = i+1)
{
printf("%d",i);
}
```

The for loop will be executed 10 times. Each time a value equal to the current value of i will be printed out.

Various forms of loop statements can be as follows

(a) *for(;condition;increment/decrement)*
  body;
A blank first statement will mean no initialization.

(b) *for (initialization;condition;)*
  body;
A blank last statement will mean no running increment/decrement.

(c) *for (initialization;;increment/decrement)*
  body;
A blank second conditional statement means no test condition to control the exit from the loop. So, in the absence of second statement, it is required to test the condition inside the loop otherwise it results in an infinite loop where the control never exits from the loop.

(d) *for (;;increment/decrement)*
  body;
Initialization is required to be done before the loop and test condition is checked inside the loop.

(e) *for (initialization;;)*
  body;
Test condition and control variable increment/decrement is to be done inside the body of the loop.

(f) *for (;condition;)*
  body;
Initialization is required to be done before the loop and control variable increment/decrement is to be done inside the body of the loop.

(g) *for (;;;)*
  body;
Initialization is required to be done before the loop, *test condition* and *control variable* increment/decrement is to be done inside the body of the loop.

Additional Features of for loop

1. More than one variable can be initialized at a time in the for statement.
   For example:
   `for (p=1,n=0; n<18; ++n)`
2. Like the initialization section, the increment section may also have more than one part.
   For example,
   ```
   for (a=2,b=30; n <= m; n=n+1,m=m-1)
   {
   p = b/a;
   printf("%d ",p);
   }
   ```
3. It is also permissible to use expressions in the assignment statements of initialization and increment sections.
   For example
   `for (w = (a+b); w>0; w=w/2)` is valid.
4. One or more sections in the for statement can be omitted, if necessary.
   For example,
   ```
   e = 2;
   for (; e != 10 ;)
   {
   printf("%d",e);
   e = e + 2;
   }
   ```

Both the initialization and increment sections are omitted in the for statement. In such cases, the sections are left blank. However, the semicolons separating the sections must remain. If the test-condition is not present, the for statement sets up an infinite loop.

Programming Problems

Problem 1

Write a program to print only the numbers from 1 to 100 with the help of for loop.

Solution

```
#include<stdio.h>
#include<conio.h>
void main()
{
int i;
for(i=1;i<=100;i++)
printf("\n%d",i);
```

```
getch();
}
```

Problem 2

Write a program to print the numbers from 100 to 1 with the help of for loop.

Solution

```
#include<stdio.h>
#include<conio.h>
void main()
{
int i;
for(i=100;i>=1;i—)
printf("\n%d",i);
getch();
}
```

Problem 3

Write a program to print only the even numbers from 1 to 100 with the help of for loop.

Solution

```
#include<stdio.h>
#include<conio.h>
void main()
{
int i;
for(i=1;i<=100;i++)
{
if(i%2==0)
printf("\n%d",i);
}
getch();
}
```

Problem 4

Write a program to print only the odd numbers from 1 to 100 with the help of for loop.

Solution

```
#include<stdio.h>
#include<conio.h>
void main()
{
int i;
for(i=1;i<=100;i++)
{
```

```
if(i%2==1)
printf("\n%d",i);
}
getch();
}
```

Problem 5

Program for generating 'N' natural numbers using for loop.

Solution

```
# include < stdio.h >
void main()
{
int n, i; //Declare and initialize the variables
printf("Enter the upper limit number"); //
Message to the user
scanf("%d", &n); //read and store the number
for(i=1;i<=n;i++)
printf("\n%d",i);
getch();
}
```

Problem 6

Write a program to input a number and print number of digits of that number with the help of for loop.

Solution

```
#include<stdio.h>
#include<conio.h>
void main()
{
int n,digit=0;
clrscr();
printf("\n enter the number");
scanf("%d",&n);
for(;n>0;n=n/10)
{
digit++;
}
printf("\n the number of digits are %d",digit);
getch ();
}
```

Problem 7

Write a program to input a number and print sum of digits of that number with the help of for loop.

Solution

```c
#include<stdio.h>
#include<conio.h>
void main()
{
int n,sum=0,r;
clrscr();
printf("\n enter the·number");
scanf("%d",&n);
for(;n>0;n=n/10)
{
r=n%10;
sum=sum+r;
}
printf("\n the sum is %d",sum);
getch();
}
```

The Nested Loop

These loops are the loops which contain another looping statement in a single loop. These types of loops are used to create matrix. Any loop can contain a number of loop statements in itself. If we are using loop within loop that is called nested loop. In this the outer loop is used for counting rows and the internal loop is used for counting columns.

The general syntax of Nested Loop is as follows:

```c
for (initializing ; test condition ; increment /
decrement) //Outer Loop
{
statement;
for (initializing ; test condition ; increment /
decrement) // Inner Loop
{
body of inner loop;
}
statement;
}
```

The execution of the nested loop is as follows:

Firstly the outer loop is executed and according to the outer loop the inner loop is executed and according to the inner loop the statements are executed. When the inner loop is terminated then the pointer is automatically goes to outer loop and when the outer loop is terminated then the complete loop is terminated.

Example

Write a program to generate the following pattern given below:

```
1
1 2
1 2 3
1 2 3 4
```

```c
/* Program to print the pattern */
#include <stdio.h>
#include<conio.h>
void main( )
{
int i,j;
clrscr();
for (i=1;i<=4;++i)
{
printf("%d\n",i);
for(j=1;j<=i;++j)
printf("%d\t",j);
}
getch();
}
```

Here, an *inner for loop* is written inside the *outer for loop*. For every value of *i*, *j* takes the value from 1 to *i* and then value of *i* is incremented and next iteration of outer loop starts ranging *j* value from 1 to *i*.

Programming Problems

Problem 1

Write a program to print the tables from 1 to 10.

Solution

```c
#include<stdio.h>
#include<conio.h>
void main()
{
int i,j;
clrscr();
for(i=2;i<=10;i++)
{
for(j=1;j<=10;j++)
{
printf("\n%d",i*j);
}
}
getch();
}
```

Problem 2

Write a program to input the initial limit and input the final limit and print the tables from initial limit to final limit.

Solution

```
#include<stdio.h>
#include<conio.h>
#include<stdlib.h>
void main()
{
int initial_limit,final_limit,i,j;
clrscr();
printf("\n Enter the initial limit");
scanf("%d",&initial_limit);
printf("\n Enter the final limit");
scanf("%d",&final_limit);
if(initial_limit>final_limit)
{
printf("\n Wrong Entries Initial Limit Cannot
be greater then final limit");
getch();
exit(1);
}
for(i=initial_limit;i<=final_limit;i++)
{
    for(j=1;j<=10;j++)
{
    printf("\n%d",i*j);
}
}
getch();
}
```

THE *GOTO* STATEMENT

The *goto* statement is used to alter the normal sequence of program instructions by transferring the control to some other portion of the program.

The syntax of goto statement is as follows:
goto label;
Here, *label* is an identifier that is used to label the statement to which control will be transferred. The targeted statement must be preceded by the unique label followed by colon.
label: statement;
Although *goto* statement is used to alter the normal sequence of program execution but its usage in the program should be avoided.

The most common applications are as follows:

i. To branch around statements under certain conditions in place of use of *if else* statement

ii. To jump to the end of the loop under certain conditions bypassing the rest of Statements inside the loop in place of *continue* statement

iii. To jump out of the loop avoiding the use of *break* statement. *goto* can never be used to jump into the loop from outside and it should be preferably used for forward jump.

Situations may arise, however, in which the *goto* statement can be useful. To the possible extent, the use of the *goto* statement should generally be avoided.

One must take care not to use too much of goto statements in their program or in other words use it only when demanded. This is because C being a highly structured programming language one must take care not to use too much of these unconditional goto branching instructions. The goto statement is warned in C, because it changes the sequential flow of logic that is the characteristic of C language. This word is redundant in C and promotes poor programming style.

goto statement allows to make an absolute jump to another point in the program. You should use this characteristic with caution since its performance causes an unconditional jump ignoring any type of nesting limitations. The destination point is described by a label, which is then used as an argument for the goto statement. A label is made of a valid symbol followed by a colon (:).

Example

Write a program to print first 10 even numbers with the help of goto statement.

Solution

```
/* Program to print 10 even numbers */
#include <stdio.h>
#include<conio.h>
void main()
{
int i=2;
while(1)
```

```
{
printf("%d ",i);
i=i+2;
if (i>=20)
goto outside;
}
outside : printf("over");
}
```

The output of the above code is as follows:
2 4 6 8 10 12 14 16 18 20 over

THE *BREAK* STATEMENT

A *break statement* lets you end an *iterative* (**do**, **for**, or **while**) statement or a **switch** statement and exit from it at any point other than the logical end. A **break** may only come along on one of these statements.

A **break** statement has the form:
break;

In an iterative statement, the **break** statement ends the loop and moves control to the next statement outside the loop. Within nested statements, the **break** instruction ends only the smallest enclosing **do**, **for**, **switch**, or **while** instruction.

In a **switch** instruction, the **break** passes control out of the **switch** body to the next statement outside the **switch** statement.

Use of Break with Switch Statement

break is used with the conditional switch statement and with the do, for, and while loop statements. In a **switch** statement, **break** induces the program to execute the next statement after the **switch**. Without a **break** statement, every statement from the matched **case** label to the end of the **switch**, admitting the **default**, is executed.

The following example demonstrates the use of break with switch statement:

Switch without Break

```
switch(c)
{
case 1:printf("\n hello world");
case 2: printf("\n Hi");
case 3: printf("OOPs");
}
```

If we define c=1 then the output will be hello world Hi and Oops because we are not using the break statement.

Switch with break

```
switch(c)
{
case 1:printf("\n hello world");
    break;
case 2: printf("\n Hi");
    break;
case 3: printf("OOPs");
    break;
}
```

If we define c=1 then the output will be hello world and switch is automatically terminated and if we define c=2 then the output will be Hi and switch is automatically terminated and if we define c=3 then the output will be Oops and switch is automatically terminated.

Use of break with loops

In loops, **break** terminates execution of the nearest enclosing **do**, **for**, or **while** statement. Control passes to the statement that follows the finished statement, if any.

Within nested statements, the **break** statement ends only the **do**, **for**, **switch**, or **while** statement that instantly encloses it. You can use a **return** or **goto** statement to transfer control from within more intensely nested structures.

Example

The following example demonstrates the use of break statement with for loop.

```
#include <stdio.h>
void main()
{
int i;
for (i = 1; i < 10; i++)
{
printf("%d\n", i);
if (i == 4)
break;
}
getch();
} // Loop exits after printing 1 through 4
```

The output of the above program is as follows:

1
2

3
4

The above code prints the value 1,2,3,4 and when the value of I is 4 then loop automatically terminates with the help of break statement.

Programming Problem

Problem 1

Write a program to calculate the first smallest divisor of a number.

Solution

```
#include <stdio.h>
#include<conio.h>
void main( )
{
int div,num,i;
clrscr();
printf("Enter any number:\n");
scanf("%d",&num);
for (i=2;i<=num;++i)
{
if ((num % i) == 0)
{
printf("Smallest divisor for number %d is
%d",num,i);
break;
}
}
getch();
}
```

The output of the above code segment is as follows:
Enter any number:
9
Smallest divisor for number 9 is 3

THE CONTINUE STATEMENT

The **continue** statement is somewhat the opposite of the break statement. It forces the next iteration of the loop to take place, skipping any code in between itself and the test condition of the loop. In while and do-while loops, a continue statement will cause control to go directly to the test condition and then continue the looping process. In the case of the for loop, the increment part of the loop continues. One good use of continue is to restart a statement sequence when an error occurs.

The *continue statement* ends the present iteration of a loop. Program control is agreed from the **continue** statement to the end of the loop body.

The general syntax of continue statement is as follows:
continue;

The **continue** statement can only appear within the body of an iterative statement.

The **continue** statement ends the dispensation of the action part of an iterative (**do, for,** or **while**) statement and moves control to the loop continuation portion of the statement. For example, if the iterative declaration is a **for** statement, control moves to the third expression in the condition part of the statement, then to the second expression (the test) in the condition part of the statement.

Within nested statements, the **continue** instruction ends only the current iteration of the **do, for,** or **while** statement immediately enclosing it.

Programming Problems

Problem 1

Write a program to display all of the numbers between 0 and 20 except 10.

Solution

```
#include<stdio.h>
#include<conio.h>
void main()
{
int i;
clrscr();
for (i = 0; i < 21; i++ )
{
if( i == 10 )
{
continue;
}
printf("\n %d",i);
}
getch();
}
```

Problem 2

Write a program to print first 20 natural numbers skipping the numbers divisible by 5.

Solution

```
#include <stdio.h>
#include<conio.h>
void main( )
{
int i;
clrscr();
for (i=1;i<=20;++i)
{
if ((i % 5) == 0)
continue;
printf("%d ",i);
}
}
```

The output of the above code segment is as follows:
1 2 3 4 6 7 8 9 11 12 13 14 16 17 18 19

MULTIPLE LOOP VARIABLES

If we define more than one variable in a single loop then that particular concept is known as multiple loop variable and we can define the multiple loop variable in for loop with the help of comma (,) operator.

Example

If we define the following loop
for(i=1,j=1;i<=10;i++)
in this loop i and j are multiple loop variables.

EXIT FUNCTION

Exit is a function defined in the stdlib.h header file.

The purpose of exit is to finish the current program with a specific exit code. Its prototype is:

void exit (int exit code);

The exit code is used by some operating systems and may be used by calling programs. By convention, an exit code of 0 means that the program completed normally and any other value means that some error or unexpected results happened.

SUMMARY

- Conditional statements are used to check any particular condition according to user's requirement.
- In C language we have if else, if else if else, nested if and switch case statements for checking any particular condition according to user's requirement.
- If else is used to check the simple conditions according to the user's requirement.
- If else if else is used to check the various conditions according to the user's requirement.
- The **switch** statement is Cs multiple selection statement. It is used to select one of several alternative paths in program execution.
- If we want to execute a statement more than once based on some particular condition then this concept is known as looping.
- In C language we have while, do…while, for and nested loop.
- In while loop the statements are executed when the given condition is true and when the result of given condition is false then the loop is automatically terminated.
- In do…while loop the statements are executed at least once.
- Nested loop means loop with in a loop. The first loop in nested loop is known as outer loop and the another loop inside the outer loop is known as inner loop.
- In nested loop firstly the outer loop is executed and on the basis of outer loop the inner loop is executed and on the basis of inner loop the statements are executed. When the inner loop is terminated then the pointer is automatically goes to outer loop and when the outer loop is terminated then the complete loop is terminated.
- Break statement is used to terminate the switch or loop based on some particular condition.
- Continue statement is used to continue the loop based on some particular condition.

Exercises

Section A

Multiple Choice Questions

1. The general syntax of if statement is
   (a) if<condition> {statement;} else {statements;}
   (b) if <condition> statement; else statement;
   (c) Both a and b
   (d) None of the above

2. Which statement is false regarding if statement
   (a) If condition is true statements in if are executed.
   (b) There needs to be a semicolon after each if statement.
   (c) After execution of if statement control shifts to else
   (d) If and else condition can never execute simultaneously

3. If we have multiple condition than we can use
   (a) if - else - if   (b) nested if
   (c) switch case   (d) All of the above

4. In switch case statement parameters can be
   (a) int
   (b) char
   (c) both a and b
   (d) logical operators

5. Default statement is executed is switch case when
   (a) All conditions are satisfied
   (b) No condition satisfy
   (c) After break statement
   (d) All of the above

6. Statements in while will execute
   (a) Until condition is false
   (b) Until condition remains true
   (c) Depends on the condition
   (d) Depends on the statements

7. This is not a looping operation
   (a) for
   (b) do...while
   (c) if else
   (d) nested while

8. Do...while statement
   (a) Executes at least once if the condition is false
   (b) Use to have a semicolon after while statement
   (c) Both a and b
   (d) None of the above

9. While using break statement
   (a) Will switch the control out of loop
   (b) Will terminate the program
   (c) Will use only one break in a program
   (d) All of the above

10. In if else if statement
    (a) Multiple statements can be executed
    (b) Multiple conditions can be tested
    (c) Both a and b
    (d) None of the above

State True/False

1. Statements in else are executed if condition is true.
2. {} Braces are not mandatory to be included after if statement.
3. Switch statement is included if we have many alternates.
4. Switch can only have int and char data type.
5. In do...while loop the condition are tested at the start of loop.
6. do...while and while loops are exactly the same.
7. In for loop initialization statement is executed each time it runs.
8. Break statement lets you come out of switch statement even.
9. In switch if no condition is satisfied control comes out of loop.
10. In while loop is executed until the condition becomes false.

Section B

Short Answer Type Questions

1. Define the concept of conditional statements and how we use these conditional statements?
2. Differentiate between if else if else and switch statement with proper explanation.
3. Differentiate between while and do while loop with proper explanation.

4. Write a program in C to check whether the given number even or odd.

5. Write a program in C to check whether the given number positive, negative or zero.

6. Write a program in C to check whether the given year is leap or not.

7. Write a program in C to check whether the given character upper, lower, numbr or not.

8. Write a program in C to change upper to lower.

9. Write a program in C to change lower to upper.

10. Write a program in C to print a word form of given number between 0 to 9.

Section C

Long Answer Type Questions

1. Write a program in C to print a word form of given number is tens between 1 and 99.

2. Write a program in C for relations operations of two given integer numbers.

3. Write a program in C for relations operations of two given float numbers.

4. Write a program in C for given mark contain which grade.

5. Write a program in C to find biggest of two given numbers.

6. Write a program in C to find smallest of two given numbers.

7. Write a program in C to find biggest of three given numbers.

8. Write a program in C to find smallest of three given numbers.

9. Write a program in C to find biggest of three given numbers using && operator.

10. Write a program in C to find smallest of three given numbers using && operator.

11. Write a program in C to show the name of the day in a week. Here given input range is 1 – 7.

12. Write a program in C to find a biggest of four given numbers.

13. Write a program in C to find a smallest of four given numbers.

14. Write a program in C to find a vowel or not of given character.

15. Write a program in C to find biggest of two numbers using ternary operator.

16. Write a program in C to find smallest of two numbers using ternary operator.

17. Write a program in C to find biggest of three numbers using ternary operator.

18. Write a program in C to find smallest of three numbers using ternary operator.

10

Modular Programming

Structured programming (sometimes known as *modular programming*) is a subset of procedural programming that enforces a logical structure on the program being written to make it more efficient and easier to understand and modify. Certain languages such as Ada, Pascal, and dBase are designed with features that encourage or enforce a logical program structure.

Structured programming frequently employs a top-down design model, in which developers map out the overall program structure into separate subsections. A defined function or set of similar functions is coded in a separate module or sub-module, which means that code can be loaded into memory more efficiently and that modules can be reused in other programs. After a module has been tested individually, it is then integrated with other modules into the overall program structure.

Program flow follows a simple hierarchical model that employs looping constructs such as "for," "repeat," and "while." Use of the "Go To" statement is discouraged.

Structured programming was first suggested by Corrado Bohm and Guiseppe Jacopini. The two mathematicians demonstrated that any computer program can be written with just three structures: decisions, sequences, and loops. Edsger Dijkstra's subsequent article, *Go To Statement Considered Harmful* was instrumental in the trend towards structured programming. The most common methodology employed was developed by Dijkstra. In this model (which is often considered to be synonymous with structured programming,

although other models exist) the developer separates programs into subsections that each has only one point of access and one point of exit.

In the modular programming concept we will discuss the concept of functions in C Language.

USE OF FUNCTIONS IN MODULAR PROGRAMMING

Almost any language can use structured programming techniques to avoid common pitfalls of unstructured languages. Unstructured programming must rely upon the discipline of the developer to avoid structural problems, and as a consequence may result in poorly organized programs. Most modern procedural languages include features that encourage structured programming. Object-oriented programming (OOP) can be thought of as a type of structured programming, uses structured programming techniques for program flow, and adds more structure for data to the model.

To make programming simple and easy to debug, we break a larger program into smaller *subprograms* which perform *'well-defined tasks'*. These subprograms are called *functions*. So far we have defined a single function *main ( )*. After reading this unit you will be able to define many other functions and the *main ( )* function can call up these functions from several different places within the program, to carry out the required processing. Functions are very important tools for *Modular Programming*, where we break large programs into small subprograms or modules (functions in case of C). The use of functions reduces complexity and makes programming simple and easy to understand.

In this chapter, we will discuss how functions are defined and how are they accessed from the main program? We will also discuss various types of functions and how to invoke them. And finally you will learn an interesting and important programming technique known as *Recursion*, in which a function calls within itself.

DEFINTION OF A FUNCTION

A *function* is a self-contained block of executable code that can be called from any other function. In many programs, a set of statements are to be executed repeatedly at various places in the program and may with different sets of data, the idea of functions comes in mind. You keep those repeating statements in a function and call them as and when required. When a function is called, the control transfers to the called function, which will be executed, and then transfers the control back to the calling function (to the statement following the function call).

When we divide a complex program into various modules then that particular various modules are called as functions. We have mentioned earlier that one of the strengths of C language is that C functions are easy to define and use. C functions can be classified into two categories, namely, library functions and user-defined functions. main is an example of user defined function. printf and scanf belong to the category of library functions. The main distinction between user-defined functions and library functions is that the former are not required to be written by user while the latter have to be developed by the user at the time of writing a program. However, the user defined function can become a part of the C program library.

A **function** is a subprogram that acts on data and often returns a value. You are already familiar with the one function that every C program possesses: **int main(void)**.

Good C programmers write programs that consist of many of these small functions. These programmers know that a program written with numerous functions is easier to maintain, update and debug than one very long program. By programming in a modular (functional) fashion, several programmers can work independently on separate functions which can be assembled at a later date to create the entire project.

Ideally, your **main( )** function should be very short and should consist primarily of function calls.

Each function has its own name. When that name is encountered in a program, the execution of the program branches to the body of that function. When the function is finished, execution returns to the area of the program code from which it was called, and the program continues on to the next line of code.

Graphically the function calling is represented as follows:

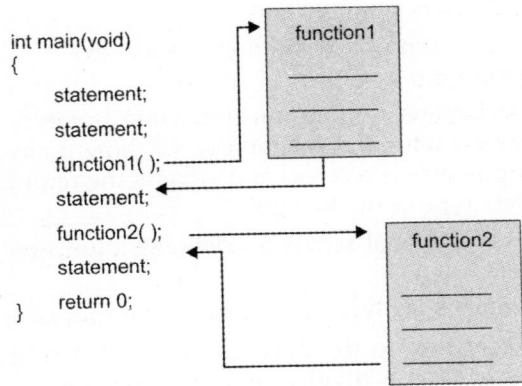

TYPES OF FUNCTION

Functions are classified into the following two categories:

(A) User defined functions

(B) Predefined functions

In this chapter we will discuss the concept of user-defined functions.

ADVANTAGES OF USING FUNCTIONS

The use of functions in C serves many advantages:

1. It facilitates top-down modular programming. In this programming style,

the high level of the overall problem is solved first while the details of each lower-level function are addressed later.

2. The length of a source program can be reduced by using functions at a appropriate places.

3. It is easy to locate and isolate a faulty function for further investigations.

4. A function may be used by many other programs. This means that a C programmer can build on what others have already done, instead of starting from scratch.

ELEMENTS OF USER-DEFINED FUNCTION

In C language user-defined function has three stages and which are as follows:

1. Declaration of a function or function prototype
2. Calling of a function
3. Body of a function

Declaration of a Function or Function Prototype

Declaration of a function or function prototype means what the function is, i.e. how many arguments it receives and what is the return data type of the function.

The general syntax for defining a function is as follows:

<return data type> functionname(arguments)

According to the declaration the function declaration is divided into four categories:

a. With arguments and with return type
b. With arguments and no return type
c. No arguments but return type
d. No arguments and no return type

Note

- If a function accepts the arguments then the inputting is done within main function and if the function does not accept the arguments then the inputting is done within the body of the function

- If a function returns the value then the printing is done within main function and if the function does not return the

value then the printing is done within the body of the function.

Calling of a Function

We have to call the function according to the arguments and return value and if the function does not return any value then we does not store that function in any variable.

Example

If we define int add(int,int) as a prototype then at the time of calling we have to call this function as follows:
c=add(a,b);
If we define void add(int,int) as a prototype then at the time of calling we have to call this function as follows:
add(a,b);
If we define int add() as a prototype then at the time of calling we have to call this function as follows:
c=add();
If we define void add() as a prototype then at the time of calling we have to call this function as follows:
add();

Body of a Function

Finally we have to write the body of the function outside of the main program, i.e. either the upper side of the main or the lower side of the main function.

WITH ARGUMENTS AND WITH RETURN TYPE

In this type of declaration we have to pass the arguments as well as we have to return the value to the main function.

For example

If we define int add (int,int). The meaning of this line is that add is a function which defines two integer arguments and according to that arguments it returns an integer type of value.

Programming Problems

Problem 1

Write a program to define two numbers and print the addition of both the numbers with the help of function and use the case of with arguments and with return value.

Solution

Program	Output

```
// function example
#include <stdio.h>
#include<conio.h>
int addition (int,int)//function prototype
void main ()
{
    int z;
    z = addition (5,3);//calling of the function
    printf("\n the Addition is %d",z);
}
int addition (int a, int b)
{
    int r;
    r=a+b;
    return (r);
}
```

The result is 8

We can see how the main function begins by declaring the variable z of type int. Right after that, we see a call to a function called addition. Paying attention we will be able to see the similarity between the structure of the call to the function and the declaration of the function itself some code lines above:

int addition   (int a, int b)
                        ↑   ↑
z = addition (5 ,   3) ;

The parameters and arguments have a clear correspondence. Within the main function we called to addition passing two values: 5 and 3, that correspond to the int a and int b parameters declared for function addition.

At the point at which the function is called from within main, the control is lost by main and passed to function addition. The value of both arguments passed in the call (5 and 3) are copied to the local variables int a and int b within the function.

Function addition declares another local variable (int r), and by means of the expression r=a+b, it assigns to r the result of a plus b. Because the actual parameters passed for a and b are 5 and 3 respectively, the result is 8.

The following line of code:

return (r);

Finalizes function addition, and returns the control back to the function that called it in the first place (in this case, main). At this moment the program follows it regular course from the same point at which it was interrupted by the call to addition. But additionally, because the return statement in function addition specified a value: the content of variable r (return (r);), which at that moment had a value of 8. This value becomes the value of evaluating the function call.

int addition   (int a, int b)
               ↓8
               ↓
z = addition (5 ,   3) ;

So being the value returned by a function the value given to the function call itself when it is evaluated, the variable z will be set to the value returned by addition (5, 3), that is 8. To explain it another way, you can imagine that the call to a function (addition (5,3)) is literally replaced by the value it returns (8).

The following line of code in main is:

printf("\n The result is %d",z);

That, as you may already expect, produces the printing of the result on the screen.

Problem 2

Write a program to input two numbers and print the maximum number between them with the help of function and use the case of with arguments and with return value.

Solution

```
#include<stdio.h>
#include<conio.h>
int maximum(int,int);//function prototype
void main()
{
int a,b,c;
clrscr();
printf("\n Enter the values of a and b");
scanf("%d %d",&a,&b);
c=maximum(a,b);//Calling of a function
printf("\n The maximum number is %d",c);
getch();
}
int maximum(int x,int y)
{
if(x>y)
return x;
```

```
else
return y;
}
```

Problem 3

Write a program to input a number and print the factorial of that number with the help of function and use the case of with arguments and with return value.

Solution

```
#include<stdio.h>
#include<conio.h>
int fact (int);//function Prototype
void main()
{
int n,f;
clrscr();
printf("\n Enter the number");
scanf("%d",&n);
f=fact(n);
printf("\n The factorial is %d",f);
getch();
}
int fact(int x)
{
int f=1;
while(x>0)
{
f=f*x;
x−;
}
return(f);
}
```

Problem 4

Write a program to input a number and print the number of digits of that particular number with the help of function and use the case of with arguments and with return value.

Solution

```
#include<stdio.h>
#include<conio.h>
int digit(int);//function prototype
void main()
{
int n,d;
clrscr();
printf("\n enter the number");
```

```
scanf("%d",&n);
d=digit(n);//calling of a function
printf("\n the number of digits are %d",d);
getch();
}
int digit(int n)
{
int d;
while(n>0)
{
n=n/10;
d++;
}
return(d);
}
```

WITH ARGUMENTS AND NO RETURN TYPE

In this type of declaration we have to pass the arguments but the function does not return any particular value. When a function does not return any particular value then the return data type must be void.

For example

If we define void add (int, int). The meaning of this line is that add is a function which accepts two integer arguments and according to that arguments it does not return any particular value that means the printing is done within the body of the function.

Programming Problems

Problem 1

Write a program to define two numbers and print the addition of both the numbers with the help of function and use the case of with arguments and no return value.

Solution

```
#include<stdio.h>
#include<conio.h>
void add (int, int);//function prototype
void main()
{
int a,b;
clrscr();
printf("\n enter the values of a and b");
scanf("%d %d",&a,&b);
add(a,b);//calling of a function
```

```
getch();
}
void add (int x, int y)
{
   int z;
   z=x+y;
   printf("\n the addition is %d", z);
}
```

Problem 2

Write a program to input two numbers and print the maximum number between them with the help of function and use the case of with arguments and no return value.

Solution

```
#include<stdio.h>
#include<conio.h>
void maximum(int,int);//function prototype
void main()
{
int a,b;
clrscr();
printf("\n enter the values of a and b");
scanf("%d %d",&a,&b);
maximum(a,b);//calling of a function
getch();
}
int maximum(int x, int y)
{
if(x>y)
printf("\n the maximum number is %d",x);
else
printf("\n the maximum number is %d",y);
}
```

Problem 3

Write a program to input a number and print the factorial of that number with the help of function and use the case of with arguments and with no return value.

Solution

```
#include<stdio.h>
#include<conio.h>
void fact (int);//function prototype
void main()
{
int n;
clrscr();
```

```
printf("\n enter the number");
scanf("%d",&n);
fact(n);
getch();
}
void fact(int x)
{
int f=1;
while(x>0)
{
f=f*x;
x—;
}
printf("\n the factorial is %d",f);
}
```

Problem 4

Write a program to input a number and print the number of digits of that particular number with the help of function and use the case of with arguments and no return value.

Solution

```
#include<stdio.h>
#include<conio.h>
void digit(int);//function prototype
void main()
{
int n,d;
clrscr();
printf("\n enter the number");
scanf("%d",&n);
digit (n);//calling of a function
getch();
}
void digit (int n)
{
int d;
while(n>0)
{
n=n/10;
d++;
}
printf ("\n the number of digits are %d",d);
}
```

NO ARGUMENTS BUT RETURN TYPE

In this type of declaration, we do not pass the arguments to the function but the function returns the value.

For example

If we define int add (). The meaning of this line is that add is a function which does not accept any argument but it returns an integer type of value.

Programming Problems

Problem 1

Write a program to input two numbers and print the addition of both the numbers with the help of function and use the case of no argument but return value.

Solution

```
#include<stdio.h>
#include<conio.h>
int add ();//function prototype
void main()
{
int c;
clrscr();
c=add();//calling of a function
printf("\n the addition is %d",c);
getch();
}
int add ()
{
   int a,b,c;
   printf("\n enter the values of a and b");
   scanf("%d %d",&a,&b);
   c=a+b;
   return(c);
}
```

Problem 2

Write a program to input two numbers and print the maximum number between them with the help of function and use the case of no arguments but return value.

Solution

```
#include<stdio.h>
#include<conio.h>
int maximum();//function prototype
void main()
{
int m;
clrscr();
m=maximum();//calling of a function
printf("\n The maximum number is %d",m);
```

```
getch();
}
int maximum()
{
int a,b,c;
clrscr();
printf("\n enter the values of a and b");
scanf("%d %d",&a,&b);
if(a>b)
return(a);
else
return(b);
}
```

Problem 3

Write a program to input a number and print the factorial of that number with the help of function and use the case of no arguments but return value.

Solution

```
#include<stdio.h>
#include<conio.h>
int fact ();//function prototype
void main()
{
int f;
clrscr();
f=fact();
printf("\n the factorial is %d",f);
getch();
}
int fact()
{
int n,f=1;
clrscr();
printf("\n enter the number");
scanf("%d",&n);
while(n>0)
{
f=f*n;
n−;
}
return(f);
}
```

Problem 4

Write a program to input a number and print the number of digits of that particular number with the help of function and use the case of no arguments but return value.

Solution

```
#include<stdio.h>
#include<conio.h>
int digit();//function prototype
void main()
{
int d;
clrscr();
d=digit();//calling of a function
printf("\n the number of digits are %d",d);
getch();
}
int digit ()
{
int n,d;
clrcsr();
printf("\n enter the number");
scanf("%d",&n);
while(n>0)
{
n=n/10;
d++;
}
return (d);
}
```

NO ARGUMENTS AND NO RETURN TYPE

In this type of declaration, we do not pass the arguments to the function and the function does not return any value.

For example

If we define void add (). The meaning of this line is that add is a function which does not accept any argument and it does not return any value.

Programming Problems

Problem 1

Write a program to input two numbers and print the addition of both the numbers with the help of function and use the case of no argument and no return value.

Solution

```
#include<stdio.h>
#include<conio.h>
void add ();//function prototype
void main()
{
add();
getch();
}
void add ()
{
    int a,b,c;
    printf("\n enter the values of a and b");
    scanf("%d %d",&a,&b);
    c=a+b;
    printf("\n the addition is %d",c);
}
```

Problem 2

Write a program to input two numbers and print the maximum number between them with the help of function and use the case of no arguments and no return value.

Solution

```
#include<stdio.h>
#include<conio.h>
void maximum();//function prototype
void main()
{
maximum();//calling of a function
getch();
}
void maximum()
{
int a,b;
clrscr();
printf("\n enter the values of a and b");
scanf("%d %d",&a,&b);
if(a>b)
printf("\n the maximum value is %d",a);
else
printf("\n the maximum value is %d",b);
}
```

Problem 3

Write a program to input a number and print the factorial of that number with the help of function and use the case of no arguments and no return value.

Solution

```
#include<stdio.h>
#include<conio.h>
void fact ();//function prototype
```

```
void main()
{
fact();
getch();
}
void fact()
{
int n,f=1;
clrscr();
printf("\n enter the number");
scanf("%d",&n);
while(n>0)
{
f=f*n;
n—;
}
printf("\n the factorial is %d",f);
}
```

Problem 4

Write a program to input a number and print the number of digits of that particular number with the help of function and use the case of no arguments and no return value.

Solution

```
#include<stdio.h>
#include<conio.h>
void digit();//function prototype
void main()
{
digit();
getch();
}
void digit ()
{
int n,d;
clrcsr();
printf("\n enter the number");
scanf("%d",&n);
while(n>0)
{
n=n/10;
d++;
}
printf("\n the number of digits are %d",d);
}
```

CALL BY VALUE AND CALL BY REFERENCE

The arguments passed to function can be of two types namely:

1. Values passed
2. Address passed

The first type refers to call by value and the second type refers to call by reference.

In the call by value only the duplicate value of the variable is passed to the function and if we perform any changes in the function then that particular changes does not reflect to the main program.

But in the call by reference the actual address is passed in the function and if perform any changes in the function then that particular changes must be reflect in the main program.

For instance consider **program1**

```
#include<stdio.h>
#include<conio.h>
void interchange(int,int);
void main()
{
int x=50, y=70;
interchange(x,y);
printf("x=%d y=%d",x,y);
}
void interchange(int x1,int y1)
{
int z1;
z1=x1;
x1=y1;
y1=z1;
printf("x1=%d y1=%d",x1,y1);
}
```

Here the value to function interchange is passed by value.

Consider **program2**

```
#include<stdio.h>
#include<conio.h>
void interchange(int *,int *);
void main()
{
int x=50, y=70;
interchange(&x,&y);
printf("x=%d y=%d",x,y);
}
void interchange(int *x1,int *y1)
{
int *z1;
*z1=*x1;
*x1=*y1;
```

```
*y1=z1;
printf("*x=%d *y=%d",*x1,*y1);
}
```

Here the function is called by reference. In other words address is passed by using symbol & and the value is accessed by using symbol *.

The main difference between them can be seen by analyzing the output of program1 and program2.

The output of program1 that is call by value is:

x1=70 y1=50 x=50 y=70

But the output of program2 that is call by reference is

*x=70 *y=50 x=70 y=50

This is because in case of call by value the value is passed to function named as interchange and there the value got interchanged and got printed as:

x1=70 y1=50

and again since no values are returned back and therefore original values of x and y as in main function namely:

x=50 y=70 got printed.

But in case of call by reference address of the variable got passed and therefore what ever changes that happened in function interchange got reflected in the address location and therefore the got reflected in original function call in main also without explicit return value. So value got printed as *x=70 *y=50 and x=70 y=50

RECURSION

Within a function body, if the function calls itself, the mechanism is known as **'Recursion'** and the function is known as **'Recursive function'**. Now let us study this mechanism in detail and understand how it works.

- As we see in this mechanism, a chaining of function calls occurs, so it is necessary for a recursive function to stop somewhere or it will result into infinite callings. So the most important thing to remember in this mechanism is that every "recursive function" should have a terminating condition.

- Let us take a very simple example of calculating factorial of a number, which we all know is computed using this formula 5! = 5*4*3*2*1
- First we will write non-recursive or iterative function for this.

Example

Write a program to find factorial of a number with the help of non-recursive function

```
#include <stdio.h>
#include<conio.h>
void main ()
{
int n, factorial;
int fact(int);
printf ("Enter any number:\n" );
scanf ("%d", &n);
factorial = fact ( n); /* function call */
printf ("Factorial is %d\n", factorial);
}
/* Non-recursive function of factorial */
int fact (int n)
{
int res = 1, i;
for (i = n; i >= 1; i—)
res = res * i;
return (res);
}
```

The output of the above code segment is as follows:

Enter any number: 5
Factorial is 120
How it works?
Suppose we call this function with n = 5
Iterations:
1. i= 5 res = 1*5 = 5
2. i= 4 res = 5*4 = 20
3. i= 3 res = 20*4 = 60
4. i= 2 res = 60*2 = 120
5. i= 1 res = 120*1 = 120

Now let us write this function **recursively**. Before writing any function recursively, we first have to examine the problem, that it can be implemented through recursion.

For instance, we know n! = n* (n – 1)! (Mathematical formula)

Or fact (n) = n*fact (n – 1)

Or fact (5) = 5*fact (4)

That means this function calls itself but with value of argument *decreased by '1'*

Now modify the above program that is to input a number and print the factorial of that number with the help of recursion

```c
/*Program to find factorial using recursion*/
#include<stdio.h>
#include<conio.h>
int fact(int);
void main ()
{
int n, factorial;
clrscr();
printf("Enter any number: \n" );
scanf("%d",&n);
factorial = fact(n); /*Function call */
printf ("Factorial is %d\n", factorial);
getch();
}
/* Recursive function of factorial */
int fact(int n)
{
int res;
if(n == 1) /* Terminating condition */
return(1);
else
res = n*fact(n-1); /* Recursive call */
return(res);
}
```

The output of the above code segment is as follows:

Enter any number: 5
Factorial is 120
How it works?

Suppose we will call this function with n = 5

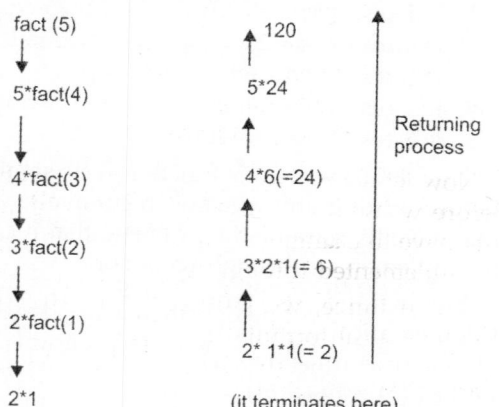

(it terminates here)

Thus, a recursive function first proceeds towards the innermost condition, which is the termination condition, and then returns with the value to the outermost call and produces result with the values from the previous return.

Note: This mechanism applies only to those problems, which repeats itself. These types of problems can be implemented either through loops or recursive functions, which one is better understood to you.

SCOPE RULES

The scope of a variable refers to that part of the program where the variable is accessible. There are two scope rules governing the scope of variables (and symbols):

- *The scope of a variable is the codeblock in which it is defined,* i.e. a variable is only valid within the block in which it is defined.
- *If two variables by the same name are defined in a block, then the closest declaration counts* and hides the original variable.

Global variables

Global variables are variables whose scope is the entire program. That means that they must be defined outside any function, typically at the top of the program. *Every function* can access and change a global variable. *Avoid using global variables without good reason.*

If we define the variable outside of the main program then that variable is known as global variable and this variable can be access any where in the program.

Example

```c
int a=34;
void main()
{
int b;
}
```

Here a is the global variable and we can access this variable at any where in the program.

Local variables

If we define the variable inside the main function then that variable is local for that main function and if we define the variable within the body of the function then that variable is a local variable for that function and we can only access that variable within that function.

Example

If we define the following

```
void main()
{
int a,b;
}
```

Here a and b are the local variables for this main() function and we can access this a and b with in this main() function.

SEPARATE COMPILATION AND LINKAGE

Separate compilation and linkage means we can divide the whole program into various modules with the help of header files and then we compile and linked that separate files according to our requirement. In C language, we can create our header files according to our requirement.

The following are the steps for creating a header file in C language.

Step 1: Create a file of .h extension and add the functions according to our requirement.

Step 2: Create a file of .c extension and include the header file according to our requirement and call the functions according to our requirement.

Example

Create a header file of calculating a factorial of a given number.

Solution

Firstly we create a file fact.h and write the following code in fact.h

```
#include<stdio.h>
#include<conio.h>
int fact(int n)
{
   int f=1;
   while(n>0)
{
   f=f*n;
   n=n-1;
}
   return(f);
}
```

Now save this file in the current directory and name it to fact.h

Now open another file and save it to try.c and write the following code.

```
#include "fact.h"
void main()
{
   int n,f;
   clrscr();
   printf("\n enter a number");
   scanf("%d",&n);
   f=fact(n);
   printf("\n the factorial is %d",f);
   getch();
}
```

Note: Here, we use #include "fact.h" in place of #include<fact.h>. The reason is that if we want to include a header file that exists in the current directory then we use #include "filename.h" in place of #include<filename.h>.

SUMMARY

- Modular programming means we can divide the whole program into various modules.
- When we divide the complex program into various modules then that various modules are called as function.
- A function has three stages: (a) declaration of the function, (b) calling of the function, (c) body of the function.
- We can declare the function with the help of four ways: (a) with arguments and with return type, (b) with arguments and no return type, (c) no arguments but return type, (d) no argument and no return type.
- We can call the function according to the arguments and return data type.
- We can write the body of the function outside of the main program.
- We can call the function either by call by value or by call by reference.
- Call by value means we call the function to the value of the variable and if we change the value of the varibale in the body of the function then that changes is not reflected in the main program.
- Call by reference means we call the function to the reference of the variable

and if we change the value of the variable in the body of the function then that changes is reflected in the main program.

- If we define the variable within the body of the main() function then that variable is the local variable to that main() function and we can only access that variable within main() function.

- If we define the variable outside of the main() function that variable is the global variable and we can directly access that variable any where of the program.

Exercises

Section A

Multiple Choice Questions

1. Functions are of types
   (a) Predefined functions
   (b) User defined functions
   (c) Both a and b
   (d) None of the above

2. We use functions because
   (a) They are reusable
   (b) Help in sorting errors
   (c) Give modular approach
   (d) All of the above

3. Various elements of user defined functions are
   (a) Declaration
   (b) Calling of function
   (c) Body of function
   (d) All of the above

4. Functions can be
   (a) With arguments and with return type
   (b) Without arguments and without return type
   (c) Both a and b
   (d) None of the above

5. When a function does not returns any value return data type must be
   (a) void
   (b) int
   (c) char
   (d) All of the above

6. void add() is a function
   (a) With argument and with data type
   (b) Without argument and without data type
   (c) With argument and without data type
   (d) Without argument and with data type

7. In call by reference function
   (a) Actual value is passed
   (b) Changes the original value
   (c) Both a and b
   (d) None of the above

8. Scope of a variable is
   (a) Codeblock in which it is defined
   (b) Farthest declaration counts
   (c) Both a and b
   (d) None of the above

9. Global variable
   (a) Defined outside any function
   (b) Accessed from anywhere
   (c) Scope is of entire program
   (d) All of the above

10. The prototype of a function must be
    (a) <return data type> function name (arguments)
    (b) <Arguments >function name(return data type)
    (c) Both a and b
    (d) None of the above

State True/False

1. Structured programming enforces a logical structure on programs.

2. A function is a subprogram that acts on data.

3. Return () statement will return the value and continue executing function.

4. Predefined functions are those functions that are defined by user.

5. main() is also a function with no parameters.

6. If a function accepts an argument then inputting is done through main.

7. There must be a data type attached with each function.

8. void add(int, int) will take two arguments and one data type.
9. We can pass address as argument to a function.
10. Global variables are those scope ends up to a function

Section B

Short Answer Type Questions

1. Define the concept of modular programming with suitable example.
2. Define the concept of function and how we create the function.
3. Define the various stages of the function.
4. Define the concept of separate compilation with proper example.
5. Write a program in C to sum of two integer numbers with the help of function.
6. Write a program in C to find the biggest of two integer numbers with the help of function.
7. Write a program in C to find the smallest of two integer numbers with the help of function.
8. Write a program in C to find the area of circle with the help of function.
9. Write a program in C to find the volume of sphere with the help of function.
10. Write a program in C to find out the given number is even or odd with the help of function.

Section C

Long Answer Type Questions

1. Write a program in C to find out the given year is leap year or not. With the help of function.
2. Write a program in C to print the single digit number into words with the help of function.
3. Write a program in C to print 11–19 range to words with the help of function?
4. Write a program in C to print the tens (10, 20, 90) into words.
5. Write a program in C to print the limit 0–99 into words with the help of function.
6. Write a program in C to print the limit 0–999 into words with the help of function.
7. Write a program in C to read and print the ten numbers with the help of function.
8. Define the term modular programming and also define the advantages of using the concept of modular programming.
9. Write a program to input the year and check whether the given year is leap year or not with the help of function.
10. Write a program to input a character and check whether the given character is small or uppercase character with the help of function.

11

Arrays

Array is the collection of similar data type and having a common name. An array is a series of elements of the same type placed in adjoining memory locations that can be individually referenced by adding an index to a unique identifier.

In programming, a sequence of objects all of which are the same size and type. Each object in an array is named an *array element*.

For example, you could have an array of integers or an array of characters or an array of everything that has a defined data type. The important features of an array are: Each element has the same data type (although they may have different values). The entire array is put in contiguously in memory (that is, there are no gaps between elements).

C language provides four basic data types—*int, char, float and double*. These basic data types are very useful; but they can handle only a limited amount of data. As programs suit larger and more complicated, it becomes increasingly difficult to manage the data. Variable names typically become longer to assure their uniqueness. And, the number of variable names makes it hard for the programmer to concentrate on the more important task of correct coding. Arrays supply a mechanism for declaring and accessing several data items with only one identifier, thereby simplifying the task of data management. Many programs require the processing of multiple, related data items that have common characteristics like *list* of numbers, marks in a course, or enrolment numbers. This could be done by creating several individual variables. But this is a hard

and tedious process. For example, suppose you want to input five numbers and print them out in reverse order. You could do it the hard way as:

```
void main()
{
int al,a2,a3,a4,a5;
scanf("%d %d %d %d %d",&a1,&a2,&a3,&a4,&a5);
printf("%d %d %d %d %d"',a5,a4,a3,a2,a1);
}
```

Does it look good if the problem is to read in 100 or more related data items and print them in reverse order? Of course, the solution is the use of the regular variable names **a1, a2** and so on. But to remember each and every variable and perform the operations on the variables is not only tedious job but also disadvantageous. One common organizing technique is to use arrays in such situations. An array is a collection of similar kind of data elements stored in adjacent memory locations and are referred to by a single array-name. In the case of C, you have to declare and define **array** before it can be used. Declaration and definition tell the compiler the name of the array, the type of each element, and the size or number of elements.

To explain it, let us consider to store marks of five students. They can be stored using five variables as follows:

int ar1, ar2, ar3, ar4, ar5;

Now, if we want to do the same thing for 100 students in a class then one will find it difficult to handle 100 variables. This can be obtained by using an array. An array declaration uses its size in [ ] brackets. For above example, we can define an array as:

int ar [100];

where *ar* is defined as an array of size 100 to store marks of integer data-type. Each element of this collection is called an *array-element* and an integer value called the *subscript* is used to denote individual elements of the array. An *ar* array is the collection of 200 consecutive memory locations referred as:

ar [0]	ar [1]		ar [99]
2001	2003		2200

In the above figure, as each integer value occupies 2 bytes, 200 bytes were allocated in the memory.

This chapter explains the use of arrays, types of arrays, declaration and initialization with the help of examples.

CLASSIFICATION OF ARRAYS

Arrays can be classified in the following categories:

1. Single dimensional array
2. Multidimensional array (2-Dimensional array, 3-Dimensional array)

SINGLE DIMENSIONAL ARRAY

If we want to store the data linearly, i.e. we want to store more than one data items of the same type then we have to use the concept of single dimensional array.

The general syntax for defining a single dimensional is as follows:

<Data type> <array name> [size];

Where data type is a valid type (like int, float...), array name is a valid identifier and the elements field (which is always enclosed in square brackets []), specifies how many of these elements the array has to contain.

The following are some of declarations for arrays:

int a[100];
float b[50];
char c[45];

There are two restrictions for using arrays in C language.

- The amount of storage for a declared array has to be specified at **compile time** before execution. This means that an array has a fixed size.

- The data type of an array applies uniformly to all the elements; for this reason, an array is called a **homogeneous** data structure.

For example, an array to contain 5 integer values of type int called billy could be represented like this:

Where each blank panel represents an element of the array, that in this case are integer values of type int. These elements are numerated from 0 to 4 since in arrays the first index is ever 0, independently of its length.

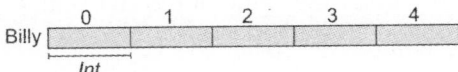

Therefore, in order to declare an array called billy as the one shown in the above diagram it is as simple as:

int billy [5];

Note: The elements field within brackets [] which represents the number of elements the array is going to hold, must be a constant value, since arrays are blocks of non-dynamic memory whose size must be determined before execution.

Size Specification of Single Dimensional Array

The size of an array should be declared using symbolic constant rather a fixed integer quantity (the subscript used for the individual element is integer quantity). The use of a symbolic constant makes it easier to modify a program that uses an array. All reference to maximize the array size can be altered simply by changing the value of the symbolic constant.

To declare size as 50 use the following symbolic constant, SIZE, defined:

#define SIZE 50

The following example shows how to declare and read values in an array to store marks of the students of a class.

Example

Write a program to declare and read values in an array and display them.

/* Program to read values in an array*/

```
# include < stdio.h >
# define SIZE 5 /* SIZE is a symbolic constant
*/
void main( )
{
int i = 0; /* Loop variable */
int stud_marks[SIZE]; /* array declaration */
clrscr();
/* enter the values of the elements */
for( i = 0;i<SIZE;i++)
{
printf ("Element no. =%d",i+1);
printf(" Enter the value of the element:");
scanf("%d",&stud_marks[i]);
}
printf("\n following are the values stored in
the corresponding array elements: \n\n");
for( i = 0; i<SIZE;i++)
{
printf("Value stored in a[%d] is %d\n"i,
stud_marks[i]);      ·
}
getch();
}
```

The output of the above code segment is as
follows:

Element no. = 1 Enter the value of the element
= 11

Element no. = 2 Enter the value of the element
= 12

Element no. = 3 Enter the value of the element
= 13

Element no. = 4 Enter the value of the element
= 14

Element no. = 5 Enter the value of the element
= 15

Following are the values stored in the
corresponding array elements.
Value stored in a [0] is 11

Value stored in a [1] is 12

Value stored in a [2] is 13

Value stored in a [3] is 14

Value stored in a [4] is 15

ARRAY INITIALIZATION

Arrays can be initialized at the time of
declaration. The initial values must appear in
the order in which they will be assigned to
the individual array elements, enclosed
within the braces and separated by commas.

Initialization of Array Elements in the Declaration

The values are assigned to individual array
elements enclosed within the braces and
separated by comma. Syntax of array initialization
is as follows:

data type array-name [size] = {val 1, val 2,
.......val n};

Where val 1 is the value for the first array
element, val 2 is the value for the second
element, and val n is the value for the n array
element. Note that when you are initializing
the values at the time of declaration, then there
is no need to specify the size.

Let us see some of the examples given below

int digits [10] = {1,2,3,4,5,6,7,8,9,10};

int digits[ ] = {1,2,3,4,5,6,7,8,9,10};

int vector[5] = {12,-2,33,21,13};

float tempe[10] ={ 3.2, 2.3, 41.4, 3.2, 2.3, 32.3,
41.1, 10.8, 11.3, 42.3};

double width[ ] = { 17.33333456, -1.212121213,
222.191345 };

int height [10] = { 60, 70, 68, 72, 68 };

Character Array Initialization

The array of characters is implemented as
strings in C. Strings are handled differently
as far as initialization is concerned. A special
character called null character ' \0', implicitly
suffixes every string. When the external or
static string character array is assigned a
string constant, the size specification is usually
omitted and is automatically assigned; it will
include the ' \0'character, added at end.

For example, consider the following two
assignment statements:

char thing [3] = "TIN";

char thing [ ] = "TIN";

In the above two statements the assignments are done differently. The first statement
is not a string but simply an array storing
three characters 'T', 'I' and 'N' and is same as
writing:

char thing [3] = {'T', 'I', 'N'};

whereas, the second one is a four character

string TIN\0. The change in the first assignment, as given below, can make it a string.
char thing [4] = "TIN";

ACCESSING THE VALUES OF AN ARRAY

In any point of a program in which an array is visible, we can access the value of any of its elements individually as if it was a normal variable, thus being able to both read and modify its value.
The format is as follows:

| Billy[0] | Billy[1] | Billy[2] | Billy[3] | Billy[4] |
Billy

name [index]
Following the previous examples in which billy had 5 elements and each of those elements was of type int, the name which we can use to refer to each element is the following:
For example, to store the value 75 in the third element of billy, we could write the following statement:
billy[2] = 75;
and, for example, to pass the value of the third element of billy to a variable called a, we could write:
a = billy[2];
Therefore, the expression billy[2] is for all purposes like a variable of type int.

Notice that the third element of billy is defined billy[2], since the first one is billy[0], the second one is billy[1], and hence, the third one is billy[2]. By this same cause, its last element is billy[4]. Hence, if we write billy[5], we would be accessing the sixth element of billy and therefore beyond the size of the array.

In C it is syntactically correct to exceed the valid range of indices for an array. This can create problems, since accessing out-of-range elements do not cause compilation errors but can cause runtime errors. The reason why this is allowed will be seen further ahead when we begin to use pointers.

At this point it is important to be able to clearly distinguish between the two uses that bracket [ ] have related to arrays. They perform two different tasks: one is to specify the size of arrays when they are declared; and the second one is to specify indices for concrete array elements. Do not confuse these two possible uses of brackets [ ] with arrays.

int billy[5]; // declaration of a new array
billy[2] = 75; // access to an element of the array.

If you read carefully, you will see that a type specifier always precedes a variable or array declaration, while it never precedes an access. Some other valid operations with arrays:
billy[0] = a;
billy[a] = 75;
b = billy [a+2];
billy[billy[a]] = billy[2] + 5;

Programming Problems

Problem 1

Write a program to input 10 numbers and then print 10 numbers with the help of array.

Solution

```
#include<stdio.h>
#include<conio.h>
void main()
{
    int a[10],i;
    clrscr();
    for(i=0;i<10;i++)
{
    printf("\n enter the element");
    scanf("%d",&a[i]);
}
    printf("\n the array elements are");
    for(i=0;i<10;i++)
{
    printf("\n%d",a[i]);
}
    getch();
}
```

Problem 2

Write a program to input 10 numbers and print the maximum number between them.

Solution

```
#include<stdio.h>
#include<conio.h>
```

```
void main()
{
    int a[10],i,max;
    clrscr();
    for(i=0;i<10;i++)
{
    printf("\n enter the element");
    scanf("%d",&a[i]);
}
    max=a[0];
    for(i=1;i<10;i++)
{
    if(max<a[i])
    max=a[i];
}
printf("\n the maximum number is %d",max);
    getch();
}
```

Problem 3

Write a program to input 10 numbers and print the minimum number between them.

Solution

```
#include<stdio.h>
#include<conio.h>
void main()
{
    int a[10],i,max;
    clrscr();
    for(i=0;i<10;i++)
{
    printf("\n enter the element");
    scanf("%d",&a[i]);
}
    max=a[0];
    for(i=1;i<10;i++)
{
    if(max<a[i])
    max=a[i];
}
printf("\n the maximum number is %d",max);
    getch();
}
```

Problem 4

Write a program to input 10 numbers and print the minimum number between them.

Solution

```
#include<stdio.h>
#include<conio.h>
void main()
{
    int a[10],i,min;
    clrscr();
    for(i=0;i<10;i++)
{
    printf("\n enter the element");
    scanf("%d",&a[i]);
}
    min=a[0];
    for(i=1;i<10;i++)
{
    if(min>a[i]])
    min=a[i];
}
printf("\n the minimum number is %d",
min);
    getch();
}
```

Problem 5

Write a program to input 10 numbers and print the sum of all the numbers.

Solution

```
#include<stdio.h>
#include<conio.h>
void main()
{
    int a[10],i,sum=0;
    clrscr();
    for(i=0;i<10;i++)
{
    printf("\n enter the element");
    scanf("%d",&a[i]);
}
    for(i=0;i<10;i++)
{
    sum=sum+a[i];
}
printf("\n the sum is %d",sum);
    getch();
}
```

Problem 6

Write a program to input 10 numbers and print the average of all the numbers.

Solution

```
#include<stdio.h>
#include<conio.h>
void main()
{
    int a[10],i,sum=0;
    float avg;
    clrscr();
    for(i=0;i<10;i++)
{
    printf("\n enter the element");
    scanf("%d",&a[i]);
}
    for(i=0;i<10;i++)
{
    sum=sum+a[i];
}
    avg=sum/10;
printf("\n the Average is %f",avg);
    getch();
}
```

Problem 7

Write a program to input 10 numbers and print the numbers in ascending order.

Solution

```
#include<stdio.h>
#include<conio.h>
void main()
{
    int a[10],i,j,temp;
    clrscr();
    for(i=0;i<10;i++)
{
    printf("\n enter the element");
    scanf("%d",&a[i]);
}
    for(i=0;i<10;i++)
{
    for(j=i+1;j<10;j++)
{
    if(a[i]>a[j])
{
    temp=a[i];
    a[i]=a[j];
    a[j]=temp;
}
}
}
```

```
printf("\n the array in ascending order is ");
    for(i=0;i<10;i++)
{
    printf("\n %d",a[i]);
}
    getch();
}
```

Problem 8

Write a program to input 10 numbers and print the numbers in descending order.

Solution

```
#include<stdio.h>
#include<conio.h>
void main()
{
    int a[10],i,j,temp;
    clrscr();
    for(i=0;i<10;i++)
{
    printf("\n enter the element");
    scanf("%d",&a[i]);
}
    for(i=0;i<10;i++)
{
    for(j=i+1;j<10;j++)
{
    if(a[i]<a[j])
{
    temp=a[i];
    a[i]=a[j];
    a[j]=temp;
}
}
}
printf("\n the array in descending order is ");
    for(i=0;i<10;i++)
{
    printf("\n %d",a[i]);
}
    getch();
}
```

MULTIDIMENSIONAL ARRAY

Suppose that you are writing a chess-playing program. A chessboard is an 8-by-8 control grid. What data structure would you use to constitute it? You could use an array that has a chessboard-like structure, i.e. a *two-dimensional array*, to store the places of the chess pieces. Two-dimensional arrays use two indices to pinpoint a single element of the array. This is very similar to what is called "algebraic notation", commonly used in chess circles to record games and chess problems.

In principle, there is no limit to the number of subscripts (or dimensions) an array can have. Arrays with more than one dimension are called *multidimensional arrays*. While humans cannot easily visualize objects with more than three dimensions, representing multidimensional arrays presents no problem to computers. In practice, however, the amount of memory in a computer tends to place limits on the size of an array. A simple four-dimensional array of double-precision numbers, merely twenty elements wide in each dimension, takes up 20^4 * 8, or 1,280,000 bytes of memory—about a megabyte.

For example, you have ten rows and ten columns, for a total of 100 elements. It is really no big deal. The first number in brackets is the number of rows; the second number in brackets is the number of columns. So, the upper left corner of any grid would be element [0][0]. The element to its right would be [0][1], and so on.

Graphically it is represented as follows:

[0][0]	[0][1]	[0][2]
[1][0]	[1][1]	[1][2]
[2][0]	[2][1]	[2][2]

Three-dimensional arrays (and higher) are stored in the same way as the two-dimensional ones. They are kept in computer memory as a linear sequence of variables, and the last index is always the one that varies fastest (then the next-to-last, and so on).

Two Dimensional Array

Two-dimensional arrays, the most common multidimensional arrays, are used to store information that we normally represent in table form. Two-dimensional arrays, like one-dimensional arrays, are homogeneous. This means that all of the data in a two-dimensional array is of the same type. Examples of applications involving two-dimensional arrays include:

- A seating plan for a room (organized by rows and columns),
- A monthly budget (organized by category and month), and
- A grade book where rows might correspond to individual students and columns to student scores.

Declaration of Two-Dimensional Arrays

The general syntax for declaration of two-dimensional array is as follows:
<datatype>
<arrayname>[sizeofrow][sizeofcolumn];

Example

The following declarations set aside storage for a two-dimensional array called *labScores* which contains 40 rows and 14 columns. Rows correspond to a particular student and columns correspond to a particular lab score.

    const int MAX_STUDENTS=40;

    const int MAX_LABS=14;

    int labScores
[MAX_STUDENTS][MAX_LABS];

Manipulation of a two-dimensional array requires the manipulation of two indices. When the two-dimensional array *labScores* is declared, enough storage is set aside to hold a table containing 40 rows and 14 columns for a total of 40 * 14 = 560 integer values. To access one particular value, we must specify the row and column. The row index ranges from 0 to MAX_STUDENTS-1 (39) and the column index ranges from 0 to MAX_LABS-1 (13). Thus the table can be visualized as follows:

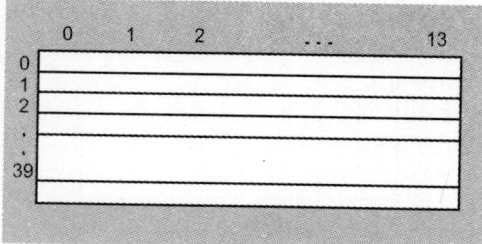

This two-dimensional array may also be visualized as a one-dimensional array of arrays. An alternative view of this array would be:

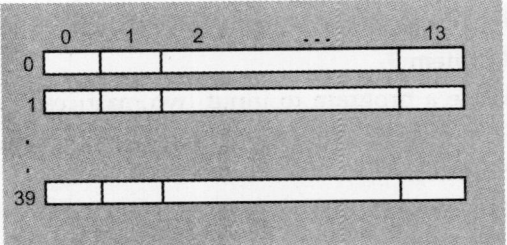

This two-dimensional array may be viewed as a one-dimensional array having 40 elements where each element is an array of 14 values.

Accessing a Two-Dimensional Array Element

In our labScores example, suppose we wish to indicate that the second student (corresponding to row 1) made a 90 on lab 10 (corresponding to column 9). We might use the statement:

labScores [1][9] = 90;

Array indices may be integer constants (as in the above example), variables, or expressions. They should be within the bounds of the array.

Two-Dimensional Array Initialization

We can declare and initialize an array A as follows:

```
//declaration
int A[3][4] = {{8, 2, 6, 5}, //row 0
{6, 3, 1, 0}, //row 1
{8, 7, 9, 6}}; //row 2
```

Memory for the array may be visualized as follows:

A	0	1	2	3
0	8	2	6	5
1	6	3	1	0
2	8	7	9	6

Programming Problems

Problem 1

Write a program to input a 3*3 matrix and print the elements in matrix form.

Solution

```
#include<stdio.h>
#include<conio.h>
void main()
{
    int a[3][3],i,j;
    clrscr();
    for(i=0;i<3;i++)
{
    for(j=0;j<3;j++)
{
    printf("\n enter the element");
    scanf("%d",&a[i][j]);
}
}

    printf("\n the matrix is");
    for(i=0;i<3;i++)
{
    for(j=0;j<3;j++)
{
    printf("\t%d",a[i][j]);
}
    printf("\n");
}
    getch();
}
```

Problem 2

Write a program to input two 3*3 matrices and print the addition of both the matrix.

Solution

The usual matrix addition is defined for two matrices of the same dimensions. The sum of two m-by-n matrices A and B, denoted by A + B, is again an m-by-n matrix computed by adding corresponding elements. For example: If we want to add the following matrices:

$$\begin{bmatrix} 0 & 1 & 2 \\ 9 & 8 & 7 \end{bmatrix} + \begin{bmatrix} 6 & 5 & 4 \\ 3 & 4 & 7 \end{bmatrix}$$

Add the pairs of entries, and simplify for the final answer:

$$\begin{bmatrix} 0 & 1 & 2 \\ 9 & 8 & 7 \end{bmatrix} + \begin{bmatrix} 6 & 5 & 4 \\ 3 & 4 & 7 \end{bmatrix} = \begin{bmatrix} 0+6 & 1+5 & 2+4 \\ 9+3 & 8+4 & 7+5 \end{bmatrix} = \begin{bmatrix} 6 & 6 & 6 \\ 12 & 12 & 12 \end{bmatrix}$$

So the answer is:

$$\begin{bmatrix} 6 & 6 & 6 \\ 12 & 12 & 12 \end{bmatrix}$$

```c
#include<stdio.h>
#include<conio.h>
void main()
{
    int a[3][3],b[3][3],c[3][3],i,j;
    clrscr();
    for(i=0;i<3;i++)
{
    for(j=0;j<3;j++)
{
    printf("\n enter the elements of ist matrix");
    scanf("%d",&a[i][j]);
}
}
    for(i=0;i<3;i++)
{
    for(j=0;j<3;j++)
{
    printf("\n enter the elements of IInd matrix");
    scanf("%d",&b[i][j]);
}
}
    for(i=0;i<3;i++)
{
    for(j=0;j<3;j++)
{
    c[i][j]=a[i][j]+b[i][j];
}
}
    printf("\n the Addition is ");
    for(i=0;i<3;i++)
{
    for(j=0;j<3;j++)
{
    printf("\t%d",c[i][j]);
}
    printf("\n");
}
    getch();
}
```

Problem 3

Write a program to input two matrices and print the multiplication of both the matrix.

Solution

The Product AB is determined as the dot products of the *i*th row in A and the *j*th column in B,

placed in *i*th row and *j*th column of the resulting *m* x *p* matrix C.

For example

To get this, we found the dot-product of each row in A with each column in B.
C[1,1] = (1 x 1) + (3 x 2) + (4 x 3) = 19, and
C[2,3] = (2 x 3) + (0 x 2) + (1 x 1) = 7

```c
#include<stdio.h>
#include<conio.h>]
#include<stdlib.h>
void main()
{
int a[10][10],b[10][10],c[10][10],i,j,k,m,n,p,q;
clrscr();
printf("Enter the rows and columns and of the first matrix:");
scanf("%d %d",&m,&n);
printf("\n enter the rows and columns and of the second matrix:");
scanf("%d %d",&p,&q);
printf("\n enter elements of the first matrix:\n");
for(i=0;i< m;i++)
{
for(j=0;j< n;j++)
```

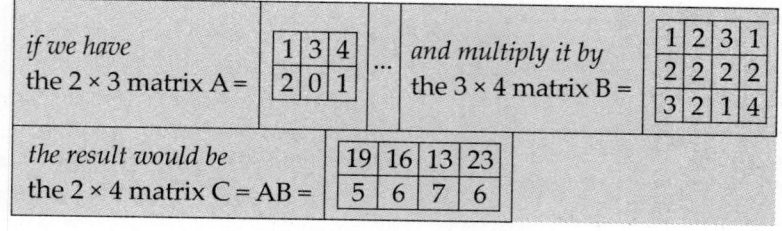

if we have the 2 × 3 matrix A =	1	3	4	...	and multiply it by the 3 × 4 matrix B =	1	2	3	1
	2	0	1			2	2	2	2
						3	2	1	4
the result would be the 2 × 4 matrix C = AB =	19	16	13	23						
	5	6	7	6						

```
scanf("%d",&a[i][j]);
}
printf("\n enter elements of the second
Matrix:\n");
for(i=0;i< p;i++)
{
for(j=0;j< q;j++)
scanf("%d",&b[i][j]);
}
printf("The first matrix is:\n");
for(i=0;i< m;i++)
{
for(j=0;j< n;j++)
printf(" %d ",a[i][j]); //print the first matrix
printf("\n");
}
printf("The second matrix is:\n");
for(i=0;i< p;i++) // print the second matrix
{
for(j=0;j< q;j++)
printf(" %d ",b[i][j]);
printf("\n");
}
if(n!=p)
{
printf("Aborting!!!!!!/n multiplication of the
above matrices not feasible.");
exit(0);
}
else
{
for(i=0;i< m;i++)
{
for(j=0;j< q;j++)
{
c[i][j] = 0;
for(k=0;k< n;k++)
{
c[i][j] = c[i][j] + a[i][k] * b[k][j];
}
}
}
printf("\n multiplication of the above two
matrices are:\n\n");
for(i=0;i< m;i++)
{
for(j=0;j< q;j++)
{
printf("\t%d ",c[i][j]);
```

```
}
printf("\n");
}
}
getch();
}
```

Problem 4

Write a program to input a matrix and print the transpose of a matrix.

Solution

A matrix which is formed by turning all the rows of a given matrix into columns and vice versa. The transpose of matrix A is written A^T.

$$\text{Example:} \begin{bmatrix} 1 & 2 & 3 \\ 4 & 5 & 6 \end{bmatrix}^T = \begin{bmatrix} 1 & 4 \\ 2 & 5 \\ 3 & 6 \end{bmatrix}$$

The program in C language is as follows:

```
#include<stdio.h>
#include<conio.h>
void main()
{
    int a[3][3],i,j;
    clrscr();
    for(i=0;i<3;i++)
{
    for(j=0;j<3;j++)
{
    printf("\n enter the element");
    scanf("%d",&a[i][j]);
}
}
    printf("\n the transpose of the matrix is");
    for(i=0;i<3;i++)
{
    for(j=0;j<3;j++)
{
    printf("\t%d",a[j][i]);
}
    printf("\n");
}
    getch();
}
```

Three-Dimensional Array

C allows arrays with more than two dimension, to the limit depending on the compiler.

The general form of a multidimensional array is

type array_name[s1][s2][s3]....[sm];
The total number of elements in any dimension of arrays is the product of all sizes included in the declaration. So it becomes s1*s2* s3*....*sm.

For most applications, two dimension arrays serve. Three dimension is maximum that you can anticipate to go in normal applications. Still higher dimension arrays are needed in the field of scientific computing, weather forecasting, time-space analysis, etc.

DYNAMIC ALLOCATION OF ARRAYS OR ARRAYS OF UNKNOWN OR VARYING SIZE

If we want to allocate the memory dynamically then we have to use the concept of dynamic allocation of arrays.

The problems with fixed size arrays
Declaring an array with a fixed size like int a[100000]; has two typical problems:

- **Exceeding maximum**. Selecting a real maximum is often impossible because the programmer has no control over the size of the data sets the user is interested in. Erroneous assumptions that a maximum will never be exceeded are the source of many programming errors. Declaring very big arrays can be extremely wasteful of memory, and if there are many such arrays, may prevent the program from running in some systems.

- **No expansion**. Using a little size may be more efficient for the typical data set, but prevents the program from running with larger data sets. If array limits are not checked, large data sets will run over the end of an array with disastrous effects. Fixed size arrays cannot elaborate as needed.

These troubles can be avoided by dynamically allocating an array of the right size, or reallocating an array when it demands to expand. Both of these are done by declaring an array as a pointer and using the malloc/ calloc function to allocate memory, and free function to free memory that is no longer needed.

The detail of this topic will be discussed in the Pointer chapter.

PASSING ARRAYS TO FUNCTIONS

We can pass the single dimensional arrays as well as multidimensional arrays to the function. We will discuss these concepts one by one.

Single Dimensional Arrays

Sometimes it is inconvenient to call a function that requires a long list of arguments.

One way around this, is to accumulate your variables into an array, then pass a **POINTER** to the array to the function. This method will be described in a great detail in the pointers section, but for now you need to know that the array is not actually passed to the function—just the array's location in the memory. This is experienced as **PASS BY REFERENCE**. The name of an array references the array's location in the memory, its **ADDRESS**.

If we write int max (int *) or int max (int [])
The meaning of this line is that max is a function in which we pass the integer type of pointer, i.e. the complete array and according to that array it returns an integer type of variable.

Programming Problems

Problem 1

Write a program to input 5 numbers and print the sum of five numbers with the help of function.

Solution

```
#include <stdio.h>
int addNumbers(int fiveNumbers[]); /* declare
function */
void main()
{
int array[5];
int i;
printf("Enter 5 integers separated by spaces:
");
for(i=0 ; i<5 ; i++)
{
scanf("%d", &array[i]);
```

```
}
printf("\n their sum is: %d\n", addNumbers
(array));
return 0;
}
int addNumbers(int fiveNumbers[])
{ /* define function */
int sum = 0;
int i;
for(i=0 ; i<5 ; i++) {
sum+=fiveNumbers[i]; /* work out the total */
}
return sum; /* return the total */
}
```

Notice that I have left the size of the array blank in both the function declaration and definition- the compiler works it out for you. Also, when I called the function in C language, I passed on the name of the array. This is the equivalent to passing &array [0]—the address of the first component. You will learn about the address-of operator afterward.

Problem 2

Write a program to input 10 numbers and print the maximum number between them with the help of function.

Solution

```
#include<stdio.h>
#include<conio.h>
int max(int []);
void main()
{
int a[10],i,m;
clrscr();
for(i=0;i<10;i++)
{
    printf("\n enter the value of a[%d]",i);
    scanf("%d",&a[i]);
}
m=max(a);
printf("\n The maximum number is %d",m);
getch();
}
int max(int p[])
{
    int m,i;
    m=p[0];
```

```
for(i=1;i<10;i++)
{
    if(m<p[i])
    m=p[i];
}
    return(m);
}
```

Multidimensional Arrays

This is similar to passing 1D arrays but, in the function declarations you must specify all the dimension sizes (but the leftmost one is optional).

The following program shows the passing of multidimensional array into the function

```
#include <stdio.h>
void printArray(int array[][4]); /* declare
function */
void main()
{
int array[3][4] = {0,1,2,3,4,5,6,7,8,9,10,11};
printArray(array);
}
void printArray(int array[][4])
{
/* define function */
int i, j;
for(i=0 ; i<3 ; i++) {
for(j=0 ; j<4 ; j++) {
printf("%2d ", array[i][j]);
}
printf("\n");
}
}
```

SUMMARY

- Array is the collection of similar data type and having a common name.
- Array can be classified into the following categories (a) single dimensional array (b) multidimensional array.
- If we want to store the data in the form of matrix then we have to use the concept of 2-D array.
- Array plays a very important role in the computer programming.
- We can pass the array into the function with the help of pointer.

- If we define int a[10] the meaning of this line is that a is an array which holds 10 integer elements and 20 bytes memory is allocated for this array.
- If we define int a[3][3] the meaning of this line is that a is an array which holds 3 rows and 3 columns and total 9 elements are stored in this array and 18 bytes memory is allocated for this array.

Exercises

Section A

Multiple Choice Questions

1. Different types of arrays includes
   (a) Single dimensional
   (b) Two dimensional
   (c) Multidimensional
   (d) All of the above

2. General syntax for defining arrays are
   (a) <data type><array name>[size]
   (b) <size><array name>[data type]
   (c) Both a and b
   (d) None of these

3. Examples of applications involving 2-D arrays are
   (a) Seating plan for a room
   (b) A monthly budget
   (c) A grade book
   (d) All of the above

4. We can declare 2-D arrays using this syntax
   (a) <data type><array name>[size of column] [size of row]
   (b) <data type><array name>[size of row] [size of column]
   (c) Both a and b
   (d) None of the above

5. Multidimensional array can have
   (a) Multiple rows
   (b) Multiple columns
   (d) Both a and b
   (d) None of the above

6. Problems with fixed size arrays are
   (a) Their size can not be expanded
   (b) Difficulty to define maximum size
   (c) Memory wastage
   (d) All of the above

7. We can pass an array to a function by
   (a) It address   (b) It name
   (c) Both a and b   (d) None of the above

8. Multidimensional array can be
   (a) Homogenous
   (b) Heterogeneous
   (c) Both a and b
   (d) Optional

9. To sum up two 2D arrays
   (a) Their sizes must be same
   (b) Number of rows must be same
   (c) Number of columns must be same
   (d) Rows and columns must be same

10. Array indices may be
    (a) integer constants
    (b) char
    (c) float
    (d) All of the above

State True/False

1. Array are collection of similar type of data.
2. The entire array is stored contiguously in memory.
3. We can access data elements of array sequentially only.
4. Multidimensional arrays are non-homogeneous.
5. Multidimensional array can be defined as array of arrays.
6. Two dimensional array is a type of multidimensional array.
7. We cannot allocate an array dynamically.
8. While passing arrays to a function we pass a pointer to it.
9. In multidimensional arrays all dimension sizes must be specified.
10. Multidimensional arrays are limited to three dimensions only.

Section B

Short Answer Type Questions

1. Define the concept of array with suitable example.
2. Define the classification of array with suitable example.
3. Why we use the concept of array define it with proper explanation?

4. Write a program in C to sum of first N numbers.

5. Write a program in C to average of M to N numbers.

6. Write a program in C to display the even or odd numbers between 1 and 100.

7. Write a program in C to display the even or odd numbers of first N numbers.

8. Write a program in C to display the even or odd numbers between two ranges.

9. Write a program in C to display the leap years between 1000 and 2000.

10. Write a program in C to display the leap years of given range.

Section C

Long Answer Type Questions

1. Write a program in C to print the numbers between two ranges.

2. Write a program in C to print the number until 1000 is given an input.

3. Write a program in C to print the positive, negative and zero until 1000 is given an input.

4. Write a program in C to count the positive, negative numbers and zero until 1000 is given an input.

5. Write a program in C to sum of the positive, negative numbers until 1000 is given an input.

6. Write a program in C to mean of the positive, negative numbers until 1000 is given an input.

7. Write a program in C to factorial of given number until 1000 is given an input.

8. Write a program in C to read and display of N numbers using array.

9. Write a program in C to find mean of ten numbers using array.

10. Write a program in C to find mean of N numbers using array.

11. Write a program in C to find the biggest of 10 numbers using array.

12. Write a program in C to find the smallest of 10 numbers using array.

13. Write a program in C to find the position of biggest number in the array.

14. Write a program in C to find the position of smallest number in the array.

15. Write a program in C to find the positions of biggest and smallest number in the array.

16. Write a program in C to interchange the biggest and smallest number in the array.

17. Write a program in C to find the second biggest of N numbers using array.

18. Write a program in C to find the second smallest of N numbers using array.

19. Write a program in C to find the second biggest and smallest of N numbers using array.

20. Write a program in C to find the positions of second biggest and smallest of N numbers using array.

21. Write a program in C to interchange the second biggest and smallest of N numbers using array.

22. Write a program in C to read and write 3x3 matrix.

23. Write a program in C to read and display the transpose of 3x3 matrix.

24. Write a program in C to find the sum of matrix.

25. Write a program in C to find the product of two matrixes.

26. Write a program in C to read and display mxm matrix.

27. Write a program in C to read and display the transpose of mxn matrix.

28. Write a program in C to sum of two mxn matrix.

29. Write a program in C to poduct of two mxn matrix.

12

Strings

In the previous chapter, we have discussed numeric arrays, a powerful data storage method that lets you group a number of same-type data items under the same group name. Individual items, or elements, in an array are identified using a subscript after the array name. Computer programming tasks that involve repetitive data processing lend themselves to array storage. Like non-array variables, arrays must be declared before they can be used. Optionally, array elements can be initialized when the array is declared. In the earlier unit, we had just known the concept of *character arrays* which are also called *strings*. String can be represented as a single-dimensional character type array. C language does not provide the intrinsic string types. Some problems require that the characters within a string be processed individually. However, there are many problems which require that strings be processed as complete entities. Such problems can be manipulated considerably through the use of special string oriented library functions. Most of the C compilers include string library functions that allow string comparison, string copy, concatenation of strings, etc. The string functions operate on null terminated arrays of characters and require the header <string.h>. The use of some of the string library functions are given as examples in this chapter.

A string is a collection of character variables. Any sequence or set of characters determined within double quotation symbols is a constant string. In C it is required to do some meaningful operations on strings they are:

- Reading string displaying strings
- Combining or concatenating strings
- Copying one string to another
- Comparing string and checking whether they are equal
- Extraction of a portion of a string.

Strings are stored in memory as ASCII codes of characters that make up the string appended with '\0'(ASCII value of null). Normally each character is store in one byte, successive characters are stored in successive bytes.

Example

If we define char s[80];
The meaning of this line is that s is a character type array which holds 80 characters and the values are stored from s[0] to s[79].

DECLARATION AND INITIALIZATION OF STRINGS

In C language strings are group of characters, digits, and symbols enclosed in quotation marks or simply we can say the string is declared as a "character array". The end of the string is marked with a special character, the '\0' (*Null character*), which has the decimal value 0. There is a difference between a *character* stored in memory and a *single character string* stored in a memory. The character requires only one byte whereas the single character string requires two bytes (one byte for the character and other byte for the delimiter).

Declaration of Strings

A string in C is simply a sequence of characters. To declare a string, specify the data type as char and place the number of characters in the array in square brackets after the string name.

The general syntax of defining a string is as follows:

char string-name[size];

Example

char name[25];

char address[30];

char city[20];

Initializing Strings

Following the discussion on characters arrays, the initialization of a string must the following form which is simpler to one dimension array.

char month1[ ]={'j','a','n','u','a','r','y'};

Then the string month is initializing to January. This is absolutely valid but C offers a special way to initialize strings. The above string can be initialized char month1[] ="January"; The characters of the string are enclosed within a part of double quotes. The compiler takes care of string closed in within a pair of a double quotes. The compiler takes care of putting in the ASCII codes of characters of the string in the memory and also stores the null terminator in the end.

```
/*String.c string variable*/
#include < stdio.h >
main()
{
char month[15];
printf ("Enter the string");
gets (month);
printf ("The string entered is %s", month);
}
```

In this example, string is stored in the character variable month the string is displayed in the statement.

printf("The string entered is %s", month");

It is one dimension array. Each character takes a byte. A null character (\0) that has the ASCII value 0 ends the string. The figure shows the computer storage of string January in the memory recall that \0 determines a single character whose ASCII value is zero.

J
A
N
U
A
R
Y
\0

Character string terminated by a null character '\0'.

A string variable is any valid C variable name & is forever declared as an array. The general form of declaration of a string variable is

char string_name[size];

The size determines the number of characters in the string name.

Example

If we define the following character type arrays

char month[10];

char address[100];

The size of the array should be one byte more than the actual space occupied by the string since the complier appends a null character at the end of the string.

STRING CONSTANTS

String constants have double quote marks around them, and can be assigned to char pointers. In lieu, you can assign a string constant to a char array—either with no size defined, or you can specify a size, but do not forget to leave a space for the null character!

Let us explain it with the help of following code fragments

```
/* Code Fragment 1 */
{
char *s;
s="Manish";
printf("%s\n",s);
}
/* Code Fragment 2 */
{
char s[100];
strcpy(s, " Manish");
printf("%s\n",s);
}
```

These two fragments produce the same output, but their internal behavior is quite different. In fragment 2, you cannot say **s = "Manish";**. To realize the differences, you have to understand how the *string constant table* works in C. When your program is compiled, the compiler fleshes the object code file, which contains your machine code and a table of all the string constants declared in the program. In fragment 1, the statement **s = "Manish";** causes **s** to point to the address of the string **Manish** in the string constant table. Since this string is in the string constant table, and therefore technically a part of the feasible code, you cannot modify it. You can only point to it and apply it in a read-only manner. In fragment 2, the string **Manish** also exists in the constant table, so you can copy it into the array of characters named **s**. Since **s** is not an address, the statement **s="Manish";** will not work in fragment 2. It will not even compile.

Example

Write a program to read a name from the keyboard and display message **Hello** onto the monitor.

Solution

```
/*Program that reads the name and display
the hello along with your name*/
#include <stdio.h>
#include<conio.h>
void main()
{
char name[10];
clrscr();
printf("\n enter your name : ");
scanf("%s", name);
printf("Hello %s\n", name);
getch();
}
```

The output of the above code segment is as follows:

Enter your name: Manish
Hello Manish

In the above example declaration char name [10] allocates 10 bytes of memory space (on 16 bit computing) to array name [ ]. We are passing the base address to scanf function and scanf() function fills the characters typed at the keyboard into array until enter is pressed. The scanf() places '\0' into array at the end of the input. The printf() function prints the characters from the array on to monitor, leaving the end of the string '\0'. The %s used in the scanf() and printf() functions is a format specification for strings.

STRINGS DISPLAY USING DIFFERENT FORMATTING TECHNIQUES

The *printf* function with %s format is used to display the strings on the screen.

For example, the below statement displays entire string: printf ("%s", name);

We can also specify the accuracy with which character array (string) is displayed.

For example, if you want to display first 5 characters from a field width of 20 characters,

You have to write as: printf ("%20.5s", name);

If you include minus sign in the format (e.g. % –10.5s), the string will be printed left justified.

printf ("% -10.5s", name);

Example

Write a program to display the string "MANI" in the following format:

```
M
MA
MAN
MANI
MANI
MAN
MA
M
```

Solution

```
/* Program to display the string in the above
shown format*/
# include <stdio.h>
#include<conio.h>
void main()
{
int x, y;
static char string[ ] = "MANI";
printf("\n");
for( x=0; x<4; x++)
{
y = x + 1;
```

/* reserves 4 character of space on to the monitor and minus sign is for left justified*/
printf ("%-4.*s \n", y, string);
/* and for every loop the * is replaced by value of y */
/* y value starts with 1 and for every time it is incremented by 1 until it reaches to 4*/
}
for(x=3; x>=0; x- -)
{
y = x + 1;
printf("%-4.*s \n", y, string);
/* y value starts with 4 and for every time it is decrements by 1 until it reaches to 1*/
}
}
The output of the above code segment is as follows:
M
MA
MAN
MANI
MANI
MAN
MA
M

ARRAY OF STRINGS

Array of strings are multiple strings, stored in the form of table. Declaring array of strings is same as strings, except it will have additional dimension to store the number of strings.
The general syntax of defining an array of string is as follow:
char array-name[size][size];
For example
char names[4][25];
where names is the name of the character array and the constant in first square brackets will gives number of string we are going to store, and the value in second square bracket will gives the maximum length of the string.

Example

char names [3][10] = {"manish", "priya", "shiva"};

It can be represented by a two-dimensional array of size[3][10] as shown below:

0	1	2	3	4	5	6	7	8	9
m	a	n	i	S	h	\0			
p	r	i	y	A	\0				
s	h	i	v	A	\0				

Programming Problem

Problem 1

Write a program to initializes 3 names in an array of strings and display them on to monitor.

Solution

```
#include <stdio.h>
#include<conio.h>
void main()
{
int n;
char names[3][10] = {"Manish", "Pragati", "Shiva" };
clrscr();
for(n=0; n<3; n++)
printf("%s \n",names[n] );
getch();
}
```
The output of the above code segment is as follows:
Manish
Pragati
Shiva

BUILT IN STRING FUNCTIONS AND ITS APPLICATIONS

The header file <string.h> contains some string manipulation functions. The following is a list of the common string managing functions in C.

strlen() function

This function counts and returns the number of characters in a string. The length does not include a null character.
The general syntax of strlen () function is as follows:
n=strlen (string);
Where n is integer variable which receives the value of length of the string.

Example

length=strlen ("Manish");
This function returns 6 and stored in the length variable.

Problem

Write a program to read a string from the keyboard and to display the length of the string onto the monitor by using strlen ( ) function.

Solution

```
#include <stdio.h>
#include <string.h>
#include<conio.h>
void main()
{
char name[80];
int length;
clrscr();
printf("Enter your name: ");
gets(name);
length = strlen(name);
printf("Your name has %d characters\n",
length);
getch();
}
```

The output of the above code is as follows
Enter your name: Manish
Your name has 6 characters

strcpy() Function

In C language, you cannot simply assign one character array to another. You have to copy element by element. The string library <string.h> contains a function called **strcpy** for this purpose. The **strcpy** function is used to copy one string to another.

The general syntax of strcpy function is as follows:

strcpy (str1, str2);

Where str1, str2 are two strings. The content of string str2 is copied on to string str1.

Problem

Write a program to read a string from the keyboard and copy the string onto the second string and display the strings on to the monitor by using strcpy( ) function.

Solution

```
#include <stdio.h>
#include <string.h>
#include<conio.h>
void main()
{
```

```
char first[80], second[80];
clrscr();
printf("Enter a string: ");
gets(first);
strcpy(second, first);
printf("\n first string is : %s, and second string
is: %s\n", first, second);
}
```

The output of the above code is as follows:
Enter a string: Manish
First string is: Manish, and second string is: Manish

strcmp () function

The **strcmp** function in the string library function which compares two strings, character by character and stops comparison when there is a difference in the ASCII value or the end of any one string and returns ASCII difference of the characters that is integer. If the return value *zero* means the two strings are equal, a negative value means that first is less than second, and a positive value means first is greater than second.

The general syntax of strcmp () function is as follows:

n = strcmp (str1, str2);

Where **str1** and **str2** are two strings to be compared and **n** is returned value of differed characters.

Example

- strcmp("Newyork","Newyork") will return zero because 2 strings are equal.
- strcmp("their","there") will return a 9 which is the numeric difference between ASCII 'i' and ASCII 'r'.
- strcmp("The", "the") will return 32 which is the numeric difference between ASCII "T" & ASCII "t".

Problem

Write a program to compare two strings using strcmp () function.

Solution

```
#include <stdio.h>
#include <string.h>
#include<conio.h>
void main()
```

```
{
char first[80], second[80];
int value;
clrscr();
printf("Enter a string: ");
gets(first);
printf("Enter another string: ");
gets(second);
value = strcmp(first, second);
if(value == 0)
puts("The two strings are equal");
else if(value < 0)
puts("The first string is smaller ");
else if(value > 0)
puts("the first string is bigger");
getch();
}
```
The output of the above code is as follows:
Enter a string: MANISH
Enter another string: VARSHNEY
The first string is smaller

strcmpi () Function
This function is same as strcmp () which compares 2 strings but not case sensitive.

Example
strcmpi("THE","the"); will return 0.

Problem
Write a program to input two strings and print the appropriate message that the strings are alphabetically equal or not whether the strings are not case sensitive.

Solution
```
#include<stdio.h>
#include<conio.h>
#include<string.h>
void main()
{
    char s[80],s1[80];
    clrscr();
    printf("\n enter first string");
    scanf("%s",s);
printf("\n enter second string");
scanf("%s",s1);
if(strcmpi(s,s1)==0)
printf("\n both the strings are alphabetically equal");
```
```
else
printf("\n both the strings are not alphabetically equal");
getch ();
}
```
The output of the above code is as follows:
Enter first string Manish
Enter second string MANISH
Both the strings are alphabetically equal.

strcat () Function
The **strcat** function is used to join one string to another. It takes two strings as arguments; the characters of the second string will be appended to the first string.

The general syntax of strcat () function is as follows:
strcat (str1, str2);
Where str1 and str2 are two string arguments, string str2 is appended to string str1.

Problem
Write a program to read two strings and append the second string to the first string.

Solution
```
#include <stdio.h>
#include <string.h>
#include<conio.h>
void main()
{
char first[80], second[80];
clrscr();
printf("Enter a string:");
gets (first);
printf("Enter another string: ");
gets (second);
strcat(first, second);
printf("\n the concated string is %s\n", first);
getch();
}
```
The output of the above code is as follows
Enter a string: Manish
Enter another string: Varshney
The concated string is Manish Varshney

strlwr () Function
The **strlwr** function converts upper case characters of string to lower case characters.
The general syntax is as follows:
strlwr (str1);

Where str1 is string to be converted into lower case characters.

Example

strlwr ("MANISH") converts to manish

Problem

Write a program to convert the string into lower case characters using predefined function.

Solution

```
#include <stdio.h>
#include <string.h>
#include<conio.h>
void main()
{
char first[80];
clrscr();
printf("Enter a string: ");
gets(first);
printf("Lower case of the string is %s",
strlwr(first));
getch();
}
```

The output of the above code segment is as follows:

Enter a string: SHIVA
Lower case of the string is shiva

strupr() Function

This function converts all characters in a string from lower case to uppercase.

The general syntax of strupr() function is as follows:

strupr(string);

Example strupr("manish") will convert the string to MANISH

Problem

Write a program to input a string in lowercase and then convert it into uppercase and print the uppercase string.

Solution

```
#include<stdio.h>
#include<conio.h>
#include<string.h>
void main()
{
    char s[80];
    clrscr();
```

```
    printf("\n enter the string in lowercase
characters");
    scanf("%s",s);
    printf("\n the uppercase string is %s",
strupr(s));
getch();
}
```

The output of the above code segment is as follows

Enter the string in lowercase characters
manish
The upper case string is MANISH

strrev() Function

The **strrev** funtion is used to reverse the given string.

The general syntax of strrev() function is as follows:

strrev (str); where string **str** will be reversed.

Example strrev("program") reverses the characters in a string into "margrop".

Problem

Write a program to input a string and print the reverse of the string with the help of predefined function.

Solution

```
#include <stdio.h>
#include <string.h>
#include<conio.h>
void main()
{
char first[80];
clrscr();
printf("Enter a string:");
gets(first);
printf("\n reverse of the given string is : %s",
strrev(first));
getch();
}
```

The output of the above code is as follows:

Enter a string: Shiva
Reverse of the given string is: avihS

strspn () Function

The **strspn** function returns the position of the string, where first string mismatches with second string.

The general syntax of strspn () function is as follows:

n = strspn (first, second);

Where **first** and **second** are two strings to be compared, **n** is the number of character from which first string does not match with second string.

Problem

Write a program, which returns the position of the string from where first string does not match with second string.

Solution

```
#include <stdio.h>
#include <string.h>
#include<conio.h>
void main()
{
char first[80], second[80];
clrscr();
printf("Enter first string: ");
gets(first);
printf("\n Enter second string: ");
gets(second);
printf("\n After %d characters there is no
match",strspn(first, second));
getch();
}
```

The output of the above code is as follows
Enter first string: HELLOBROTHER
Enter second string: HELLOSISTER
After 5 characters there is no match

OTHER STRING FUNCTIONS

Now we will discuss some other important string functions.

strncpy () Function

The **strncpy** function same as strcpy. It copies characters of one string to another string up to the specified length.

The general syntax of strncpy () function is as follows:

strncpy (str1, str2, No. of characters);

Example

strncpy (str1, str2, 10);

Where str1 and str2 are two strings. The **10** characters of string **str2** are copied onto string **str1**.

strncmp Function

The **strncmp** function is same as *strcmp*, except it compares two strings up to a specified length.

The general syntax of the above function is as follows:

n = strncmp(str1, str2, 10);

Where **10** characters of **str1** and **str2** are compared and **n** is returned value of differed characters.

strstr () Function

The **strstr** function takes two arguments address of the string and second string as inputs. And returns the address from where the second string starts in the first string.

The general syntax of the above function is as follows:

cp = strstr (first, second);

where **first** and **second** are two strings, **cp** is character pointer.

Programming Problems

Problem 1

Write a program to concatenate two strings without using the strcat () function.

Solution

```
# include<string.h>
# include <stdio.h>
#include<conio.h>
void main ()
{
char str1 [10];
char str2 [10];
char output_str [20];
clrscr();
int i=0, j=0, k=0;
printf("\n enter first string: ");
gets(str1);
printf("\n enter second string: ");
gets (str2);
while (str1[i] != '\0')
output_str[k++] = str1[i++];
while(str2[j] != '\0')
output_str[k++] = str2[j++];
output_str[k] = '\0';
puts (output_str);
getch ();
}
```

The output of the above program is as follows:
Enter first string Manish
Enter second string Varshney
ManishVarshney

Problem 2

Write a program to find the string length without using the strlen() function.

Solution

```
# include<stdio.h>
# include<string.h>
#include<conio.h>
void main ()
{
char string [60];
int len=0, i=0;
printf ("\n enter the string: ");
gets (string);
while (string [i++]! = '\0')
len ++;
printf("Length of input string = %d", len);
getch ();
}
```

The output of the above program is as follows:
Enter the string Manish
Length of input string = 6

Problem 3

Write a program to convert the lowercase letters to uppercase in a given string without using strupr() function.

Solution

```
#include<stdio.h>
#include<conio.h>
void main ()
{
int i= 0; char source[10], destination[10];
printf ("\n enter the source string");
gets (source);
while (source[i]! = '\0')
{
if ((source[i]>='a') && (source[i]<='z'))
destination[i] =source[i]-32;
else
destination[i] =source[i];
i++;
}
destination[i]= ' \0 ';
puts (destination);
getch ();
}
```

The output of the above program is as follows:
Enter the source string manish
MANISH

SUMMARY

- Strings are sequence of characters.
- Strings are to be null-terminated if you want to use them properly.
- Remember to take into account null-terminators when using dynamic memory allocation.
- The string.h library has many useful functions.
- strlen () is a predefined function in string.h which is used to calculate the length of the string.
- strcmp () is a predefined function in string.h which is used to compare two strings.
- strcat () is a predefined function in string.h which is used to concatanated two strings.
- strrev () is a predefined function which is used to reverse the string.
- A string is nothing but an array of characters terminated by '\0'.
- Being an array, all the characters of a string are stored in contiguous memory locations.
- Though **scanf( )** can be used to receive multi-word strings, **gets( )** can do the same job in a cleaner way.
- Both **printf( )** and **puts( )** can handle multi-word strings.
- Strings can be operated upon using several standard library functions like **strlen( )**, **strcpy( )**, **strcat( )** and **strcmp( )** which can manipulate strings. More importantly we imitated some of these functions to learn how these standard library functions are written.
- Though in principle a 2-D array can be used to deal several strings, in practice an array of pointers to strings is preferred since it takes less space and is effective in processing strings.
- **malloc ( )** function can be used to allocate space in memory on the fly throughout execution of the program.

The following table gives the basic idea of all the functions used in string.h header file.

String functions	Its use
strlen	Returns number of characters in string
str11vr	Converts all the characters in the string into lower case characters
strcat	Adds one string at the end of another string
strcpy	Copies a string into another
strcmp	Compares two strings and returns zero if both are equal.
strdup	Duplicates a string
strchr	Finds the first occurrence of given character in a string
strstr	Finds the first occurrence of given string in another string
strset	Sets all the characters of string to given character or symbol
strrev	Reverse a string

Exercises

Section A

Multiple Choice Questions

1. In char *strcpy(char *s1, const char *s2)
   (a) Copies the string **s2** into **s1**, the value of **s1** is returned.
   (b) Copies the string **s1** into **s2**, the value of **s1** is returned.
   (c) Copies the string **s2** into **s1**, the value of **s2** is returned.
   (d) Copies the string **s1** into **s2**, the value of **s2** is returned.

2. char *strcat(char *s1, const char *s2)
   (a) Appends the string **s1** to the end of character array **s2**,the value of **s1** is returned.
   (b) Appends the string **s2** to the end of character array **s1**,the value of **s1** is returned.
   (c) Appends the string **s2** to the end of character array **s1**,the value of **s2** is returned.
   (d) Appends the string **s1** to the end of character array **s2**,the value of **s2** is returned.

3. int strcmp(const char *s1, const char *s2)
   (a) The function returns 0 if they are the same, a number < 0 if **s1** < **s2**, a number > 0 if **s1** > **s2**.
   (b) The function returns 1 if they are the same
   (c) The function returns <1 if they are the same
   (d) None of the above

4. strlen() function returns
   (a) Length of string including null character
   (b) Length of string excluding null character
   (c) Value of string in ASCII
   (d) None of the above

5. If we have to compare first x characters of a string we will use
   (a) strcpy
   (b) strncpy
   (c) strxcpy
   (d) None of the above

6. In strcat function
   (a) The null value at the end of s2 gets deleted
   (b) The null value at the end of s1 gets deleted
   (c) Both a and b
   (d) None of the above

7. In strchr() if the char is not found in the string
   (a) 0 is returned
   (b) char is returned
   (c) null is returned
   (d) No value is returned

8. The library used to store all the string functions is
   (a) string.h
   (b) conio.h
   (c) stdio.h
   (d) All of the above

9. Strings are stored in memory as
   (a) Pointer to address
   (b) Array of characters
   (c) Array of addresses
   (d) All of the above

10. If we want to copy a string from one memory location to another
    (a) strcpy()
    (b) strncpy()
    (c) memcpy()
    (d) None of the above

State True/False

1. A string is an array of characters ended by a null.
2. The functions expect a pointer to the first character in the string.
3. All of the comparing functions compare full string simultaneously.
4. char *strcat(char *s1, const char *s2) statement concatenates s1 in s2.
5. memmove() is identical to memcpy.
6. strncmp compares atmost n characters.
7. strchr ()Returns a pointer to the first occurrence of (char)c in *s.
8. The header <string.h> declares string functions.
9. strcpy() returns an integer value.
10. strcmp() function returns a char pointer.

Section B

Short Answer Type Questions

1. Define the term string and how we handle the string in C language.
2. Define the working of strlen() function with suitable example.
3. Define the working of strcat() function with suitable example.
4. Define the working of strcpy() function with suitable example.
5. Define the working of strrev() function with suitable example.
6. Define the working of strlwr() function with suitable example.
7. Define the working of strupr() function with suitable example.

8. Write a program to input a string and print the string in Title Case with the help of user defined function.
9. Write a program to input a string and print the string in sentence case with the help of user defined function.
10. Write a program to input a string and print how many characters are there in that string.

Section C

Long Answer Type Questions

1. Write a program to input a string and calculate the length of that string with the help of user defined function.
2. Write a program to input a string and then copy that string into another string with the help of user defined function.
3. Write a program to input two strings and then print the concated string with the help of user defined function.
4. Write a program to input a string and check whether the given string is palindrome or not with the help of user defined function.
5. Write a program to input a string and print the string in upper case with the help of user defined function.
6. Write a program to input a string and print the string in lowercase with the help of user defined function.
7. Write a program to input a text and input the string and check whether the string is available in the text or not.
8. Write a program to input a string and print the total number of vowels in that string.
9. Define the working of strncmp() function with suitable example.
10. Define the working of strncpy() function with suitable example.

13

Structure and Union

We have seen so far how to store numbers, characters, strings, and even large sets of these primitives using arrays, but what if we want to store collections of different kinds of data that are somehow related. For example, a file about an employee will probably have his/her name, age, the hours of work, salary, etc. Physically, all of that is usually stored in someone's filing cabinet. In programming, if you have lots of related information, you group it together in an organized fashion. Let us say you have a group of employees, and you want to make a database! It just would not do to have tons of loose variables hanging all over the place. Then we need to have a single data entity where we will be able to store all the related information together. But this cannot be achieved by using the arrays alone, as in the case of arrays, we can group multiple data elements that are of the same data type, and is stored in consecutive memory locations, and is individually accessed by a subscript. That is where the user defined data type *Structures* come in.

Structure is commonly referred to as a user-defined data type. C's *structures* allow you to store multiple variables of any type in one place (the structure). A structure can contain any of C's data types, including arrays and other structures. Each variable within a structure is named a *member* of the structure. They can hold any number of variables, and you can make arrays of structures. This flexibility makes structures ideally useful for creating databases in C. Similar to the structure there is another user defined data type called *Union* which allows the programmer to view

a single storage in more than one way, i.e. a variable declared as union can store within its storage space, the data of different types, at different times. In this unit, we will be discussing the user-defined data type structures and unions.

DEFINITION OF STRUCTURE

A structure is a collection of variables under a single name. These variables can be of different types, and each has a name which one can select it from the structure. A structure is a suitable way of grouping several pieces of related information together.

A structure can be defined as a new named type, thus extending the number of available types. It can use other structures, arrays or pointers as some of its members, though this can get complicated unless you are careful. That means we can say that structures is the collection of different data types and having a common name. A struct (short for STRUCTURE), is a collection of variables of different types. Structures are sometimes referred to as abstract data types or ADTs for short. We already know how to store a group of variables of the same type using arrays. Variables in a struct are called members or fields, and are accessed by their name. Variables in an array are called elements, and are accessed using square brackets and an index.

DECLARATION OF STRUCTURES

To declare a structure you must start with the keyword **struct** followed by the structure name or structure tag and within the braces the list of the structure's member variables. Note that the structure declaration does not actually create any variables.

Method No 1

The general syntax of structure declaration is as follows:

```
struct structure-tag
{
datatype variable1;
datatype variable2;
dataype variable 3;
...
};
```

Example 1

Consider the student database in which each student has a roll number, name and course and the marks obtained. Hence to group this data with a structure tag as **student**, we can have the declaration of structure as follows:

```
struct student
{
int roll_no;
char name[20];
char course[20];
int marks_obtained ;
};
```

The point you need to remember is that, till this time no memory is allocated to the structure. This is only the definition of structure that tells us that there exists a user defined data type by the name of student which is composed of the following members. Using this structure type.

We have to create the structure variables as follows:

struct student stud1, stud2 ;

At this point, we have created two instances or structure variables of the user-defined data type student. Now memory will be allocated. The amount of memory allocated will be the sum of all the data members which form part of the structure template.

In the above example 44 bytes memory is allocated for stud1 and stud2.

Method No. 2

The second method of declaring a structure is as follows:

```
struct
```

```
{
int roll_no;
char name[20];
char course[20];
int marks_obtained ;
} stud1, stud2 ;
```

In this case, a tag name *student* is missing, but still it happens to be a valid declaration of structure. In this case the two variables are allocated memory equivalent to the members of the structure.

The advantage of having a tag name is that we can declare any number of variables of the tagged named structure later in the program as per requirement.

If you have a small structure that you just want to define in the program, you can do the definition and declaration together as shown below.

This will define a structure of type *struct telephone* and declare three instances of it.

Example 2

Consider the example for declaring and defining a structure for the telephone billing with three instances:

```
struct telephone
{
int tele_no;
int cust_code;
char cust_address[40];
int bill_amt;
} tele1, tele2, tele3;
```

Here tele1, tele2 and tele3 are structure variables which hold 46 bytes memory.

SHORTHAND STRUCTURE WITH *TYPEDEF* KEYWORD

To make your source code more concise, you can use *typedef* keyword to create a synonym for a structure. This is an example of using *typedef* keyword to define address structure so when you want to create an instance of it you can omit the keyword *struct*

```
typedef struct{
    unsigned int house_number;
    char street_name[50];
    int zip_code;
    char country[50];
    } address;
```

Now here address will become a user-defined data type and you can define the variable of user defined data type address as follows:
address billing_addr;
address shipping_addr;
The typedef statement does not occupy storage: it simply defines a new type.

ACCESSING THE MEMBERS OF A STRUCTURE

We can directly access all the data members of the structure with the help of dot operator. The general syntax for accessing the member of the structure is as follows:
structurevariable. member-name;

Example
Let us define a structure student which has sno and name are the data members.
struct student
{
    int sno;
    char name[25];
}s;
And we want to access the sno and name then we can directly access the sno and name as follows:
s.sno and s.name
But if the structure variable is not a simple variable that is a pointer type variable then we use -> sign in place of. Operator to access the members of the structure.
struct student
{
    int sno;
    char name[25];
}*s;
Now we want to access the sno and name then we can directly access the sno and name as follows:
s->sno and s->name

STRUCTURE AND SIZEOF FUNCTION

sizeof is used to get the size of any data types even with any structures. Let us take a look at simple program:

Example
#include <stdio.h>
typedef struct __address{

int house_number;// 2 bytes
char street[50]; // 50 bytes
int zip_code; // 2 bytes
char country[20];// 20 bytes
} address;//74 bytes in total
void main()
{
// it returns 80 bytes
printf("Size of address is %d bytes\n",sizeof (address));
}

You will never get the size of a structure exactly as you think it must be. The sizeof function returns the size of structure larger than it is because the compiler pads struct members so that each one can be accessed faster without delays. So you should be careful when you read the whole structure from file which was written from other programs.

Programming Problems

Problem 1
Write a program to input the record of a student which contains sno, name, age and city and then print this record with the help of structure.

Solution
#include<stdio.h>
#include<conio.h>
struct student
{
int sno,age;
char name[25];
}s;
void main()
{
    clrscr();
    printf("\n enter the enrollment number");
    scanf("%d",&s.sno);
    printf("\n enter the name");
    scanf("%s",s.name);
    printf("\n enter the age");
    scanf("%d",&s.age);
    printf("\n the enrollment number is %d", s.sno);
    printf("\n the name is %s",s.name);
    printf("\n the age is %d",s.age);

```
getch();
}
```
The output of the above code segment is as follows:

Enter the enrollment number 0001
Enter the name Manish
Enter the age 30
The enrollment number is 0001
The name is Manish
The age is 30

Problem 2

Write a program to store and retrieve the values from the student structure.

Solution

```
#include<stdio.h>
#include<conio.h>
struct student
{
int roll_no;
char name[20];
char course[20];
int marks_obtained ;
};
void main()
{
student s1 ;
clrscr();
printf ("Enter the student roll number:");
scanf ("%d",&s1.roll_no);
printf ("\n enter the student name: ");
scanf ("%s",s1.name);
printf ("\n enter the student course");
scanf ("%s",s1.course);
printf ("Enter the student percentage\n");
scanf ("%d",&s1.marks_obtained);
printf ("\n data entry is complete");
printf ( "\n the data entered is as follows:\n");
printf ("\n the student roll no is %d",s1.roll_no);
printf ("\n the student name is %s",s1.name);
printf ("\n the student course is %s",s1.course);
printf ("\n the student percentage is %d",s1.
marks_obtained);
getch();
}
```
The output of the above code segment is as follows:

Enter the student roll number: 1234
Enter the student name: MANISH
Enter the student course: MSC(COMPUTER)
Enter the student percentage: 84
Date entry is complete
The data entered is as follows:
The student roll no is 1234
The student name is MANISH
The student course is MSC(COMPUTER)
The student percentage is 84.

INITIALIZING STRUCTURES

Like other C variable types, structures can be initialized when they are declared. This procedure is similar to that for initializing arrays. The structure declaration is followed by an equal sign and a list of initialization values is separated by commas and enclosed in braces.

Example 1

Let us define a structure student which have sno, name and age and then initialized the valued at the time of declaration.

Solution

```
struct student
{
int sno,age;
char name[25]
} s= {1001,23,"manish"};
```
In a structure that contains structures as members, list the initialization values in order. They are placed in the structure members in the order in which the members are listed in the structure definition.

Programming Problems

Problem 1

Write a program to access the values of the structure initialized with some initial values.

Solution

```
#include<stdio.h>
#include<conio.h>
struct telephone
{
int tele_no;
int cust_code;
char cust_name[20];
char cust_address[40];
int bill_amt;
```

```
};
void main()
{
struct telephone tele =
{
2314345,
5463,
"Manish",
"Bareilly",
2435
};
printf("The values are initialized in this
program.");
printf("\n the telephone number is %d",tele.
tele_no);
printf("\n the customer code is %d",tele.cust_
code);
printf("\n the customer name is %s",tele.
cust_name);
printf("\n the customer address is %s",tele.
cust_address);
printf("\n the bill amount is %d",tele.bill_
amt);
getch();
}
```

The output of the above code is as follows
The values are initialized in this program.
The telephone number is 2314345
The customer code is 5463
The customer name is Manish
The customer address is Bareilly
The bill amount is 2435

ARRAYS OF STRUCTURE

Thus far we have studied as to how the data of heterogeneous nature can be grouped together and be referenced as a single unit of structure. Now we come to the next step in our real world problem. Let us consider the example of students and their marks. In this case, to avoid declaring various data variables, we grouped together all the data concerning the student's marks as one unit and call it student. The problem that arises now is that the data related to students is not going to be of a single student only. We will be required to store data for a number of students. To solve this situation one way is to declare a structure and then create sufficient number of variables

of that structure type. But it gets very cumbersome to manage such a large number of data variables, so a better option is to declare an array. So, revising the array for a few moments we would refresh the fact that an array is simply a collection of homogeneous data types.

Hence, if we make a declaration as:

int a [20];

If we want to store more than one record of any employee/student or anybody then we have to consider the concept of arrays of structure.

Like with the usual data types, you can store a group of similar typed variables in an array.

The general syntax for defining any array of structure is as follows

struct <structure name>

{

<data members>;

} <structure variable> [size];

Example

struct student stud[20], we can access the *roll_no* of this array as

stud[0].roll_no;

stud[1].roll_no;

stud[2].roll_no;

stud[3].roll_no;

...

...

stud[19].roll_no;

Please remember the fact that for an array of twenty elements the subscripts of the array will be ranging from 0 to 19 (a total of twenty elements).

Programming Problems

Problem 1

Write a program to input 10 records of students which will consists sno, name, age and city and then print that record with the help of structure.

Solution

```
#include<stdio.h>
#include<conio.h>
struct student
{
int sno, age;
```

```
char name[25],city[10];
}s[10];
void main()
{
int i;
clrscr();
for(i=0;i<10;i++)
{
    printf("\n enter the student number");
    scanf("%d",&s[i].sno);
    printf("\n enter the name");
    scanf("%s",s[i].name);
    printf("\n enter the age");
    scanf("%d",&s[i].age);
    printf("\n enter the city");
    scanf("%s",s[i].city);
}
printf("\n the students records are as
follows");
for(i=0;i<10;i++)
{
    printf("\n the student number is %d",s
[i].sno);
    printf("\n the name is %s",s[i].name);
    printf("\n the age is %d",s[i].age);
    printf("\n the city is %s",s[i].city);
}
getch();
}
```

Problem 2

Write a program to read and display data for
20 students.

Solution

```
#include <stdio.h>
#include<conio.h>
struct student
{
int roll_no;
char name[20];
char course[20];
int marks_obtained ;
};
void main( )
{
struct student stud [20];
int i;
clrscr();
```

```
printf ("Enter the student data one by
one\n");
for(i=0; i<=19; i++)
{
printf ("Enter the roll number of %d student",
i+1);
scanf ("%d",&stud[i].roll_no);
printf ("Enter the name of %d student",i+1);
scanf ("%s",stud[i].name);
printf ("Enter the course of %d student",i+1);
scanf ("%d",stud[i].course);
printf ("Enter the marks obtained of %d
student",i+1);
scanf ("%d",&stud[i].marks_obtained);
}
printf ("the data entered is as follows\n");
for (i=0;i<=19;i++)
{
printf ("The roll number of %d student is
%d\n",i+1,stud[i].roll_no);
printf ("The name of %d student is %s\n",i+1,
stud[i].name);
printf ("The course of %d student is %s\n",
i+1,stud[i].course);
printf ("The marks of %d student is %d\n",
i+1,stud[i].marks_obtained);
}
getch();
}
```

Problem 3

Write a program to read and print data related
to five students having marks of three subjects
each using the concept of arrays.

Solution

```
#include<stdio.h>
#include<conio.h>
struct student
{
int roll_no;
char name [20];
char course [20];
int subject [3];
};
void main( )
{
struct student stud[5];
int i,j;
printf ("Enter the data for all the students:\n");
```

```
for (i=0;i<=4;i++)
{
printf ("Enter the roll number of %d student",
i+1);
scanf ("%d",&stud[i].roll_no);
printf("Enter the name of %d student",i+1);
scanf ("%s",stud[i].name);
printf ("Enter the course of %d student",i+1);
scanf ("%s",stud[i].course);
for (j=0;j<=2;j++)
{
printf ("Enter the marks of the %d subject of
the student %d:\n",j+1,i+1);
scanf ("%d",&stud[i].subject[j]);
}
}
printf ("The data you have entered is as
follows:\n");
for (i=0;i<=4;i++)
{
printf ("The %d th student's roll number is
%d\n",i+1,stud[i].roll_no);
printf ("The %d the student's name is %s\n",
i+1,stud[i].name);
printf ("The %d the student's course is
%s\n",i+1,stud[i].course);
for (j=0;j<=2;j++)
{
printf ("The %d the student's marks of %d I
subject are %d\n",i+1, j+1,stud[i].subject[j]);
}
}
getch();
}
```

STRUCTURE WITHIN STRUCTURE OR NESTING STRUCTURE

Nested structure means structure within a structure.

The general syntax for defining a nested structure is as follows:

```
struct<structurename1>
{
    <data_members1>;
struct<structurename2>
{
    <data_members2>;
}<structure_variable2>;
}<structure_variable1>;
```

If we want to access the data members of structurename1 then we can directly access with the help of <structurevariable1>.<data_ members1 > and if we want to access the data members of <structurename2> then we can directly access with the help of <structure_ variable1>. <structure_variable2>.<data_members2>.

Example

If we define the following structure:

```
struct date
{
    int day, month, year;
struct time
{
    int hour,min,sec;
} t;
}d;
```

If we want to access the day, month and year then we can directly access with the help of d.day, d.month, d.year and if we want to access the hour, min and sec then we can directly access with the help of d.t.hour, d.t.min and d.t.sec.

The following example demonstrates the concept of nesting structure.

```
#include<stdio.h>
#include<conio.h>
struct course
{
int couno;
int coufees;
};
struct student
{
int studno;
course sc;
course sc1;
};
void main( )
{
student s1;
int x;
s1.studno=100;
s1.sc.couno=123;
s1.sc.coufees=5000;
s1.sc1.couno=200;
s1.sc1.coufees=5000;
x = s1.sc.coufees + s1.sc1.coufees;
printf(:\n the student number is
```

```
%d",s1.studno);
printf("\n the total fees is %d",x);
getch();
}
```

The output of the above program is as follows:

Student Number: 100

Total Fees: Rs.10000

In the above example, the structure course is nested inside the structure student. To access such nested structure members, the programmer must use dot operator in the above case twice to access the nested structure members. In the above example:

s1.sc.couno

s1 is the name of the structure variable. sc is the member in the outer structure student. couno is the member in the inner structure course. This is how nested structure members are accessed.

PURPOSE AND USES OF STRUCTURES

As we have seen, a structure is a good way of storing related data together. It is also a good way of representing certain types of information. Complex numbers in mathematics inhabit a two dimensional plane (stretching in real and imaginary directions). These could easily be represented here by

```
typedef struct {
double real;
double imag;
} complex;
```

doubles have been used for each field because their range is greater than floats and because the majority of mathematical library functions deal with doubles by default.

In a similar way, structures could be used to hold the locations of points in multi-dimensional space. Mathematicians and engineers might see a storage efficient implementation for sparse arrays here.

Apart from holding data, structures can be used as members of other structures. Arrays of structures are possible, and are a good way of storing lists of data with regular fields, such as databases.

Another possibility is a structure whose fields include pointers to its own type. These can be used to build chains (programmers call these linked lists), trees or other connected structures. These are rather daunting to the new programmer, so we will not deal with them here.

UNION

Structures are a way of grouping homogeneous data together. But it often happens that at any time we require only one of the member's data. For example, in case of the support price of shares you require only the latest quotations. And only the ones that have changed need to be stored. So if we declare a structure for all the scripts, it will only lead to crowding of the memory space. Hence it is beneficial if we allocate space to only one of the members. This is achieved with the concepts of the *unions*. *Unions* are similar to *structures* in all respects but differ in the concept of storage space.

A union is a collection of variables of different types, just like a structure. However, with unions, you can only store information in one field at any one time.

You can picture a union as like a chunk of memory that is used to store variables of different types. Once a new value is assigned to a field, the existing data is wiped over with the new data.

DECLARING A UNION

Declaring a union is exactly the same as declaring a struct, except you use the union keyword:

A *union* is declared and used in the same way as the structures. Yet another difference is that only one of its members can be used at any given time. Since all members of a union occupy the same memory and storage space, the space allocated is equal to the largest data member of the union. Hence, the member who has been updated last is available at any given time.

The general syntax for defining a union as follows:

```
union union-tag
{
datatype variable1;
datatype variable2;
...
};
```

Example 1

This example demonstrates what happens if you initialize all the data for a union variable, all at once:

```c
#include <stdio.h>
#include<conio.h>
struct robot1 {
int ammo;
int energy;
};
union robot2 {
int ammo;
int energy;
};
typedef struct robot1 ROBOT1;
typedef union robot2 ROBOT2;
void main()
{
ROBOT1 red = {10,200};
/* ROBOT2 blue = {15,100}; does not work with unions*/
ROBOT2 blue;
clrscr();
blue.ammo = 15;
blue.energy = 100;
printf ("The red robot has %d ammo ", red.ammo);
printf("and %d units of energy.\n\n", red.energy);
printf("The blue robot has %d ammo ", blue.ammo);
printf("and %d units of energy\n.", blue.energy);
getch();
}
```

The output of the above program is as follows:

The red robot has 10 ammo and 200 units of energy. The blue robot has 100 ammo and 100 units of energy.

The explanation of the above code segment is as follows:

First notice how similar the declaration of a union is to a struct.

Things get a little different inside main. You cannot initialize fields of a union variable all at once—try removing the correct initialization above and see what your compiler says. So I initialized each field separately...

The first printf statement displays the expected for the struct version of our robot.

The second printf shows you that the program has overwritten data in the ammo field with data of the energy field—this is how unions work. All fields share the same address, where as with structs, each field has its own address.

Example 2

This example is similar to the last, except I used printf before initializing the energy field of the union:

```c
#include <stdio.h>
#include<conio.h>
struct robot1
{
int ammo;
int energy;
};
union robot2
{
int ammo;
int energy;
};
typedef struct robot1 ROBOT1;
typedef union robot2 ROBOT2;
void main()
{
ROBOT1 red = {10,200};
ROBOT2 blue;
clrscr();
printf("The red robot has %d ammo ", red.ammo);
printf("and %d units of energy.\n\n", red.energy);
blue.ammo = 15;
printf("The blue robot has %d ammo", blue.ammo);
blue.energy = 100;
printf(" and %d units of energy.", blue.energy);
getch();
}
```

The output of the above program is as follows:

The red robot has 10 ammo and 200 units of energy.

The blue robot has 15 ammo and 100 units of energy.

Data in the energy field has still overwritten the ammo field though—rearranging the printfs has changed made the outcome different from that of the last example.

ARRAYS OF UNION VARIABLES

The concept is the same as that of structs. However, because of the problem of new data overwriting existing data in the other fields, the struct version of the following example will not work if you simply replace all instances of struct with union.

```c
#include <stdio.h>
#include<conio.h>
typedef union robot ROBOT;
union robot
{
char *name;
int energy;
};
void main()
{
int i;
ROBOT robots[3];
robots[0].name = "Lunar Lee";
robots[0].energy = 50;
robots[1].name = "Planetary Pete";
robots[1].energy = 20;
robots[2].name = "Martian Matt";
robots[2].energy = 30;
for(i=0 ; i<3 ; i++)
{
/*printf("Robot %d is called %s ", i, robots[i].name);*/
printf("and has %d units of energy.\n", robots[i].energy);
}
getch();
}
```

The output of the above program is as follows:

and has 50 units of energy.
and has 20 units of energy.
and has 30 units of energy.

The reason why I commented out the printf that displays the name is because I initialized everything at once. As a consequence, by the time the program reaches the for loop the union holds an int variable, rather than a char *. In general, my program crashes whenever an int is passed to the string format specifier, %s—other compilers might produce a different result.

UNIONS INSIDE STRUCTURES

I cannot see any obvious reasons why one might use unions, but here is a case when they could be useful. Suppose we want a field of a structure to contain a string or an integer, depending on what the user specifies. Take this example:

```c
#include <stdio.h>
#include <conio.h>
struct choice
{
union {
char *name;
int age;
};
int number;
};
void main()
{
struct choice example;
char d[4];
printf("You can enter a name OR an age. \n");
printf("Would you like to enter a name? (Yes or No) : ");
gets(d);
if(d[0]=='y' || d[0]=='Y') {
printf("\n what's your name? ");
gets(example.name);
}
else {
printf("\n what's your age? ");
scanf("%d", &example.age);
}
printf("\n what's you favourite number? ");
scanf("%d", &example.number);
if(d[0]=='y' || d[0]=='Y') {
printf("\nYour name is %s ", example.name);
}
else {
printf("\n your age is %d ", example.age);
}
printf("and your favourite number is %d.\n", example.number);
```

printf("The size of the struct is %d\n", sizeof (example));
getch();
}

The output of the above program is as follows:

Output varies on what you entered, but I shall run through the program:

You can have an **INLINE** declaration of union within a struct declaration (inline is a term meaning "embedded within"). So the first field of choice can either be a char * or an int. Once inside main, a struct choice variable called example is declared, as well as a sized char array called d. The used is told to make a Yes / No choice—the response gets stored into d. Now the program checks to see if the first character of d is Y or y—if so, the user can enter a name, else an age. Entering a favorite number is compulsory. Then the program prints out what the user has entered. The last printf displays the size of the struct: 8 bytes in all cases.

DIFFERENCE BETWEEN UNION AND STRUCTURE

The difference between structure and union in C are as follows:

1. union allocates the memory equal to the maximum memory required by the member of the union but structure allocates the memory equal to the total memory required by the members.
2. In union, one block is used by all the member of the union but in case of structure, each member have their own memory space.

SUMMARY

- Structure is the collection of different data items and having a common name.
- The general syntax for defining the structure is as follows:
  - Struct <structure name>
  - {
  - <data members>;
  - }<structure variable>;
- We can directly access all the data members of the structure with the help of <structure variable> and .(dot) operator.

- When we want to store more than one record with the help of structure then we have to use the concept of arrays of structure.
- Nested structure means structure within structure.
- A union, is a collection of variables of different types, just like a structure. However, with unions, you can only store infor-mation in one field at any one time.

Section A

Multiple Choice Questions

1. Declaring a structure requires
   (a) Struct keyword followed by name
   (b) Structure can be defined before main
   (c) Homogenous data types to be taken
   (d) None of the above

2. If we wants to access data members of a field, we can use
   (a) . operator
   (b) –> operator
   (c) Any one of a and b
   (d) None of the above

3. struct student
   {
   int sno;
   char name[25]
   }s; means
   (a) Student is a structure having sno and name as members
   (b) s is a pointer to structure variable
   (c) s is a structure having sno and name as members
   (d) None of the above

4. struct student
   {int sno,age;
   Char name[29];
   }*s;
   Is a structure how do we access students age
   (a) s.age   (b) s–>age
   (c) Both a and b   (d) None of the above

5. If we want to consider more data to be stored we can use
   (a) Arrays of structures
   (b) Structures of array

(c) Both a and b
(d) None of the above

6. Pointer to structures holds
   (a) Memory address to structure variable
   (b) Memory address to structure
   (c) Both a and b (d) None of the above

7. In the function declaration you need to specify
   (a) Type, names and arguments
   (b) Type, address, argument,
   (c) Both a and b (d) None of the above

8. We can access a member of second structure
   (a) struct_var1.struct_var2.data_member
   (b) struct_var2.struct_var1.data_member
   (c) Both a and b
   (d) Cannot be done

9. Structures can be used for
   (a) Hold the locations of points in multidimensional space
   (b) Members of other structures
   (c) Used to create link list
   (d) All of the above

10. In unions
    (a) One block of memory is used by each member
    (b) Allocates memory equals to maximum memory of member
    (c) Both a and b
    (d) None of the above

State True/False
1. A structure is a collection of variables of different types.
2. Variables in a structure can only be accessed by their address.
3. Size of struct is the sum of size of variables it holds.
4. sizeof() function returns the size smaller than actually it is.
5. The pointer to structure stores its memory address.
6. We cannot declare structures within a structure.
7. In union you can store one information at a time.
8. A structure cannot hold a field which is pointer to it.

9. In case of union fields share a same memory space.
10. We can have an inline declaration of union in a structure.

Section B
Short Answer Type Questions
1. Define the concept of structure with proper explanation.
2. Define the concept of union with proper explanation.
3. Differentiate between structure and union with proper explanation.
4. Write a program in C to read and display the bio-data using structure.
5. Write a program in C to area of circle using structure.
6. Write a program in C to find the biggest of three numbers using structure.
7. Write a program in C to find the smallest of three numbers using structure.
8. Write a program in C to read and display the bio-data using structure.
9. Write a program in C to read and display the bio-data with date of birth using structure.
10. Write a program in C to read and display the bio-data of 10 students using structure.

Section C
Long Answer Type Questions
1. Write a program in C to read and display the numbers using typedef.
2. Write a program in C to read and display the bio-data using typedef.
3. Write a program in C to read and display the bio-data using typedef.
4. Write a program in C to read and display the complete bio-data using typedef.
5. Write a program in C to read and display the complex number.
6. Write a program in C to read and display the two complex number.
7. Write a program in C to read and display the sum of two complex numbers.
8. Write a program in C to read and display the difference of two complex numbers.
9. Differentiate between structure and union with suitable example.
10. Define the term union and why we use the concept of union explain in detail.

Pointer

If you want to be proficient in the writing of code in the C programming language, you must have a thorough working knowledge of how to use pointers. One of those things, beginners in C find difficult is the concept of pointers. The purpose of this chapter is to provide an introduction to pointers and their efficient use in the C programming. Actually, the main difficulty lies with the C's pointer terminology than the actual concept.

C uses pointers in three main ways. First, they are used to create *dynamic data structures*: data structures built up from blocks of memory allocated from the heap at run-time. Second, C uses pointers to handle *variable parameters* passed to functions. And third, pointers in C provide an alternative means of accessing information stored in arrays, which is especially valuable when you work with strings.

A normal variable is a location in memory that can hold a value. For example, when you declare a variable *i* as an integer, four bytes of memory is set aside for it. In your program, you refer to that location in memory by the name *i*. At the machine level, that location has a memory address, at which the four bytes can hold one integer value. A *pointer* is a variable that points to another variable. This means that it holds the memory address of another variable. Put another way, the pointer does not hold a value in the traditional sense; instead, it holds the address of another variable. It points to that other variable by holding its address. Because a pointer holds an address rather than a value, it has two parts. The pointer itself holds the address.

That addresses points to a value. There is the pointer and the value pointed to. As long as you are careful to ensure that the pointers in your programs always point to valid memory locations, pointers can be useful, powerful, and relatively trouble-free tools.

CONCEPT BEHIND POINTER

Computer's memory is made up of a sequential collection of storage cells called bytes. Each byte has a number called an address associated with it. When we declare a variable in our program, the compiler immediately assigns a specific block of memory to hold the value of that variable. Since every cell has a unique address, this block of memory will have a unique starting address. The size of this block depends on the range over which the variable is allowed to vary. For example, on 32 bit PC's the size of an integer variable is 4 bytes. On older 16 bit PC's integers were 2 bytes. In C the size of a variable type such as an integer need not be the same on all types of Machines. If you want to know the size of the various data types on your system, running the following code given in the example will give you the information.

Example
Write a program to know the size of the various data types on your system.

Solution
```
# include <stdio.h>
#include<conio.h>
void main( )
{
```

```
clrscr();
printf ("n size of a int = %d bytes", sizeof (int));
printf ("\n size of a float = %d bytes", sizeof (float));
printf ("\n size of a char = %d bytes", sizeof (char));
getch();
}
```

The output of the above code segment is as follows

Size of int = 2 bytes

Size of float = 4 bytes

Size of char = 1 byte

An *ordinary variable* is a location in memory that can hold a value. For example, when you declare a variable *num* as an integer, the compiler sets aside 2 bytes of memory (depends on the PC) to hold the value of the integer. In your program, you refer to that location in memory by the name *num*. At the machine level that location has a memory address.

int num = 100;

We can access the value 100 either by the name num or by its memory address. Since, addresses are simply digits, they can be stored in any other variable.

Such variables that hold addresses of other variables are called *Pointers*.

The general syntax of defining a pointer is as follows:

<datatype> * <pointername>;

In other words, a *pointer* is simply a variable that contains an address, which is a location of another variable in memory. A pointer variable "points to" another variable by holding its address. Since a pointer holds an address rather than a value, it has two parts. The pointer itself holds the address. That addresses points to a value. There is a pointer and the value pointed to. This fact can be a little confusing until you get comfortable with it, but once you get familiar with it, then it is extremely easy and very powerful.

One good way to visualize this concept is to examine in the following figure.

CHARACTERISTIC FEATURES OF POINTERS

The following are the features of pointers

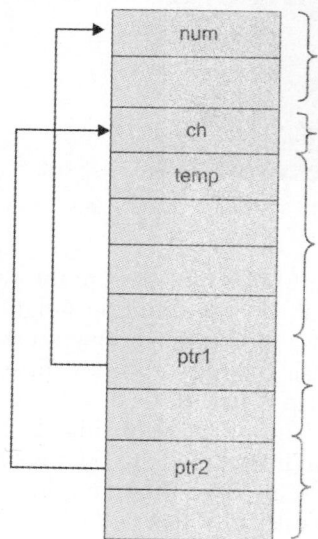

- With the use of pointers in programming, the program execution time will be faster as the data is manipulated with the help of addresses directly.
- With the use of pointers in programming, will save the memory space.
- With the use of pointers in programming, the memory access will be very efficient.
- With the use of pointers in programming, dynamic memory is allocated.

THE ADDRESS AND INDIRECTION OPERATORS

Now we will consider how to determine the address of a variable. The operator that is available in C for this purpose is "&" (*address of*) operator. The operator & and the immediately preceding variable returns the address of the variable associated with it. C's other unary pointer operator is the "*", also called as *value at address* or indirection operator. It returns a value stored at that address.

This can be illustrated with the help of following examples.

Example

Write a program to print the address associated with a variable and value stored at that address.

Solution

```
# include <stdio.h>
#include<conio.h>
void main( )
{
int qty = 5;
clrscr();
printf ("Address of qty = %u \n",&qty);
printf ("Value of qty = %d \n",qty);
printf("Value of qty = %d",*(&qty));
getch();
}
```

The output of the above code is as follows:

Address of qty = 65524
Value of qty = 5
Value of qty = 5

The explanation of the above output if as follows:

Look at the *printf* statement carefully. The format specifier *%u* is taken to increase the range of values the address can possibly cover. The system-generated address of the variable is not fixed, as this can be different the next time you execute the same program. Remember unary operator operates on single operands. When & is preceded by the variable qty, has returned its address. Note that the & operator can be used only with simple variables or array elements. It cannot be applied to expressions, constants, or register variables. Observe the third line of the above program. *(&qty) returns the value stored at address 65524, i.e. 5 in this case. Therefore, *qty* and *(&qty)* will both evaluate to 5.

POINTER TYPE DECLARATION AND ASSIGNMENT

We already know that &*qty* returns the address of *qty* and this address can be stored in a variable as shown below:

ptr = &qty;

In C programming language, every variable must be declared for its data type before it is used. Even this holds good for the pointers too. We know that *ptr* is not an ordinary variable like any integer variable. We declare the data type of the pointer variable as that of the type of the data that will be stored at the address to which it is pointing to. Since *ptr* is a variable, which contains the address of an integer variable *qty*, it can be declared as follows:

int *ptr;

Where *ptr* is called a *pointer variable*. In C, we define a pointer variable by preceding its name with an asterisk(*). The "*" informs the compiler that we want a pointer variable, i.e. to set aside the bytes that are required to store the address in memory. the *int* says that we intend to use our pointer variable to store the address of an integer.

Let us consider the following memory map as shown in the figure below.

Example 1

Write a program to demonstrates the relationship between & and * operator.

Solution

```
# include <stdio.h>
#include<conio.h>
void main( )
{
int qty = 5;
int *ptr; /* declares ptr as a pointer variable
that points to an integer variable */
clrscr();
ptr = &qty; /* assigning qty's address to ptr ->
Pointer Assignment */
printf ("Address of qty = %u \n", &qty);
printf ("Address of qty = %u \n", ptr);
printf ("Address of ptr = %u \n", &ptr);
```

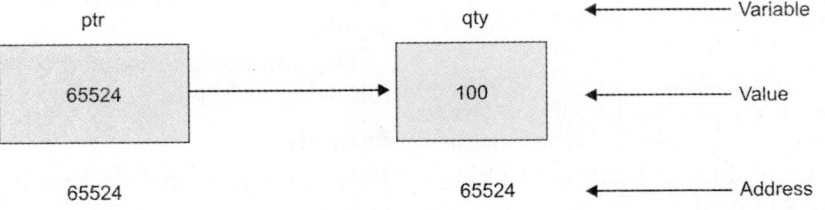

```
printf ("Value of ptr = %d \n", ptr);
printf ("Value of qty = %d \n", qty);
printf ("Value of qty = %d \n", *(&qty));
printf ("Value of qty = %d", *ptr);
getch();
}
```

The output of the above code segment is as follows:

Address of qty = 65524
Address of ptr = 65522
Value of ptr = 65524
Value of qty = 5
Value of qty = 5
Value of qty = 5

Example 2

Write a program that tries to reference the value of a pointer even though the pointer is uninitialized.

Solution

```
# include <stdio.h>
#include<conio.h>
void main()
{
int *p; /* a pointer to an integer */
clrscr();
*p = 10;
printf("the value is %d", *p);
printf("the value is %u",p);
getch();
}
```

This gives you an error. Pointer p is uninitialized and points to a random location in memory when you declare it. It could be pointing into the system stack, or the global variables, or into the program's code space, or into the operating system. When you say *p=10; the program will simply try to write a 10 to whatever random location p points to. The program may explode immediately. It may subtly corrupt data in another part of your program and you may never realize it. Almost always, an uninitialized pointer or a bad pointer address causes the fault.

This can make it difficult to track down the error. Make sure you initialize all pointers to a valid address before dereferencing them.

Within a variable declaration, a pointer variable can be initialized by assigning it the address of another variable. Remember the variable whose address is assigned to the pointer variable must be declared earlier in the program.

This concept is illustrated in the following example

Example

```
# include <stdio.h>
#include<conio.h>
void main( )
{
int *p; /* a pointer to an integer */
int x;
clrscr();
p = &x;
*p=10;
printf("The value of x is %d",*p);
printf("\n the address in which the x is stored is %d",p);
getch();
}
```

The Output of the above code segment is as follows:

The value of x is 10
The address in which x is stored is 52004

This statement puts the value of 20 at the memory location whose address is the value of px. As we know that the value of px is the address of x and so the old value of x is replaced by 20. This is equivalent to assigning 20 to x. Thus, we can change the value of a variable *indirectly* using a pointer and the *indirection operator*.

POINTER TO A POINTER

The concept of pointer can be extended further. As we have seen earlier, a pointer variable can be assigned the address of an ordinary variable. Now, this variable itself could be another pointer. This means that a pointer can contain address of another pointer.

This concept is illustrated with the help of following example.

Example

Write a program that declares a pointer to a pointer.

Solution

```
# include<stdio.h>
#include<conio.h>
void main( )
{
int i = 100;
int *pi;
int **pii;
clrscr();
pi = &i;
pii = &pi;
printf ("Address of i = %u \n", &i);
printf ("Address of i = %u \n", pi);
printf ("Address of i = %u \n", *pii);
printf ("Address of pi = %u \n", &pi);
printf ("Address of pi = %u \n", pii);
printf ("Address of pii = %u \n", &pii);
printf ("Value of i = %d \n", i);
printf ("Value of i = %d \n", *(&i));
printf ("Value of i = %d \n", *pi);
printf ("Value of i = %d", **pii);
getch();
}
```

The output of the above code segment is as follows:

```
Address of i = 65524
Address of i = 65524
Address of i = 65524
Address of pi = 65522
Address of pi = 65522
Address of pii = 65520
Value of i = 100
Value of i = 100
Value of i = 100
Value of i = 100
```

This can be graphically represented with the help of following figure.

Null Pointer Assignment

It does make sense to assign an integer value to a pointer variable. An exception is an assignment of 0, which is sometimes used to indicate some special condition. A macro is used to represent a null pointer. That macro goes under the name *NULL*. Thus, setting the value of a pointer using the *NULL*, as with an assignment statement such as *ptr = NULL*, tells that the pointer has become a *null* pointer. Similarly, as one can test the condition for an integer value as zero or not, like *if (i == 0)*, as well we can test the condition for a null pointer using *if (ptr == NULL)* or you can even set a pointer to *NULL* to indicate that it is no longer in use.

This can be illustrated with the help of following example.

Example

Write a program to illustrate the concept of NULL pointer.

Solution

```
# include<stdio.h>
#include<conio.h>
# define NULL 0
void main()
{
int *pi = NULL;
clrscr();
printf("The value of pi is %u", pi);
getch();
}
```

The output of the above code segment is as follows:
The value of pi is 0

POINTERS vs ARRAY

When an array is declared, the compiler allocates a base address and sufficient amount of storage to contain all the elements of the array in contiguous memory locations. The base address is the location of the first element (index 0) of the array. The compiler also defines the array name as a constant pointer to the first element suppose we declare an

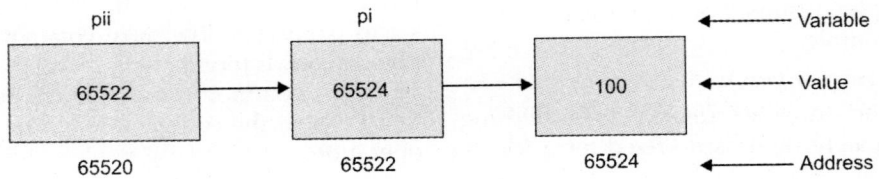

array X as follows:

static int X [6] = {1, 2, 3, 4, 5, 6};

Suppose the base address of X is 1000 and assuming that each integer requires two bytes, the five elements will be stored as follows:

ELEMENTS x[0] x[1] x[2] x[3] x[4] x[5]

VALUE 1 2 3 4 5 6

Address 1000 1002 1004 1006 1008 1010

BASE ADDRESS

The name X is defined as a constant pointer pointing to the first element, x[0] and therefore the value of X is 1000, the location whose X[0] is stored. That is

$X = \& \, x[0] = 1000$

If we declare P as an integer pointer, then we can make the pointer P to point to the array X by the following assignment:

$P = X$;

This is equivalent to $P = \& \, X$ [0];

Now we can access every value of x using P+ + to more from one element to another. The relationship between P and X is shown below:

$P = \& \, x$ [0] (= 1000)

$P+1 = \& \, x[1]$ (= 1002)

$P+2 = \& \, x[2]$ (= 1004)

$P+3 = \& \, x[3]$ (= 1006)

$P+4 = \& \, x[4]$ (= 1008)

$P+5 = \& \, x[5]$ (= 1010)

The address of an element is calculated using its index and the scale factor of the data type. For instance, address of $X[3]$ = base address + $(3 \times$ Scale factor of int$) = 1000 + (3 \times 2) = 1006$.

When handling array, instead of using array indexing, we can use pointers to access array elements. Note that $X(P+3)$ gives the value of $X[3]$. The pointer accessing method is faster than array indexing.

POINTER RULES

One of the nice things about pointers is that the rules that govern how they work are pretty simple. The rules can be layered together to get complex results, but the individual rules remain simple.

Pointers and Pointees

A **pointer** stores a reference to something. Unfortunately there is no fixed term for the thing that the pointer points to, and across different computer languages there is a wide variety of things that pointers point to.

We use the term **pointee** for the thing that the pointer points to, and we stick to the basic properties of the pointer/pointee relationship which are true in all languages.

The term "reference" means pretty much the same thing as "pointer" — "reference" implies a more high-level discussion, while "pointer" implies the traditional compiled language implementation of pointers as addresses. For the basic pointer/pointee rules covered here, the terms are effectively equivalent.

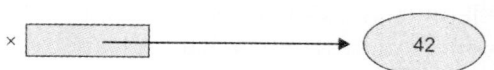

The above drawing shows a pointer named x pointing to a pointee which is storing the value 42. A pointer is usually drawn as a box, and the reference it stores is drawn as an arrow starting in the box and leading to its pointee.

Allocating a pointer and allocating a pointee for it to point to are two separate steps. You can think of the pointer/pointee structure as operating at two levels. Both the levels must be set up for things to work. The most common error is concentrating on writing code which manipulates the pointer level, but forgetting to set up the pointee level. Sometimes pointer operations that do not touch the pointees are called "shallow" while operations on the pointees are called "deep".

Dereferencing

The **dereference** operation starts at the pointer and follows its arrow over to access its pointee. The goal may be to look at the pointee state or to change the pointee state.

The dereference operation on a pointer only works if the pointer has a pointee—the pointee must be allocated and the pointer must be set to point to it. The most common error in pointer code is forgetting to set up the pointee. The most common runtime crash because of that error in the code is a failed dereference operation.

In Java the incorrect dereference will be flagged politely by the runtime system. In compiled languages such as C, C++, and Pascal, the incorrect dereference will sometimes crash, and other times corrupt memory in some subtle, random way. Pointer bugs in compiled languages can be difficult to track down for this reason.

Pointer Assignment

Pointer assignment between two pointers makes them point to the same pointee. So the assignment $y = x$; makes y point to the same pointee as x.

Pointer assignment does not touch the pointees. It just changes one pointer to have the same reference as another pointer. After pointer assignment, the two pointers are said to be "sharing" the pointee.

POINTER ARITHMETIC

Pointer variables can also be used in arithmetic expressions. The following operations can be performed on pointers.

1. Pointers can be incremented or decremented to point to different locations like

    ptr1 = ptr2 + 4;

    ptr ++;

    — ptr;

    However, *ptr++* will cause the pointer *ptr* to point the next address value of its type.

    For example, if *ptr* is a pointer to float with an initial value of 65526, then after the operation *ptr ++* or *ptr = ptr+1*, the value of *ptr* would be 65530. Therefore, if we increment or decrement a pointer, its value is increased or decreased by the length of the data type that it points to.

2. If *ptr1* and *ptr2* are properly declared and initialized pointers, the following operations are valid:

    res = res + *ptr1;

    *ptr1 = *ptr2 + 5;

    prod = *ptr1 * *ptr2;

    quo = *ptr1 / *ptr2;

    Note that there is a blank space between / and * in the last statement because if you write /* together, then it will be considered as the beginning of a comment and the statement will fail.

3. Expressions like *ptr1 == ptr2*, *ptr1 < ptr2*, and *ptr2 != ptr1* are permissible provided the pointers *ptr1* and *ptr2* refer to same and related variables. These comparisons are common in handling arrays.

Operations that cannot be used with respect to pointers

Suppose *p1* and *p2* are pointers to related variables. The following operations cannot work with respect to pointers:

1. Pointer variables cannot be added. For example, *p1 = p1 + p2* is not valid.

2. Multiplication or division of a pointer with a constant is not allowed.

    For example, *p1 * p2* or *p2 / 5* are invalid.

3. An invalid pointer reference occurs when a pointer's value is referenced even though the pointer does not point to a valid block. Suppose *p* and *q* are two pointers. If we say, *p = q*; when *q* is uninitialized. The pointer *p* will then become uninitialized as well, and any reference to **p* is an invalid pointer reference.

Note: Once we have created a pointer that points to a certain variable, we can either reassign it another variable's address, or we can move it around by performing **Pointer Arithmetic.**

Suppose we had an int pointer, ptr. If we say ptr++, ++ptr or ptr+=1, we are actually moving the pointer forward by 2 bytes (i.e. the size of its data type). So it will point somewhere else in the computer's memory.

Similarly, ptr—, —ptr and ptr-=1 will move the int pointer "back" 2 bytes.

And you are not restricted to move in steps of 1 either: ptr+=n will move it "forward" by n*2 bytes, (assuming ptr is an int pointer).

The larger the data type, the larger the step size. So if ptr was a double pointer, ptr-=3 will move it "back" by 3*8 = 24 bytes.

The arithmetic assignment operators you can use with pointers are += and -= ptr*=2 and ptr/=2 cannot be used!

Examples

The following are the examples based on pointer arithmetic.

Example 1

```
char s[100];
char *ptr = s;
ptr += 1;
```

Solution

I assume ptr is increased by 1 byte, pointing to the 2nd element in the array, because character type array holds 1 bye per character.

Example 2

```
int i [100]; // suppose the size of an int is 4 bytes
int *ptr = i;
ptr += 1;
```

Solution

Since the type of data type is integer and it holds 4 bytes then ptr increased by 4 bytes in here

Example 3

```
int i[100];
void *ptr = i;
ptr += 1; // So how many bytes is the ptr increased by here?
// Will it point to the 2nd element of the array.
```

Solution

If I know the byte-size of elements in the array, but I do not know the actual type.

How can I make the void * pointer points to the next element of the array?

PASSING POINTERS TO FUNCTIONS

We know that in Functions, argument can generally be passed to functions in one of the two following ways:

1. Pass by value method
2. Pass by reference method

In the first method, when arguments are passed by value, a copy of the *values* of actual arguments is passed to the calling function. Thus, any changes made to the variables inside the function will have no effect on variables used in the actual argument list.

However, when arguments are passed by reference (i.e. when a pointer is passed as an argument to a function), the *address* of a variable is passed. The contents of that address can be accessed freely, either in the called or calling function. Therefore, the function called by reference can change the value of the variable used in the call.

The following program illustrates the difference.

Problem 1

Write a program to swap the values using the pass by value and pass by reference methods.

Solution

```
/* Program that illustrates the difference
between ordinary arguments, which are passed
by value, and pointer arguments, which are
passed by reference */
# include <stdio.h>
#include<conio.h>
void swapVal ( int, int ); /* function prototype */
void swapRef ( int *, int * ); /*function prototype*/
void main ()
{
int x = 10;
int y = 20;
printf("Pass by Value Method\n");
printf ("Before calling function swapVal x=%d y=%d",x,y);
swapVal (x, y); /* copy of the arguments are passed */
printf ("\n after calling function swapVal x=%d y=%d",x,y);
printf("\n\n pass by Reference Method");
printf ("\n before calling function swapRef x=%d y=%d",x,y);
swapRef (&x,&y); /*address of arguments are passed */
printf("\n after calling function swapRef x=%d y=%d",x,y);
getch();
}
/* Function using the pass by value method*/
void swapVal (int x, int y)
{
int temp;
```

```
temp = x;
x = y;
y = temp;
printf ("\nWithin function swapVal x=%d
y=%d",x,y);
}
/*Function using the pass by reference method*/
void swapRef (int *px, int *py)
{
int temp;
temp = *px;
*px = *py;
*py = temp;
printf ("\nWithin function swapRef *px=%d
*py=%d",*px,*py);
}
```

The output of the above code segment is as follows:

Pass by Value Method
Before calling function swapVal x=10 y=20
Within function swapVal x=20 y=10
After calling function swapVal x=10 y=20
Pass by Reference Method
Before calling function swapRef x=10 y=20
Within function swapRef *px=20 *py=10
After calling function swapRef x=20 y=10

The explanation of the above program segment is as follows:

This program contains two functions, *swapVal* and *swapRef*.

In the function *swapVal*, arguments x and y are passed by *value*. So, any changes to the arguments are local to the function in which the changes occur. Note the values of x and y remain unchanged even after exchanging the values of x and y inside the function **swapVal**.

Now consider the function *swapRef*. This function receives two *pointers* to integer variables as arguments identified as pointers by the indirection operators that appear in argument declaration. This means that in the function *swapRef*, arguments x and y are passed by *reference*. So, any changes made to the arguments inside the function *swapRef* are reflected in the function *main ( )*. Note the values of x and y is interchanged after the function call *swapRef*.

A Function Returning more than One Value

Using *call by reference* method we can make a function return more than one value at a time, which is not possible in the *call by value* method.

Example

If we write the following function prototype
int * sort(int *);
The meaning of this line is that sort is a function which accepts an integer type of pointer, i.e. an integer type of array and according to that array it returns an integer type of pointer, i.e. the complete array.

This concept is illustrated with the help of following problem.

Problem

Write a program to input 10 numbers and print the numbers in ascending order with the help of function and use the with argument and with return type method.

Solution

```
#include<stdio.h>
#include<conio.h>
int * sort (int *);
void main()
{
int a[10],*p,i;
clrscr();
for(i=0;i<10;i++)
{
    printf("\n enter the value of a[%d]",i);
    scanf("%d",&a[i]);
}
p=sort(a);//Calling of the function
printf("\n the sorted array is ");
for(i=0;i<10;i++)
{
    printf("\n %d",*(p+i));
}
getch();
}
int *sort(int *p)
{
    int temp,i,j;
    for(i=0;i<10;i++)
    {
```

```
for(j=i+1;j<10;j++)
{
    if(*(p+i)>*(p+j))
    {
        temp=*(p+i);
        *(p+i)=*(p+j);
        *(p+j)=temp;
    }
}
return(p);
}
```

ARRAYS AND POINTERS

Pointers and arrays are closely related. An array declaration such as *int arr [5]* will lead the compiler to pick an address to store a sequence of 5 integers, and *arr* is a name for that address. The array name in this case is the *address* where the sequence of integers starts. Note that the value is not the first integer in the sequence, nor is it the sequence in its entirety. The value is just an address.

Now, if *arr* is a one-dimensional array, then the address of the first array element can be written as *&arr[0]* or simply *arr*. Moreover, the address of the second array element can be written as *&arr[1]* or simply *(arr+1)*. In general, address of array element *(i+1)* can be expressed as either *&arr[i]* or as *(arr+ i)*. Thus, we have two different ways for writing the address of an array element. In the latter case, i.e. expression *(arr+ i)* is a symbolic representation for an address rather than an arithmetic expression. Since *&arr[i]* and *(ar+ i)* both represent the address of the *ith* element of *arr*, so *arr[i]* and **(ar + i)* both represent the contents of that address, i.e. the value of *ith* element of *arr*.

Note that it is not possible to assign an arbitrary address to an array name or to an array element. Thus, expressions such as *arr*, *(arr+ i)* and *arr[i]* cannot appear on the left side of an assignment statement. Thus we cannot write a statement such as:

&arr[0] = &arr[1]; /* Invalid */

However, we can assign the value of one array element to another through a pointer, For example,

ptr = &arr[0]; /* ptr is a pointer to arr[0] */

arr[1] = *ptr; /* Assigning the value stored at address to arr[1] */

The following program illustrates the above concept

Problem

Write a program that accesses array elements of a one-dimensional array using pointers.

Solution

```
# include<stdio.h>
#include<conio.h>
void main()
{
int arr[5] = {10, 20, 30, 40, 50};
clrscr();
int i;
for (i = 0; i < 5; i++)
{
printf ("i=%d\t arr[i]=%d\t *(arr+i)=%d\t", i, arr[i], *(arr+i));
printf ("&arr[i]=%u\t arr+i=%u\n", &arr[i], (arr+i)); }
}
getch();
}
```

The output of the above code segment is as follows:

i=0 arr[i]=10 *(arr+i)=10 &arr[i]=65516 arr+i=65516

    i=1 arr[i]=20 *(arr+i)=20 &arr[i]=65518 arr+i=65518

    i=2 arr[i]=30 *(arr+i)=30 &arr[i]=65520 arr+i=65520

    i=3 arr[i]=40 *(arr+i)=40 &arr[i]=65522 arr+i=65522

    i=4 arr[i]=50 *(arr+i)=50 &arr[i]=65524 arr+i=65524

Pointers and Multidimensional Arrays

C allows multidimensional arrays, lays them out in memory as contiguous locations, and does more behind the scenes address arithmetic. Consider a 2-dimensional array.

int arr[3][3] = {{1, 2, 3}, {4, 5, 6}, {7, 8, 9}};

The compiler treats a 2 dimensional array as an array of arrays. As you know, an array name is a pointer to the first element within the array. So, **arr** points to the first 3-element array, which is actually the first row (i.e. row 0) of the two-dimensional array. Similarly, (*arr* + 1) points to the second 3-element array (i.e. row 1) and so on. The value of this pointer, *(arr + 1), refers to the entire row. Since row 1 is a one-dimensional array, (*arr* + 1) is actually a pointer to the first element in row 1. Now add 2 to this pointer. Hence, (*(*arr* + 1) + 2) is a pointer to element 2 (i.e. the third element) in row 1. The value of this pointer, *(*(*arr* + 1) + 2), refers to the element in column 2 of row 1.

These relationships are illustrated as follows:

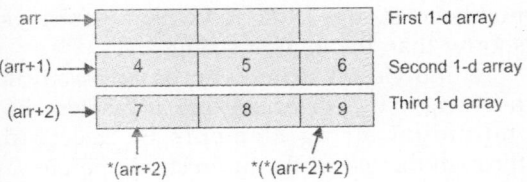

ARRAY OF POINTERS

The way there can be an array of integers, or an array of float numbers, similarly, there can be array of pointers too. Since a pointer contains an address, an array of pointers would be a collection of addresses. For example, a multidimensional array can be expressed in terms of an array of pointers rather than a pointer to a group of contiguous arrays.

Two-dimensional array can be defined as a one-dimensional array of integer pointers by writing:
int *arr[3];
Rather than the conventional array definition,
int arr[3][5];
Similarly, an *n*-dimensional array can be defined as (*n* − 1)-dimensional array of pointers by writing
data-type *arr[*subscript 1*] [*subscript 2*]....
[*subscript n − 1*];
The subscript1, subscript2 indicate the maximum number of elements associated with each subscript.

POINTERS AND STRINGS

As we have seen in strings, a string in C is an array of characters ending in the null character

(written as '\0'), which specifies where the string terminates in memory. Like in one-dimensional arrays, a string can be accessed via a pointer to the first character in the string. The value of a string is the (constant) address of its first character. Thus, it is appropriate to say that a string is a constant pointer. A string can be declared as a character array or a variable of type *char* *.
The declarations can be done as shown below:
char country[ ] = "INDIA";
char *country = "INDIA";
Each initialize a variable to the string "INDIA". The second declaration creates a pointer variable *country* that points to the letter I in the string "INDIA" somewhere in memory.

Once the base address is obtained in the pointer variable *country*, *country would yield the value at this address, which gets printed through,
printf ("%s", *country);
Here is a program that dynamically allocates memory to a character pointer using *the* library function *malloc* at run-time. We will discuss malloc function later on.

An advantage of doing this way is that a fixed block of memory need not be reserved in advance, as is done when initializing a conventional character array.

Problem 1
Write a program to test whether the given string is a palindrome or not.

Solution
/* *Program tests a string for a palindrome using pointer notation* */
include <stdio.h>
include <conio.h>
include <stdlib.h>
short int palindrome(char,int); /*Function Prototype */
void main()
{
char *palin, c;
int i, count;
palin = (char *) malloc (20 * sizeof(char));
printf("\nEnter a word: ");
do

```
{
c = getchar( );
palin[i] = c;
i++;
}while (c != '\n');
i = i-1;
palin[i] = '\0';
count = i;
if (palindrome(palin,count) == 1)
printf ("\n entered word is not a palindrome.");
else
printf ("\n entered word is a palindrome");
getch();
}
short int palindrome(char *palin, int len)
{
short int i = 0, j = 0;
for(i=0, j=len-1; i < len/2;i++,j—)
{
if (palin[i] == palin[j])
continue;
else
return(1);
}
return(0);
}
```

The output of the above code segment is as follows:

Enter a word: dalda
Entered word is not a palindrome
Enter a word: abccba
Entered word is a palindrome.

Array of pointers to strings

Arrays may contain pointers. We can form an array of strings, referred to as a string array. Each entry in the array is a string, but in C a string is essentially a pointer to its first character, so each entry in an array of strings is actually a pointer to the first character of a string. Consider the following declaration of a string array:

char *country[ ] =
{

"INDIA", "CHINA", "BANGLADESH", "PAKISTAN", "US"
};

The *country[ ] of the declaration indicates an array of five elements. The char* of the declaration indicates that each element of array country is of type "pointer to char".

Thus, *country* [0] will point to INDIA, *country* [1] will point to CHINA, and so on.

Thus, even though the array *country* is fixed in size, it provides access to character strings of any length. However, a specified amount of memory will have to be allocated for each string later in the program, for example, country[i] = (char *) malloc(15 * sizeof (char));

The *country* character strings could have been placed into a two-dimensional array but such a data structure must have a fixed number of columns per row, and that number must be as large as the largest string. Therefore, considerable memory is wasted when a large number of strings are stored with most strings shorter than the longest string.

As individual strings can be accessed by referring to the corresponding array element, individual string elements be accessed through the use of the indirection operator.

For example, * (* *country* + 3) + 2) refers to the third character in the fourth string of the array *country*.

Problem 2

Write a program to enter a list of strings and rearrange them in alphabetical order, using a one-dimensional array of pointers, where each pointer indicates the beginning of a string:

Solution

```
/* Program to sort a list of strings in alphabetical
order using an array of pointers */
# include <stdio.h>
# include <conio.h>
# include <stdlib.h>
# include <string.h>
void readinput (char *[ ], int);
void writeoutput (char *[ ], int);
void reorder (char *[ ], int);
void main( )
{
char *country[ 5 ];
int i;
clrscr();
for (i = 0; i < 5; i++)
{
country[ i ] = (char *) malloc (15 * sizeof (char));
}
printf ("Enter five countries on a separate
line\n");
```

```
readinput (country, 5);
reorder (country, 5);
printf ("\nReordered list\n");
writeoutput (country, 5);
getch( );
}
void readinput (char *country[ ], int n)
{
int i;
for (i = 0; i < n; i++)
{
scanf ("%s", country[ i ]);
}
}
void writeoutput (char *country[ ], int n)
{
int i;
for (i = 0; i < n; i++)
{
printf ("%s", country[ i ]);
printf ("\n");
}
}
void reorder (char *country[ ], int n)
{
int i, j;
char *temp;
for (i = 0; i < n-1; i++)
{
for (j = i+1; j < n; j++)
{
if (strcmp (country[ i ], country[ j ]) > 0)
{
temp = country[ i ];
country[ i ] = country[ j ];
country[ j ] = temp;
}
}
}
}
```

The output of the above code segment is as follows:

Enter five countries on a separate line
INDIA
BANGLADESH
PAKISTAN
CHINA
SRI LANKA
Reordered list

BANGLADESH
CHINA
INDIA
PAKISTAN
SRI LANKA

The limitation of the string array concept is that when we are using an array of pointers to strings we can initialize the strings at the place where we are declaring the array, but we cannot receive the strings from keyboard using *scanf( )*.

SUMMARY

- Pointers are the variables which stores the address of another variable.
- If we want to pass the array into the function then we can pass the array into the function with the help of pointer.
- If we want to define a function which returns more than one value, i.e. the array then the return data type must be pointer type.
- The most important application of pointer is to allocate the memory dynamically.
- If we define int *p the meaning of this line is that p is a pointer of integer type that means it always holds the address of integer type variable.
- Pointers play a very important role in arrays and strings.
- Variables declared in a function are not available to other functions in a program. So, there will not be any clash even if we give same name to the variables declared in different functions.
- Pointers are variables which hold addresses of other variables.
- A function can be called either by value or by reference.
- Pointers can be used to make a function return more than one value simultaneously.

Exercises

Section A
Multiple Choice Questions

1. a= *b signifies
   (a) b holds the address of a
   (b) a holds the address of b
   (c) a = a*b   (d) Both a and c

2. We declare a pointer as (type *pointer name) where type signifies
   (a) Type of pointer
   (b) Type of data that it points to
   (c) Has no significance
   (d) Optional

3. The address of variable a can be obtained by using
   (a) &a
   (b) *a
   (c) address(a)
   (d) address(*a)

4. If a points to a char string than we can access 5th character just
   (a) *(a+5)
   (b) a[5]
   (c) Both a and b
   (d) None of the above

5. *a++ can be taken as
   (a) (*a)++
   (b) *(a++)
   (c) Both a and b
   (d) None of the above

6. The general syntax of defining a pointer is
   (a) <data type> * <pointer name>;
   (b) <data type> <pointer name>;
   (c) <data type> &<pointer name>;
   (d) None of the above

7. & operator is known as
   (a) Reference operator
   (b) Address of operator
   (c) Both a and b
   (d) None of the above

8. * operator is known as
   (a) Value of operator
   (b) Multiplication operator
   (c) Both a and b
   (d) None of the above

9. The output of the following code is void
   main()
   {
   int i = 5, j = 2 ;
   junk (i, j) ;
   printf ("\n%d %d", i, j) ;
   }
   junk (int i, int j)
   {
   i = i * i ;
   j = j * j ;
   }

(a) 52
(b) 254
(c) 54
(d) None of the above

10. The output of the following code is.
    main( )
    {
    int i = 5, j = 2 ;
    junk (&i, &j) ;
    printf ("\n%d %d", i, j) ;
    }
    junk (int *i, int *j)
    {
    *i = *i * *i ;
    *j = *j * *j ;
    }

(a) 52
(b) 254
(c) 54
(d) None of the above

State True/False

1. &x returns the "address of x" in hexadecimal format.
2. %p is the format specifier for address.
3. * identifier acts as dereference operator can be translated to "value pointed by".
4. Base address is the address of last element of array.
5. Array indexing method is faster than pointer method.
6. A pointer can store a reference to a pointer.
7. you cannot modify the variable by accessing it through pointers.
8. A function may contain more than one **return** statements.
9. Each **return** statement in a function may return a different value.
10. A function can still be useful even if you do not pass any arguments to it and the function does not return any value back.

Section B

Short Answer Type Questions

A. Point out the errors, if any, in the following programs
   1. void main( )
      {
      int i = 135, a = 135, k ;
      k = pass (i, a) ;
      printf ("\n%d", k) ;
      }
      pass (int j, int b)

```
int c ;
{
c = j + b ;
return ( c ) ;
}
```

2. ```
void main( )
{
int p = 23, f = 24 ;
manish (&p, &f) ;
printf ("\n%d %d", p, f) ;
}
manish (int q, int g)
{
q = q + q ;
g = g + g ;
}
```

3. ```
void main( )
{
int k = 35, z ;
z = vineet_check ( k ) ;
printf ( "\n%d", z ) ;
}
vineet_check ( m )
{
int m ;
if ( m > 40 )
return ( 1 ) ;
else
return ( 0 ) ;
}
```

4. ```
void main( )
{
int i = 35, *z ;
z = function (&i) ;
printf ("\n%d", z) ;
}
function (int *m)
{
return (m + 2) ;
}
```

B. What would be the output of the following programs?

1. ```
void main( )
{
int i = 0 ;
i++ ;
if ( i <= 5 )
{
printf ( "\nC adds wings to your
thoughts" ) ;
exit( ) ;
main( ) ;
}
}
```

2. ```
void main( )
{
int i = 0 ;
i++ ;
if (i <= 5)
{
printf ("\nC adds wings to your
thoughts") ;
main( ) ;
}
}
```

3. ```
void main( )
{
int i = 0 ;
i++ ;
if ( i <= 5 )
{
printf ( "\nC adds wings to your
thoughts" ) ;
exit( ) ;
}
}
```

4. ```
main( )
{
static int i = 0 ;
i++ ;
if (i <= 5)
{
printf ("\n%d", i) ;
main( ) ;
}
else
exit( ) ;
}
```

## Section C

*Long Answer Type Questions*

1. Write a program in C to change upper to lower with the help of pointer.
2. Write a program in C to change lower to upper with the help of pointer.
3. Write a program in C to print a word form of given number between 0 and 9 with the help of pointer.
```

4. Write a program in C to print a word form of given number is tens between 1 and 99 with the help of pointer.

5. Write a program in C for relations operations of two given integer numbers with the help of pointer.

6. Write a program in C for relations operations of two given float numbers with the help of pointer.

7. Write a program in C for given mark contain which grade with the help of pointer.

8. Write a program in C to find biggest of two given numbers with the help of pointer.

9. Write a program in C to find smallest of two given numbers with the help of pointer.

10. Write a program in C to find biggest of three given numbers with the help of pointer.

11. Write a program in C to find smallest of three given numbers with the help of pointer.

12. Write a program in C to find biggest of three given numbers using && operator with the help of pointer.

13. Write a program in C to find smallest of three given numbers using && operator with the help of pointer.

14. Write a program in C to display the name of the day in a week .here given input range is 1–7. with the help of pointer.

15. Write a program in C to find the biggest of four given numbers with the help of pointer.

16. Write a program in C to find the smallest of four given numbers with the help of pointer.

17. Write a program in C to find a vowel or not of given character with the help of pointer.

18. Write a program in C to print the numbers 1 to 100 with the help of pointer.

19. Write a program in C to sum of first 100 numbers with the help of pointer.

20. Write a program in C to sum of first N numbers with the help of pointer.

21. Write a program in C to sum of M to N numbers with the help of pointer.

22. Write a program in C to find the average of 1 to 100 with the help of pointer.

23. Write a program in C to sum of first N numbers with the help of pointer.

24. Write a program in C to average of M to N numbers with the help of pointer?

25. Write a program in C to display the even or odd numbers between 1 and 100 with the help of pointer.

26. Write a program in C to display the even or odd numbers of first N numbers with the help of pointer.

27. Write a program in C to display the even or odd numbers between to ranges with the help of pointer.

28. Write a program in C to display the leap years between 1000 and 2000 with the help of pointer.

29. Write a program in C to display the leap years of given range with the help of pointer.

30. Write a program in C to print the numbers between two ranges with the help of pointer.

31. Write a program in C to print the number until 1000 is given an input with the help of pointer.

32. Write a program in C to print the positive, negative and zero until 1000 is given an input with the help of pointer.

33. Write a program in C to count the positive, negative numbers and zero until 1000 is given an input with the help of pointer.

34. Write a program in C to sum of the positive, negative numbers until 1000 is given an input with the help of pointer.

35. Write a program in C to mean of the positive, negative numbers until 1000 is given an input with the help of pointer.

36. Write a program in C to factorial of given number until 1000 is given an input with the help of pointer.

15 Dynamic Memory Allocation

Memory allocation means reserving memory for specific purposes. Operating systems and applications generally reserve fixed amounts of memory at start up and allocate more when the processing requires it. If there is not enough free memory to load the core kernel of an application, it cannot be launched. Although a virtual memory function will simulate an almost unlimited amount of memory, there is always a certain amount of "real" memory that is needed.

In C language memory allocation is divided into two parts:

(a) Static memory allocation
(b) Dynamic memory allocation

(a) Static Memory Allocation

Static memory allocation means the memory is allocated at the time of compilation. The compiler allocates the required memory space for a declared variable. By using the address of operator, the reserved address is obtained and this address may be assigned to a pointer variable. Since most of the declared variable has static memory, this way of assigning pointer value to a pointer variable is known as static memory allocation. Memory is assigned during compilation time. In C language the static memory allocation is done with the help of array and simple variables.

For example

If we define int a[100]. The meaning of this line is that a is an array which holds 100 integer elements and 200 bytes memory is allocated for this array.

(b) Dynamic Memory Allocation

Dynamic memory allocation means the memory is allocated at the time of running of the program. It uses functions such as malloc( ) or calloc( ) to get memory dynamically. If these functions are used to get memory dynamically and the values returned by these functions are assingned to pointer variables, such assignments are known as dynamic memory allocation. Memory is assigned during run time.

FUNCTIONS FOR DYNAMIC MEMORY ALLOCATION

In C language the following are the functions which are used for dynamic memory allocation and deallocation. We will discuss the following function

- malloc() function
- calloc() function
- realloc() function
- free() function

All these functions are predefined in stdlib.h or alloc.h header file.

(a) malloc() Function

The malloc function is one of the functions in standard C to allocate memory dynamically. Its function prototype is as follows:

void *malloc(size_t size);

We can also write the following syntax for the malloc function.

<Pointer Variable>=<datatype *> malloc(No of elements * sizeof(data type));

malloc requires one argument—the number of bytes you want to allocate dynamically. If the memory allocation was successful, malloc will return a void pointer—you can assign this

247

to a pointer variable, which will store the address of the allocated memory.

If memory allocation failed (for example, if you are out of memory), malloc will return a NULL pointer.

Example

The standard method of creating an array of ten int objects:

int array[10];

To allocate a similar array dynamically, the following code could be used:

```
/* Allocate space for an array with ten elements of type int. */
int *ptr = malloc(10 * sizeof (int));
if (ptr == NULL) {
/* Memory could not be allocated, the program should handle the error here as appropriate. */
} else {
/* If ptr is not NULL, allocation succeeded. */
}
```

malloc returns NULL to indicate that no memory is available, or that some other error occurred which prevented memory being allocated.

Programming Problem

Problem 1

Write a program to input how many numbers and then input the numbers and then print the maximum number between them with the help of malloc function.

Solution

```
#include<stdio.h>
#include<conio.h>
#include<stdlib.h>
void main()
{
    int *p, n, i,max;
    clrscr();
    printf("\n enter how many elements");
    scanf("%d",&n);
    p=(int *) malloc(n*sizeof(int));
    for(i=0;i<n;i++)
{
    printf("\n enter the element");
    scanf("%d",&p[i]);
}
```

```
    max=p[0];
    for(i=1;i<n;i++)
{
    if(max<p[i])
    max=p[i];
}
    printf("\n the maximum number is %d",max);
    getch();
}
```

Problem 2

Write a program to input how many numbers and then input the numbers and then print the minimum number between them with the help of malloc function.

Solution

```
#include<stdio.h>
#include<conio.h>
#include<stdlib.h>
void main()
{
    int *p, n, i,min;
    clrscr();
    printf("\n enter how many elements");
    scanf("%d",&n);
    p=(int *) malloc(n*sizeof(int));
    for(i=0;i<n;i++)
{
    printf("\n enter the element");
    scanf("%d",&p[i]);
}
    min=a[0];
    for(i=1;i<n;i++)
{
    if(min>p[i])
    min=p[i];
}
    printf("\n the minimum number is %d", min);
    getch();
}
```

Problem 3

Write a program to input how many numbers and then input the numbers and then print the sum of all the numbers with the help of malloc function?

Solution

```
#include<stdio.h>
```

```
#include<conio.h>
#include<stdlib.h>
void main()
{
    int *p, n, i,sum=0;
    clrscr();
    printf("\n enter how many elements");
    scanf("%d",&n);
    p=(int *) malloc(n*sizeof(int));
    for(i=0;i<n;i++)
{
    printf("\n enter the element");
    scanf("%d",&p[i]);
}
    for(i=0;i<n;i++)
{
    sum=sum+p[i];
}
printf("\n the sum is %d",sum);
    getch();
}
```

(b) calloc() Function

The *calloc* function allocates space for an array of nmemb objects, each of whose size is size. The space is initialized to all bits zero
The standard C library declares the function calloc() in as follows:
void *calloc(size_t elements, size_t sz);
We can also define the syntax of calloc() function as follows:
<Pointer variable>= (<data type*) calloc (no of elements, sizeof (data type));
calloc () allocates space for an array of elements, each of which occupies sz bytes of storage. The space of each element is initialized to binary zeros. In other words, calloc() is similar to malloc(), except that it handles arrays of objects rather than a single chunk of storage and that it initializes the storage allocated. The following example allocates an array of 100 int's using calloc():
int * p = (int*) calloc (100, sizeof(int));

Programming Problem

Problem 1

Write a program to input how many numbers and then input the numbers and then print the maximum number between them with the help of calloc function.

Solution

```
#include<stdio.h>
#include<conio.h>
#include<stdlib.h>
void main()
{
    int *p, n, i,max;
    clrscr();
    printf("\n enter how many elements");
    scanf("%d",&n);
    p=(int *) calloc(n,sizeof(int));
    for(i=0;i<n;i++)
{
    printf("\n enter the element");
    scanf("%d",&p[i]);
}
    max=p[0];
    for(i=1;i<n;i++)
{
    if(max<p[i])
    max=p[i];
}
printf("\n the maximum number is %d", max);
    getch();
}
```

Problem 2

Write a program to input how many numbers and then input the numbers and then print the minimum number between them with the help of calloc function.

Solution

```
#include<stdio.h>
#include<conio.h>
#include<stdlib.h>
void main()
{
    int *p, n, i,min;
    clrscr();
    printf("\n enter how many elements");
    scanf("%d",&n);
    p=(int *) calloc(n,sizeof(int));
    for(i=0;i<n;i++)
{
    printf("\n enter the element");
    scanf("%d",&p[i]);
}
```

```
   min=a[0];
   for(i=1;i<n;i++)
{
   if(min>p[i])
   min=p[i];
}
printf("\n the minimum number is %d",
min);
   getch();
}
```

Problem 3

Write a program to input how many numbers and then input the numbers and then print the sum of all the numbers with the help of calloc function.

Solution

```
#include<stdio.h>
#include<conio.h>
#include<stdlib.h>
void main()
{
   int *p, n, i,sum=0;
   clrscr();
   printf("\n enter how many elements");
   scanf ("%d", &n);
   p=(int *) calloc (n, sizeof(int));
   for(i=0;i<n;i++)
{
   printf("\n enter the element");
   scanf("%d",&p[i]);
}
   for(i=0;i<n;i++)
{
   sum=sum+p[i];
}
printf("\n the Sum is %d",sum);
   getch();
}
```

Difference between malloc () and calloc () Function

Both the malloc() and the calloc() functions are used to allocate dynamic memory. Each operates slightly different from the other. malloc() takes a size and returns a pointer to a chunk of memory at least that big:
void *malloc(size_t size);
calloc() takes a number of elements, and the size of each, and returns a pointer to a chunk of memory at least big enough to hold them all:
void *calloc(size_t numElements, size_t sizeOfElement);

There is one major difference and one minor difference between the two functions. The major difference is that malloc() does not initialize the allocated memory. The first time malloc() gives you a particular chunk of memory, the memory might be full of zeros. If memory has been allocated, freed, and reallocated, it probably has whatever junk was left in it. That means, unfortunately, that a program might run in simple cases (when memory is never reallocated) but break when used harder (and when memory is reused). calloc() fills the allocated memory with all zero bits. That means that anything there you're going to use as a char or an int of any length, signed or unsigned, is guaranteed to be zero. Anything you are going to use as a pointer is set to all zero bits. That is usually a null pointer, but it is not guaranteed. Anything you are going to use as a float or double is set to all zero bits; that is a floating-point zero on some types of machines, but not on all.

The minor difference between the two is that calloc() returns an array of objects; malloc() returns one object. Some people use calloc() to make clear that they want an array.

(c) free() Function

When you allocate memory with either malloc() or calloc(), it is taken from the dynamic memory pool that is available to your program. This pool is sometimes called the *heap*, and it is finite. When your program finishes using a particular block of dynamically allocated memory, you should deallocate, or free, the memory to make it available for future use. To free memory that was allocated dynamically, use free(). Its prototype is
void free(void *ptr);

The free() function releases the memory pointed to by ptr. This memory must have been allocated with malloc(), calloc(), or realloc(). If ptr is NULL, free() does nothing.

The following program demonstrates the free() function.

Example

This example will ask you how many integers you would like to store in an array. It will then allocate the memory dynamically using malloc and store a certain number of integers, print them out, then releases the used memory using free.

```c
#include <stdio.h>
#include <stdlib.h> /* required for the malloc and free functions */
void main()
{
int number;
int *ptr;
int i;
printf("How many ints would you like store? ");
scanf("%d", &number);
ptr = malloc(number*sizeof(int)); /* allocate memory */
if(ptr!=NULL) {
for(i=0 ; i<number ; i++) {
*(ptr+i) = i;
}
for(i=number ; i>0 ; i—) {
printf("%d\n", *(ptr+(i-1))); /* print out in reverse order */
}
free(ptr); /* free allocated memory */
return 0;
}
else {
printf("\n memory allocation failed—not enough memory.\n");
return 1;
}
}
```

The output of the above code is as follows:
Output if I entered 3:

How many ints would you like store? 3
2
1
0

When I first wrote the example using a Borland compiler, I had to cast the returned pointer like this:

```c
ptr = (int *)malloc(number*sizeof(int));
```

The above example was tested in MSVC but try casting the pointer if your compiler displays an error.

(d) realloc () Function

The size of the memory block pointed to by the *ptr* parameter is changed to the *size* bytes, expanding or reducing the amount of memory available in the block.

The function may move the memory block to a new location, in which case the new location is returned. The content of the memory block is preserved up to the lesser of the new and old sizes, even if the block is moved. If the new *size* is larger, the value of the newly allocated portion is indeterminate.

In case that *ptr* is NULL, the function behaves exactly as malloc, assigning a new block of *size* bytes and returning a pointer to the beginning of it.

In case that the *size* is 0, the memory previously allocated in *ptr* is deallocated as if a call to free was made, and a NULL pointer is returned.

The general syntax for realloc () function is as follows:

```c
void * realloc ( void * ptr, size_t size );
```

The ptr argument is a pointer to the original block of memory. The new size, in bytes, is specified by *size*. There are several possible outcomes with realloc():

- If sufficient space exists to expand the memory block pointed to by ptr, the additional memory is allocated and the function returns ptr.
- If sufficient space does not exist to expand the current block in its current location, a new block of the size for size is allocated, and existing data is copied from the old block to the beginning of the new block. The old block is freed, and the function returns a pointer to the new block.
- If the ptr argument is NULL, the function acts like malloc(), allocating a block of size bytes and returning a pointer to it.
- If the argument size is 0, the memory that ptr points to is freed, and the function returns NULL.

- If memory is insufficient for the re-allocation (either expanding the old block or allocating a new one), the function returns NULL, and the original block is unchanged.

Example

This example uses calloc to allocate enough memory for an int array of five elements. Then realloc is called to extend the array to hold seven elements.

```
#include<stdio.h>
#include <stdlib.h>
void main()
{
int *ptr;
int i;
ptr = calloc(5, sizeof(int));
if(ptr!=NULL)
{
*ptr = 1;
*(ptr+1) = 2;
ptr[2] = 4;
ptr[3] = 8;
ptr[4] = 16;
/* ptr[5] = 32; wouldn't assign anything */
ptr = realloc(ptr, 7*sizeof(int));
if(ptr!=NULL) {
printf("Now allocating more memory... \n");
ptr[5] = 32; /* now it's legal! */
ptr[6] = 64;
for(i=0 ; i<7 ; i++) {
printf("ptr[%d] holds %d\n", i, ptr[i]);
}
realloc(ptr,0); /* same as free(ptr); —just fancier! */
return 0;
}
else {
printf("Not enough memory—realloc failed.\ n");
return 1;
}
}
else {
printf("Not enough memory—calloc failed.\n");
return 1;
}
}
```

The output of the above code is as follows:

Now allocating more memory...
ptr[0] holds 1
ptr[1] holds 2
ptr[2] holds 4
ptr[3] holds 8
ptr[4] holds 16
ptr[5] holds 32
ptr[6] holds 64

Notice the two different methods I used when initializing the array: ptr[2] = 4; is the equivalent to *(ptr+2) = 4; (just easier to read!).

Before using realloc, assigning a value to ptr[5] would not cause a compile error. The program would still run, but ptr[5] would not hold the value you assigned.

STACK

A *stack* is an ordered list in which all insertions and deletions are made at one end, called the *top*. Stack is a linear data structure which holds the property of LIFO (last in first out).

Stack is an abstract data type and data structure based on the principle of *last in first out (LIFO)*.

The following are the important points related to stack data structure.

- The last element put on the stack is the first element to be taken off.
- Insertion and deletion can be takes place at one end called TOP.
- It looks like one side closed tube.
- The insertion operation of the stack is called *push* operation
- The deletion operation is called as *pop* operation.
- Push operation on a full stack causes **overflow**.
- Pop operation on an empty stack causes **underflow**.
- Stack pointer is a pointer, which is used to access the top element of the stack.
- If you push an element that is added at the top of the stack; in the same way when we pop the element the item at the top of the stack is deleted.

For example, the following figures give an idea about the stack.

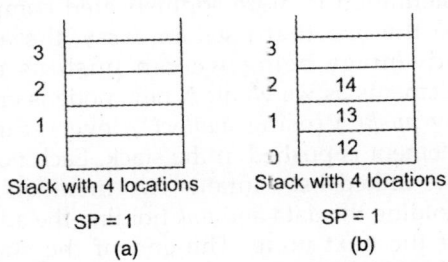

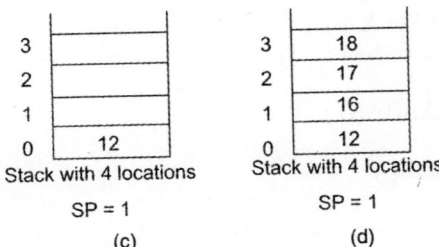

Stack with 4 locations   Stack with 4 locations

SP = 1   SP = 1

(a)   (b)

Stack with 4 locations   Stack with 4 locations

SP = 1   SP = 1

(c)   (d)

From the above Fig. (a), it is clear that the size of the stack is 4 locations and sp = –1 indicates that the stack is empty and we can push the new items into the stack. Now Fig. (b) shows how the stack grows, when we push an element 12, 13, 14 into the stack. Figure (c) shows the status of the stack after popping 2 times. Now performing (one more) pop operation one more time on the above stack causes an error known as "stack under flow". Figure (d) shows the content of the stack after pushing 16, 17, 18, and 19. If you try to add one more element stack generate error "stack over flow".

Note: We can also refer top as SP.

OPERATIONS OF STACK

There are two operations applied on stack, they are
1. push
2. Pop

1. Push

Push operation is used to insert a new element in to the stack. At the time of insertion first check the stack is full or not. If the stack is full it generates an error message "stack overflow".

2. Pop

Pop operation is used to delete an element from the stack. At the time of deletion first check the stack is empty or not. If the stack is empty it generates an error message "stack underflow".

Algorithms for Push and Pop Function

Assumptions

sp stack pointer whose initial value is -1
max_stack is the size of the queue
stack[] is an array
Element is the element to be added or deleted

Algorithm to push an element into the stack

Step: 1) start
Step: 2) take the element to push into stack
step: 3) If (Sp == max_stack)
display "The stack is full"
else
{Sp = Sp+1
stack [Sp] = item}
step: 4) return to main program

Algorithm to pop an element from the stack

Step: 1) start
Step: 2) if (Sp == -1)
display "stack is empty"
else
{element = stack[Sp]
Sp = Sp -1}
Step: 3) return element

Programming Problems

Problem 1

The program given below implements the stack data structure using an array. In this program the elements are pushed into array stack [ ] through push ( ) function. The parameters passed to push ( ) are the base address of the array, the position in the stack at which the element is to be placed and the element itself. Care is taken by the push ( ) function that the user does not try to place the element beyond the bounds of the stack. This is done by checking the value stored in pos. pop ( ) function pops out the last element stored in the stack[ ]. Because, top holds the position which has the last element in the stack.

This problem is also called as static implementation of stack or array implementation of stack.

```
/* To pop and push items in a stack */
#include<stdio.h>
```

```
#include<conio.h>
#define MAX 10
void push (int) ;
int pop( ) ;
int stack[MAX] ;
int top ;
void main( )
{
    int n ;
    clrscr( ) ;
    top = -1 ; /* stack is empty */
    push (10) ;
    push (20) ;
    push (30) ;
    push (40) ;
    n = pop( ) ;
    printf ("\n item popped out is : %d", n) ;
    n = pop( ) ;
    printf ("\n item popped out is : %d", n) ;
}
/* pushes item on the stack */
void push (int data)
{
    if ( top == MAX - 1 )
        printf ("\n Stack is full");
    else
    {
        top++ ;
        stack[top] = data ;
    }
}
/* pops off the items from the stack */
int pop( )
{
    int data ;
    if (top == -1)
    {
        printf ("\n stack is empty") ;
        return (-1) ;
    }
    else
    {
        data = stack[ top ] ;
        top- ;
        return (data) ;
    }
}
```

Problem 2

In this program stack is implemented and maintained using linked list. This imple-mentation is more sophisticated compared to the one that uses an array, the added advantage being we can push as many elements as we want. A new node is created by *push( )* (using *malloc( )*) every time an element is pushed in the stack. Each node in the linked list contains two members, *data* holding the data and *link* holding the address of the next node. The end of the stack is identified by the node holding NULL in its link part. The *pop( )* function pops out the last element inserted in the stack and frees the memory allocated to hold it. stack_display( ) displays all the elements that stack holds. *count( )* counts and returns the number of elements present in the stack.

This problem is also known as dynamic implementation of stack or linked imple-mentation of Stack.

```
#include<stdio.h>
#include<conio.h>
#include <alloc.h>
struct node
{
    int data ;
    struct node *next ;
} *top=NULL;
void push (struct node *, int) ;
int pop (struct node *) ;
void main( )
{
    int item ;
    push (top, 11) ;
    push (top, 12) ;
    push (top, 13) ;
    push (top, 14) ;
    push (top, 15) ;
    push (top, 16) ;
    push (top, 17) ;
    clrscr( ) ;
    stack_display (top ) ;
    printf ("No. of items in stack = %d", count
    (top)) ;
    printf ("\n Items extracted from stack : ") ;
    item = pop (top) ;
    printf ("%d ", item) ;
    item = pop (top) ;
    printf ("%d ", item) ;
```

```
item = pop (top) ;
printf ("%d ", item) ;
stack_display (top) ;
printf ("No. of items in stack = %d", count
(top)) ;
}
/* adds a new element on the top of stack */
void push (struct node *top, int item)
{
    struct node *q ;
    q = (struct node *) malloc (sizeof (struct
    node)) ;
    q –> data = item ;
    q –> next = top ;
    top = q ;
}
/* removes an element from top of stack */
int pop (struct node *top)
{
    int item ;
    struct node *q ;
    /* if stack is empty */
    if ( top == NULL )
    printf (" stack is empty") ;
    else
    {
        q=top;
        item = q –> data ;
        top = q –> next ;
        free ( q ) ;
        return (item) ;
    }
}
/* displays whole of the stack */
void stack_display (struct node *top)
{
    printf ("\n") ;
    /* traverse the entire linked list */
    while (top ! = NULL)
    {
        printf ("%2d ", q –> data) ;
        q = q –> next ;
    }
    printf ("\n") ;
}
/* counts the number of nodes present in the
linked list representing a stack */
int count (struct node * q)
{
    int c = 0 ;
```

```
    /* traverse the entire linked list */
    while (q != NULL)
    {
        q = q –> next ;
        C++;
    }
    return c ;
}
```

LINKED LIST

A linked list is the collection of nodes and every node is divided into two parts (a) information field and (b) address of the next node.

A linked list is called so because each of items in the list is a part of a structure, which is linked to the structure containing the next item. This type of list is called a linked list since it can be considered as a list whose order is given by links from one item to the next.

The graphical representation of the linked list is as follows:

Item →

Each item has a node consisting two fields one containing the variable and another consisting of address of the next item (i.e. pointer to the next item) in the list. A linked list is therefore a collection of structures ordered by logical links that are stored as the part of data.

Consider the following example to illustrate the concept of linking. Suppose we define a structure as follows:

```
struct linked_list
{
float age;
struct linked_list *next;
}
struct linked_list node1,node2;
```

This statement creates space for nodes each containing 2 empty field's

```
node1
    node1.age
    node1.next
node2
    node2.age
    node2.next
```

The next pointer of node1 can be made to point to the node 2 by the same statement.

```
node1.next=&node2;
```

This statement stores the address of node 2 into the field node1.next and this establishes a link between node1 and node2. Similarly, we can combine the process to create a special pointer value called null that can be stored in the next field of the last node.

Advantages of Linked List

A linked list is a dynamic data structure and therefore the size of the linked list can grow or shrink in size during execution of the program. A linked list does not require any extra space therefore it does not waste extra memory. It provides flexibility in rearranging the items efficiently.

The limitation of linked list is that it consumes extra space when compared to a array since each node must also contain the address of the next item in the list to search for a single item in a linked list is cumbersome and time consuming.

Types of Linked List

There are different kinds of linked lists they are as follows:

- Linear singly linked list
- Circular singly linked list
- Two way or doubly linked list
- Circular doubly linked list.

Applications of Linked Lists

Linked lists concepts are useful to model many different abstract data types such as queues stacks and trees. If we restrict the process of insertions to one end of the list and deletions to the other end then we have a mode of a queue that is we can insert an item at the rear end and remove an item at the front end obeying the discipline first in first out. If we restrict the insertions and deletions to occur only at one end of list the beginning then the model is called stacks. Stacks are all inherently one-dimensional. A tree represents a two dimension linked list. Trees are frequently encounters in every day life one example are organization chart and the other is sports tournament chart.

We only discuss the concept of linear linked list and other linked lists are discussed in the data structure paper.

LINEAR LINKED LIST

A **list** that displays the relationship of adjacency between elements is said to be linear list. It also can be defined to consist of an ordered set of elements. Linear linked list is the most commonly used data structure of linked lists. A simple way to represent a linear list is to expand each node to contain a link or pointer to the next node

Graphically it is represented as follows:

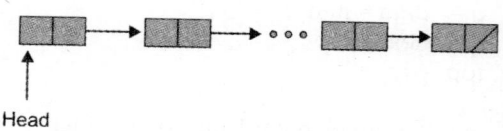

Head

In the above figure, the variable *Head* contains an address or pointer that gives the location of the first node of the list. The last node of the list does not have a successor node, and consequently, no actual address is stored in the pointer field. In such case, a null value is stored as the address. The arrow emanating from the link field of a particular node indicates its successor node in the structure. The basic operations of linear linked list include insert, delete, and search. We will discuss them in detail in the following sections.

BASIC OPERATIONS

In the previous section, we have already given the definition of the linear linked list. Now we will discuss some basic operations that can be performed on a linear linked list.

1. Insert

This operation can add a new node in the first, last or interior of the list.

2. Delete

This operation can delete a node in the first, last or interior of the list.

3. Search

This operation is to search a node containing particular value in the linked list.

To perform these operations, the first and last nodes in the linked list must be marked in some way. See the following figure, *head* is a pointer to the first node, while a slash in the link field signals the terminal node.

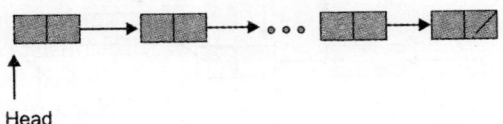

Head

For convenience, we also need two pointers, *cur* and *prev*, that indicate the current and previous nodes. Now, we can discuss these operations in detail.

1. Insertion

Insertion is to add a new node into a linked list. It can take place anywhere—the first, last, or interior of the linked list. Here, we only give the details of insertfirst and insertlast.To perform the insertion, we need a temporary pointer, *new*, that points to a node that will be added into the list.

(a) Insert First

To add a new node to the head of the linear linked list, we need to construct a new node that is pointed by pointer *new*. Assume there is a global variable *head* which points to the first node in the list. The new node points to the first node in the list. The *head* is then set to point to the new node.

The following figures show the procedure step by step:

Step 1. Create a new node that is pointed by pointer *new*. See the following figure.

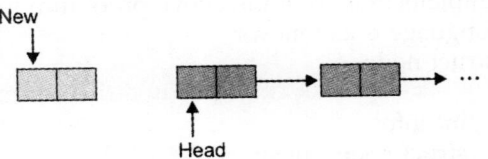

New

Head

Step 2. Link the new node to the first node of the linked list. See the following figure.

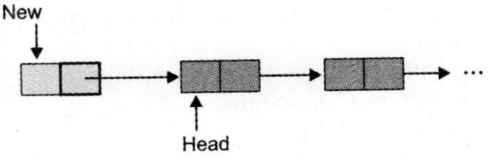

New

Head

Sep 3. Set the pointer *head* to the new node. See the following figure.

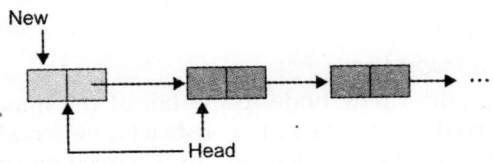

New

Head

Procedure insert head
{Procedure insert a node at the beginning of the linked list.}
Procedure Insert Head (var head: pointer; new: pointer);
begin
new^.info := Info;
new^.link := null; {create a new node.}
if (head = null) {the list is empty}
then
head := new
else
begin
new^.link := head;
head := new
end;
end;
Implementation in C language
Implementation of the above procedure in C language is as follows:

```
struct node
{
    int info;
    struct node * next;
}*start=NULL;
void insertatbegin()
{
    struct node * temp;
    temp=(struct node *) malloc(sizeof(struct node));
    printf("\n enter the information");
    scanf("%d",&temp->info);
    temp->next=NULL;
    if(start==NULL)
    start=temp;
    else
    {
        temp->next=start;
```

```
    start=temp;
}
}
```

(b) Insert Last

To add a new node to the tail of the linear linked list, we need to construct a new node and set it's link field to "null". Assume the list is not empty, locate the last node and change it's link field to point to the new node.

The following figures show the procedure step by step:

Step 1. Create the new node. Initializing pointer *cur* points to the first node of the list, while the pointer *prev* has a value of null. See the following figure.

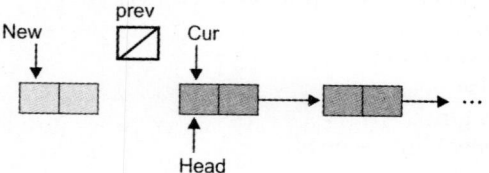

Step 2. Traverse the entire list until pointers *prev* and *cur* advanced to the end of list. To advance *cur* and *prev* one node forward, we first advance the pointer, *prev*, and then move *cur*. See the following figure.

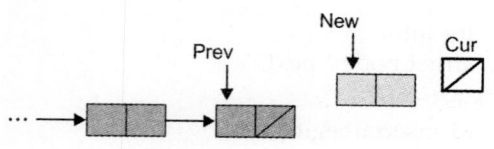

Step 3. Assign the value of the LINK part of new node to null. See the following figure.

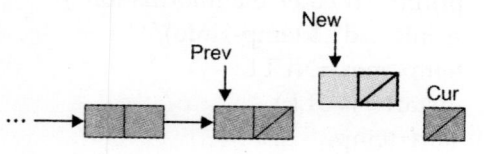

Step 4. Link the node that is pointed by pointer *prev* to the new node. See the following figure.

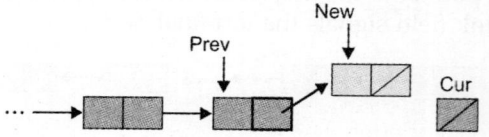

Procedure insert last

{Procedure to insert a node at the end of the linked. Assume the list is not empty.}
Procedure InsertLast (prev, new: pointer);
begin
var
head: pointer;
cur: pointer;
begin
new^.info := Info; {create a new node.}
new^.link := null;
prev := null; {initialize the pointers *prev* and *cur*.}
cur := head;
while (cur != null) do {traverse the entire list until find the last node}
begin
prev :=cur;
cur := prev^.link;
end; {end of while}
new^.link := cur; {link the new node at the end of list}
prev^.link := new
end;
end;

Implementation in C language

Implementation of the above procedure in C language is as follows:

```
struct node
{
    int info;
    struct node * next;
}*start=NULL;
void insertatlast()
{
    struct node * p,*temp;
    p=(struct node *) malloc(sizeof(struct node));
    printf("\n enter the information");
    scanf("%d",&p->info);
    p->next=NULL;
    if(start==NULL)
```

```
start=p;
else
{
    temp=start;
    while(temp->next!=NULL)
    temp=temp->next;
    temp->next=p;
}
}
```

2. Deletion

Deletion is a common operation of linear lined list. It can remove the first node, the last node, or a node containing particular value in the linked list.

We will discuss the following deletion operation of the linked list

- Deletion at the first node of linked list
- Deletion at the end of a linked list
- Deletion within the linked list

(a) *Delete First*

To delete the first node of the linked list, we not only want to advance the pointer *head* to the second node, but we also want to release the memory occupied by the abandoned node. To accomplish this removal, we use pointer *cur* to hold the position of the first node.

The following figures show the procedure step by step.

Step 1. Initializing the pointer *cur* point to the first node of the list. See the following figure.

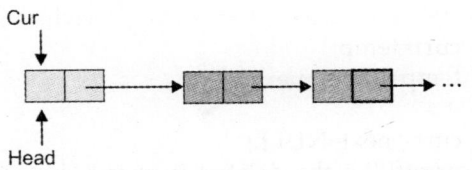

Step 2. Moving the pointer *head* to the second node of the list. See the following figure.

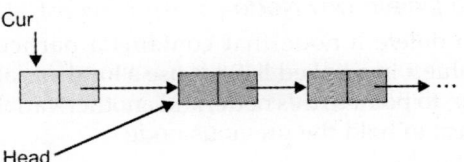

Step 3. Remove the node that is pointed by the pointer *cur*. See the following figure.

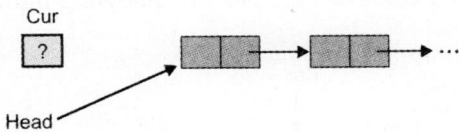

Procedure delete first

{ Procedure to delete the first node in the linked list. Assume the linked list is not empty.}

```
Procedure Delete First (head: pointer);
var cur: pointer;
begin
cur := head;
head := cur^.link;
cur := null
end;
```

Implementation in C language

Implementation of the above procedure in C language is as follows:

```
struct node
{
    int info;
    struct node * next;
}*start=NULL;
void deletefirst()
{
    struct node * p,*temp;
    if(start==NULL)
{
    printf("\n list is empty");
    getch();
    exit(1);
}
    temp=start;
    start=start->next;
    printf("\n the deleted item is %d".temp->info);
    free(temp);
}
```

(b) *Delete Last*

To delete the last node in a linked list, we use a local variable, *cur*, to point to the last node. We also use another variable, *prev*, to point to the last second node in the linked list.

The following figures show the procedure step by step:

Step 1. Initializing pointer *cur* points to the first node of the list, while the pointer *prev* has a value of null. See the following figure.

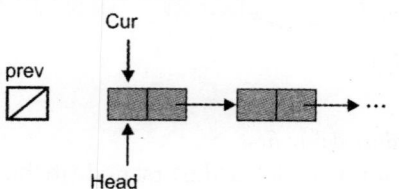

Step 2.Traversing the entire list until the pointer *cur* points to the last node of the list. See the following figure.

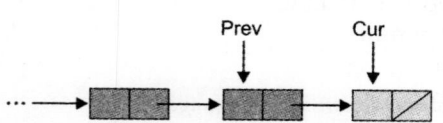

Step 3. Assign the LINK field of the node that is pointed by pointer *prev* to nil. See the following figure.

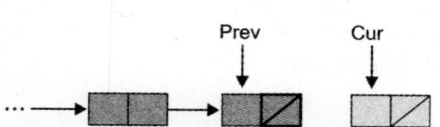

Step 4. Remove the last node that is pointed by pointer *cur*. See the following figure.

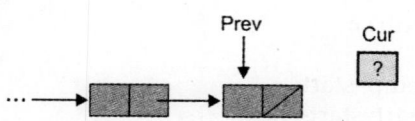

Procedure delete last
{ Procedure to delete the last node in the linked list. Assume the linked list is not empty.}
procedure DeleteLast (prev: pointer);var cur: pointer;
head: pointer;
begin

cur := head; {initialize the pointers *prev* and *cur*.}
prev := null;
while (cur^.link != null) do {traverse the entire list until the pointer *cur* points to the last node}
begin
prev :=cur;
cur := prev^.link;
end; {end of while}
prev^.link := null; {set the LINK portion of the node pointed by *prev*}
cur := null; {dispose the node pointed by *cur*}
end;

Implementation in C language
Implementation of the above procedure in C language is as follows:
```
struct node
{
    int info;
    struct node * next;
}*start=NULL;
void deletelast()
{
    struct node * curr,*temp;
    if(start==NULL)
{
    printf("\n list is empty");
    getch();
    exit(1);
}

    temp=start;
    while(temp->next!=NULL)
{
    curr=temp;
    temp=temp->next;
}

    curr->next=NULL;
    printf("\n the deleted item is %d",temp->info);
    free(temp);
}
```

(c) *Delete any Node*

To delete a node that contains a particular value x in a linked list, we use a local variable, *cur*, to point to this node, and another variable, *prev*, to hold the previous node.

The following figures show the procedure step by step:

Step 1. Initializing pointer *cur* points to the first node of the list, while the pointer *prev* has a value of null. See the following figure.

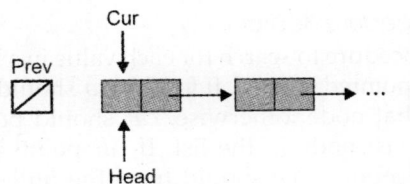

Step 2. Traversing the entire list until the pointer *cur* points to the node that contains value of x, and *prev* points to the previous node. See the following figure.

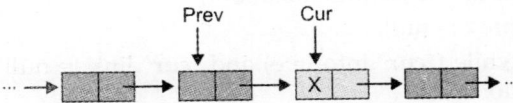

Step 3. Link the node pointed by pointer *prev* to the node after the *cur's* node. See the following figure.

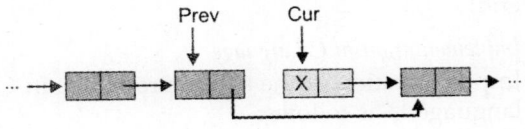

Step 4. Remove the node pointed by *cur*. See the following figure.

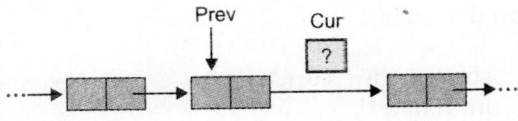

Procedure delete inside
{ Procedure to delete the node that contains value x in the linked list. Assume the linked list is not empty.}
procedure DeleteInside (x: datatype; prev: pointer);
var cur: pointer;

head: pointer;
begin
cur := head; {initialize the pointers *prev* and *cur*.}
prev := null;
{traverse the entire list until the pointer *cur* points to the node that contains value of x}
while (cur^.info != x) do
begin
prev :=cur;
cur := prev^.link;
end; {end of while
prev^.link := cur^.link; {linked the node pointed by *prev* to the node after *cur's* node}
cur := null {dispose the node pointed by *cur*}
end;

Implementation in C language
Implementation of the above procedure in C language is as follows:

```
struct node
{
    int info;
    struct node * next;
}*start=NULL;
void deleteatpos()
{
    struct node * curr,*temp;
    int pos,i;
    if(start==NULL)
{
    printf("\n list is empty");
    getch();
    exit(1);
}
    printf("\n enter the position");
    scanf("%d",&pos);
    temp=start;
for(i=0;i<pos;i++)
{
    curr=temp;
    temp=temp->next;
    if (temp==NULL)
{
    printf("\n wrong position");
}
}
    curr->next=temp->next;
    printf("\n the deleted item is %d", temp-
```

```
>info);
    free(temp);
}
```

3. Search

As long as the linked list is not empty, we can search the list for the node containing particular value *e*. To accomplish this operation, we need two pointers, *prev* and *cur*. After completing the searching, the pointer *cur* will point to the node containing the value *e*, and *prev* will hold the previous node of *cur'*.

The following figures show the procedure step by step.

Step 1. Initializing pointer *cur* points to the first node of the list, while the pointer *prev* has a value of null. See the following figure.

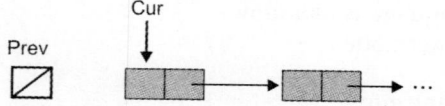

Step 2. Set *prev* point to the node pointed by *cur*. See the following figure.

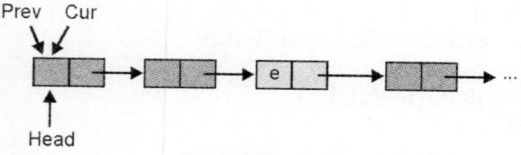

Step 3. Set *cur* point to the node after *prev'*. See the following figure.

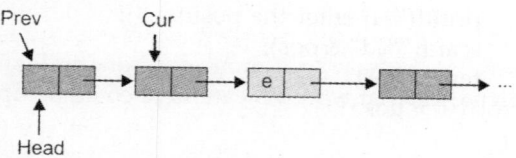

Step 4. Repeat step 2, step 3, until *cur* is pointing to the node that contains value. See the following figures.

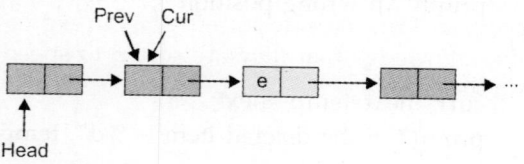

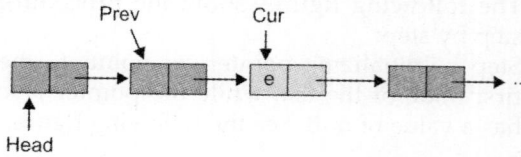

Procedure Search

{Procedure to search for each value in a linked list pointed by *head*. If found, *cur* should point to that node; otherwise, *cur* should point to the last node in the list. If *cur* points to the first node, *prev* should have the null value; otherwise, *prev* should point to the previous node. }

```
Procedure Search (head, cur, prev: pointer; e:
datatype);
begin
cur := head; {initialization}
prev := null;
while ( cur^.info != e) and (cur^.link != null)
do
begin
prev := cur;
cur := cur^.link
end;
end;
```

Implementation in C language

Implementation of the above procedure in C language is as follows:

```
struct node
{
    int info;
    struct node * next;
}*start=NULL;
void search()
{
    struct node * curr,*temp;
    int n,flag=0;
    if(start==NULL)
{
    printf("\n list is empty");
    getch();
    exit(1);
}

    printf("\n enter the number which you
want to search");
```

```
scanf("%",&n);
temp=start;
while(temp->next!=NULL)
{
    if(n==temp->info)
    {
    flag=1;
    break;
    }
    temp=temp->next;
}
    if(flag==1)
    printf("\n the given number is found in the
linked list");
    else
    printf("\n the given number is not found
in the linked list");
}
```

Sequential Search

The sequential search is best used if the array you are searching is unsorted. This method of searching is usually used on small arrays of less than 16 elements. We start the sequential search by first declaring a target to be found. The search initiates at the beginning of the array until it finds the target.

In the following example we will find a target value of 23 within a one dimensional array. At index 0, 32 is not equal to 23 so we proceed on to the next element.

a[0]	a[1]	a[2]	a[3]	a[4]
32	431	-34	23	12

At index 1, 431 is not equal to 23 so we proceed.

a[0]	a[1]	a[2]	a[3]	a[4]
32	431	-34	23	12

At index 2, -34 is not equal to 23 so we proceed.

a[0]	a[1]	a[2]	a[3]	a[4]
32	431	-34	23	12

Finally at index 3, 23 is equal to 23 and we have found our target.

a[0]	a[1]	a[2]	a[3]	a[4]
32	431	-34	23	12

Now we will implement this example of a sequential search into C code. The program below asks the user for a target to be found, then uses a for loop to analyze each element of the array. If the array element is equal to the target it will display that the target was found. Whenever a target is found the variable "flag" will be incremented by 1. At the end of the program if the variable "flag" is less than one, then the target was obviously not found.

```
#include <stdio.h>
#include<conio.h>
void main()
{
const int arraySize = 5;
int target,cntr;
int array[arraySize] = {32, 431, -34, 23, 12};
int flag;
// flag is used to log how many times the target
is encountered.
flag = 0;
printf( "Enter a target to be found: ");
scanf("%d",&target);
for(cntr = 0; cntr < arraySize; cntr++)
{
if (array[cntr] == target)
{
printf("Target found in array index %d",cntr);
flag += 1;
}
}
// Test to see if target was found.
if(flag < 1)
{
printf("\n the target not found");
}
}
```

The sequential search does have a pitfall. It is very slow and its performance rating is low. If a person had an array of one million elements, that would mean there could be up to one million comparisons, and that takes time! The sequential search method would be advisable to use only if the array you were searching was unsorted and small.

Binary Search

The binary search algorithm **can only be applied if the data are sorted**. You can exploit the knowledge that they are sorted to speed up the search.

The idea is analogous to the way people look up an entry in a dictionary or telephone book. You do not start at page 1 and read every entry! Instead, you turn to a page somewhere about where you expect the item to be. If you are lucky you find the item straight away. If not, you know which part of the book will contain the item (if it is there), and repeat the process with just that part of the book.

If you always split the data in half and check the middle item, you halve the number of remaining items to check each time. This is much better than linear search, where each unsuccessful comparison eliminates just one item.

Algorithm

The algorithm of binary search is as follows:
bottom := first element
top := last element
while ((top>=bottom) and (not found)) loop
mid := (top+bottom)/2
if (list(mid) = item to find) then
found := true
elsif (item to find > list(mid)) then
bottom := mid +1
else
top := mid - 1
end if
end loop
if found = true
wanted item is in database
else
wanted item is NOT in database
end if

Implementation in C language

The implementation of binary search in C language is as follows:

```
include<stdio.h>
#include<conio.h>
void main()
{
int a[10],n,i,j,temp;
int beg,end,mid,target;
clrscr();
printf("enter the array elements:? );
for(i=0;i<10;i++)
scanf("%d",&a[i]);
//Sorting process
for(i=0;i<n-1;i++)
{
for(j=0;j<n-i-1;j++)
{
if(a[j+1]<a[j])
{
temp=a[j];
a[j]=a[j+1];
a[j+1]=temp;
}
}
}
printf("the sorted numbers are:?);
for(i=0;i<n;i++)
printf("%4d",a[i]);
beg=a[0];
end=a[9];
mid=(beg+end)/2;
printf("\n enter the number to be searched:?);
scanf("%d",&target);
while(beg<=end && a[mid]!=target)
{
if(target<a[mid])
end=mid-1;
else
beg=mid+1;
mid=(beg+end)/2;
}
If(a[mid]==target)
{
printf("\n the number is found at position %2d",mid);
}
else
{
printf("\n the number is not found");
}
getch();
}
```

Example

The array is given and we want to search 123 value with the help of binary search method then it is illustrated as follows

Sorting is the process of putting data in order; either numerically or alphabetically.

It is often necessary to arrange the elements in an array in numerical order from highest to lowest values (descending order) or vice versa (ascending order). If the array contains

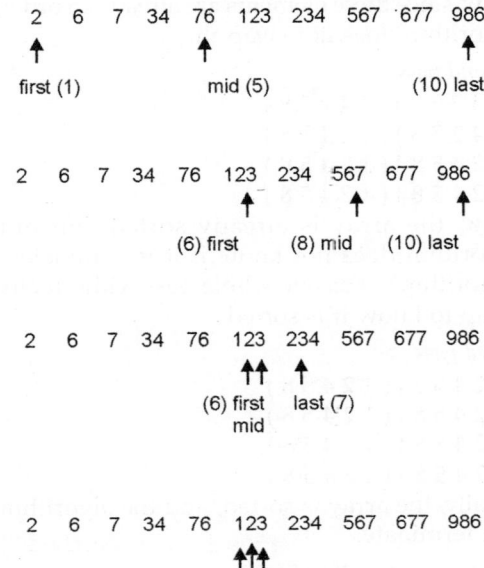

string values, alphabetical order may be needed (which is actually ascending order using ASCII values).

The process of sorting an array requires the exchanging of values. While this seems to be a simple process, a computer must be careful that no values are lost during this exchange. Consider the following dilemma:

Suppose that grade [1] = 10 and grade [2] = 8 and you want to exchange their values so that grade [1] = 8 and grade [2] = 10. You could **NOT** just do this:

grade[1] = grade[2];
grade [2] = grade [1]; // **DOES NOT WORK!!!**

In the first step, the value stored in grade[1] is erased and replaced with grade[2].

The result is that both grade[1] and grade[2] now have the same value. Oops! Then what happened to the value in grade[1]? It is lost!!! In order to swap two values, you must use a **third variable,**
(a "temporary holding variable"), to temporarily hold the value you do not want to lose:
//**swapping variables**
temp = grade[1]; // **holding variable**
grade[1] = grade[2];
grade[2] = temp;

This process successfully exchanges, "swaps", the values of the two variables (without the loss of any values).

WAYS TO SORT ARRAYS

There are literally hundreds of different ways to sort arrays. The basic goal of each of these methods is the same: to compare each array element to another array element and swap them if they are in the wrong position. The bubble sort is one of the easiest algorithms to understand and we will begin our in-vestigation with this sort.

Bubble Sort

Bubble sort is a simple sorting algorithm. It works by repeatedly stepping through the list to be sorted, comparing two items at a time and swapping them if they are in the wrong order. The pass through the list is repeated until no swaps are needed, which indicates that the list is sorted. The algorithm gets its name from the way smaller elements "bubble" to the top of the list. Because it only uses comparisons to operate on elements, it is a comparison sort.

Bubble sorting is a simple sorting technique in sorting algorithm. In bubble sorting algorithm, we arrange the elements of the list by forming pairs of adjacent elements. It means we repeatedly step through the list which we want to sort, compare two items at a time and swap them if they are not in the right order. Another way to visualize the bubble sort algorithm is as its name, the smaller element bubble to the top. Here is the source code implements bubble sorting algorithm in C which sorts an unordered list of integer.

The following program implements the concept of bubble sort for an array.

```c
#include <stdio.h>
#include <stdlib.h>
void swap(int *x,int *y)
{
int temp;
temp = *x;
*x = *y;
*y = temp;
```

```
}
void bublesort(int list[], int n)
{
int i,j;
for(i=0;i<(n-1);i++)
for(j=0;j<(n-(i+1));j++)
if(list[j] > list[j+1])
swap(&list[j],&list[j+1]);
}
void printlist(int list[],int n)
{
int i;
for(i=0;i<n;i++)
printf("%d\t",list[i]);
}
void main()
{
const int MAX_ELEMENTS = 10;
int list[MAX_ELEMENTS];
int i = 0;
// generate random numbers and fill them to
the list
for(i = 0; i < MAX_ELEMENTS; i++ ){
    list[i] = rand();
}
printf("The list before sorting is:\n");
printlist(list, MAX_ELEMENTS);
// sort the list
bublesort(list,MAX_ELEMENTS);
// print the result
printf("The list after sorting using bubble
sorting algorithm:\n");
printlist(list, MAX_ELEMENTS);
}
```

Step by step Example

Let us take the array of numbers "5 1 4 2 8", and sort the array from lowest number to greatest number using bubble sort algorithm. In each step, elements written in **bold** are being compared.

First pass

(**5 1** 4 2 8) (**1 5** 4 2 8)

Here, algorithm compares the first two elements, and swaps them.

(1 **5 4** 2 8) (1 **4 5** 2 8)
(1 4 **5 2** 8) (1 4 **2 5** 8)
(1 4 2 **5 8**) (1 4 2 **5 8**)

Now, since these elements are already in order, algorithm does not swap them.

Second pass

(**1 4** 2 5 8) (**1 4** 2 5 8)
(1 **4 2** 5 8) (1 **2 4** 5 8)
(1 2 **4 5** 8) (1 2 **4 5** 8)
(1 2 4 **5 8**) (1 2 4 **5 8**)

Now, the array is already sorted, but our algorithm does not know if it is completed. Algorithm needs one **whole** pass without **any** swap to know it is sorted.

Third pass

(**1 2** 4 5 8) (**1 2** 4 5 8)
(1 **2 4** 5 8) (1 **2 4** 5 8)
(1 2 **4 5** 8) (1 2 **4 5** 8)
(1 2 4 **5 8**) (1 2 4 **5 8**)

Finally, the array is sorted, and the algorithm can terminate.

Programming Problems

Problem 1

Write a program to input 10 numbers and then sort that numbers in ascending order.

Solution

```
#include<stdio.h>
#include<conio.h>
void main()
{
    int a[10],i,j,temp;
    clrscr();
    for(i=0;i<10;i++)
{
    printf("\n enter the element");
    scanf("%d",&a[i]);
}
for(i=0;i<10;i++)
{
for(j=i+1;j<10;j++)
{
    if(a[i]>a[j])
{
    temp=a[i];
    a[i]=a[j];
    a[j]=temp;
}
}
printf("\n the sorted array in ascending order
is");
```

```
for(i=0;i<10;i++)
{
    printf("\n%d",a[i]);
}
getch();
}
```

Problem 2

Write a program to input 10 numbers and then sort that numbers in descending order.

Solution

```
#include<stdio.h>
#include<conio.h>
void main()
{
    int a[10],i,j,temp;
    clrscr();
    for(i=0;i<10;i++)
{
    printf("\n enter the element");
    scanf("%d",&a[i]);
}
for(i=0;i<10;i++)
{
for(j=i+1;j<10;j++)
{
    if(a[i]<a[j])
{
    temp=a[i];
    a[i]=a[j];
    a[j]=temp;
}
}
printf("\n the sorted array in descending order is");
for(i=0;i<10;i++)
{
    printf("\n%d",a[i]);
}
getch();
}
```

SUMMARY

- The most important application of pointer is dynamic memory allocation.
- If we want to allocate the memory at the run time then this concept is known as dynamic memory allocation.

- malloc(), calloc, free and realloc() are the predefined functions for dynamic memory allocation.
- The general syntax of malloc() function is as follows:
  <pointer variable>=(<data type *) malloc (number of elements * sizeof(data type));
- The general syntax of calloc() function is as follows:
  <pointer variable>=(<data type *) malloc (number of elements, sizeof(data type));
- Stack is a data structure which follow the concept of last in first out (LIFO).
- Push is a function which is used to insert an element into the stack.
- Pop is a function which is used to delete the top most element from the stack.
- Linked list is the collection of nodes and every node is divided into two parts (a) information part, (b) address of the next node.
- Sequential search means we want to search the data linearly
- Binary search is only implemented when the data is sorted either in ascending order or in descending order.

Exercises

Section A

Multiple Choice Questions

1. We can allocate memory at run time by using
   (a) malloc
   (b) calloc
   (c) realloc
   (d) All of the above

2. In stack insertion and deletion can be done at
   (a) Top
   (b) Bottom
   (c) Both a and b
   (d) None of the above

3. The syntax of calloc function is
   (a) void *calloc(size_t size);
   (b) void *calloc(size_t numElements, size_t sizeOfElements);

(c) void *calloc(void*ptr,size_t size);

(d) All of the above

4. Push operation on a full stack cause

(a) Underflow

(b) Overflow

(c) Insertion of elements

(d) None of the the above

5. Different types of linked lists are

(a) Linear linked list

(b) Circular linked list

(c) Circular doubly linked list

(d) All of the above

6. push operation means

(a) To insert an element in the stack

(b) To insert the element in the queue

(c) To insert the element in the linked list

(d) None of the above

7. pop operation means

(a) To delete the top most element from the stack

(b) To delete the top most element from the queue

(c) To delete the element from the linked list.

(d) None of the above

8. LIFO property is followed by

(a) Stack

(b) Queue

(c) Linked List

(d) None of the above

9. malloc() function is predefined in

(a) malloc.h

(b) alloc.h

(c) Both a and b

(d) None of the above

10. calloc() function is predefined in

(a) calloc.h

(b) alloc.h

(c) Both a and b

(d) None of the above

State True/False

1. In static memory allocation compiler allocates memory for variable.

2. malloc function is used to allocate dynamic memory allocation.

3. While pushing an element we check whether a stack is empty or not.

4. push is a function which is used to insert the element to the top of the stack

5. pop is a function which is used to delete the top most element from the stack.

6. In dynamic memory allocation memory is allocated at the time of running of the program.

7. calloc function is predefined in alloc.h header file.

8. Sorting means to sort the data in ascending or descending order.

9. Binary search is only implemented when the data is sorted either in ascending order or in descending order.

10. Linked list is implemented with the help of pointer and malloc() function

Section B

Short Answer Type Questions

1. Define the concept of dynamic memory allocation with proper explanation.

2. Differentiate between static and dynamic memory allocation.

3. Define the concept of stack and how dynamic memory allocation implements with stack.

4. Define the concept of linked list and how dynamic memory allocation implements with linked list.

5. Write a program to push the element onto the stack with the help of array.

6. Write a program to pop the element from the stack with the help of array.

7. Write a program to push the element onto the stack with the help of linked list.

8. Write a program to pop the element from the stack with the help of linked list.

9. Write a program to insert the element at the beginning of the linked list.

10. Write a program to delete the element from the beginning of the linked list.

Section C

Long Answer Type Questions

1. Differentiate between static and dynamic memory allocation with suitable example.

2. Define the term searching and differentiate between linear search and binary search with suitable example.

3. Write a program to input 10 numbers and then sort that numbers in ascending order.

4. Write a program to input 10 numbers and then sort that numbers in descending order.

5. Define linked list with suitable example and also classify the linked list with suitable example.

6. Write a program to implement the stack with the help of linked list.

7. Write a program to input 10 names and then sort that names in ascending order.

8. Write a program to input 10 names and then sort that names in descending order.

9. Write a program to sort the linear linked list.

10. Write a program to search an element in the linear linked list.

16 Preprocessor Directives

Preprocessors in C as name signifies is something which is done before processing. Before your compiler compiles your code you can give your compiler various instruction on how to handle the source code and its compilation. Major aim of C preprocessor is to expand the scope of the programming environment and to make your code more portable and manageable. Also code developed using preprocessor directives tends to be more maintainable than normal because changes can be introduced with minimum fuss and side effects.

Theoretically, the "preprocessor" is a translation phase that is applied to the source code before the compiler gets its hands on it. The C preprocessor is not part of the compiler, but is a separate step in the compilation process. C preprocessor is just a text substitution tool, which filters your source code before it is compiled. The preprocessor more or less provides its own language, which can be a very powerful tool for the programmer. All preprocessor directives or commands begin with the symbol #.

The preprocessor makes programs easier to develop, read and modify. The preprocessor makes C code portable between different machine architectures and customizes the language.

The preprocessor performs textual substitutions on your source code in three ways:
1. *File inclusion:* Inserting the contents of another file into your source file, as if you had typed it all in there.
2. *Macro substitution:* Replacing instances of one piece of text with another.

3. *Conditional compilation:* Arranging that, depending on various circumstances, certain parts of your source code are seen or not seen by the compiler at all.

TYPES OF PREPROCESSORS

There are only handfull of preprocessor defined in Standard C and few of them work together in combination yet they allow you to achieve much more without much less effort. In addition, your compiler must be providing many more preprocessor to exploit your programming environment and programming potential to maximum but keep in mind they produce very specific targeted code when used which is not very portable.

Preprocessor can be broadly classified into following groups on the basis of functionality they provide.

1. **#define and #undef**—Used for defining and undefining MACROS.
2. **#error**—Used for debugging
3. **#include**—Used for combining source code files
4. **#if, #else, #else if, and #endif**—Used for conditional compilation similar to if-else statement.
5. **#ifdef and #ifndef**—Conditional Compilation on basis of #define and #undef
6. **#line**—Controls the program line and file macros.
7. **#pragma**—Used for giving compiler instruction. Highly specific to the compiler being used.

8. # **and** ##—Operators used for stringize and concating operation respectively.

As you can see all types of preprocesor begins with a #.

One important property is that a preprocessor must be in its own line, i.e. code below will produce an error—

#include <stdio.h> #include <iostream.h>

Also the preprocessor doesnot need a semicolon like other statement does the reason is that semicolon is placed after every executable statement but since preprocessor is not a executable statement but an instruction to the compiler hence they do not need a semicolon.

DEFINE TO IMPLEMENT CONSTANTS

#define is used to define constants.

The general syntax is as follows:

define <literal> <replacement-value>

Where literal is identifier which is replaced with *replacement-value* in the program.

Example

The preprocessor allows us to customize the language. For example to replace {and} of C language to *begin* and *end* as block-statement delimiters (as like the case in PASCAL) we can achieve this by writing:

define begin {

define end }

During compilation all occurrences of *begin* and *end* get replaced by corresponding {and}. So the subsequent C compilation stage does not know any difference!

The syntax of defining a constant with the help of # *define* is as follows:

define <constant-name> <replacement-value>

Example

#define MAXSIZE 256

#define PI 3.142857

The C preprocessor simply searches through the C code before it is compiled and replaces every instance of *MAXSIZE* with 256.

define FALSE 0

define TRUE !FALSE

The literal *TRUE* is substituted by! FALSE and *FALSE* is substituted by the value 0 at every occurrence, before compilation of the program.

Since, the values of the literal are constant throughout the program, they are called as constant.

Illustrated Examples

define M 10

define SUBJECTS 10

define PI 3.142857

define COUNTRY INDIA

Note that no semicolon (;) need to be placed as the delimiter at the end of a # define line. This is just one of the ways that the syntax of the preprocessor is different from the rest of C statements (commands). If you unintentionally place the semicolon at the end as below:

#define MAXLINE 100; /* WRONG */

And if you declare as shown below in the declaration section,

char line [MAXLINE];

The preprocessor will expand it to:

char line[100;]; /* WRONG */

Which gives you the syntax error? This shows that the preprocessor does not know much of anything about the syntax of C.

DEFINE TO CREATE FUNCTIONAL MACROS

Macros are inline code, which are substituted at compile time. The definition of a macro is that which accepts an argument when referenced.

Problem 1

Write a program to find the square of a given number using macro.

Solution

```
/* Program to find the square of a number
using macro*/
#include <stdio.h>
#include<conio.h>
# define SQUARE(x) (x*x)
void main()
{
int v,y;
clrscr();
printf("Enter any number to find its square:
");
scanf("%d", &v);
y = SQUARE(v);
printf("\nThe square of %d is %d", v, y);
```

getch();
}
The output of the above code segment is as follows:

Enter any number to find its square: 10
The square of 10 is 100

The explanation of the above code segment is as follows:

In this case, v is equated with x in the macro definition of *square*, so the variable y is assigned the square of v. The brackets in the macro definition of *square* are necessary for correct evaluation. The expansion of the macro becomes: $y = (v * v)$;

Note

Macros can make long, ungainly pieces of code into short words. Macros can also accept parameters and return values. Macros that do so are called *macro functions*. To create a macro, simply define a macro with a parameter that has whatever name you like, such as *my_val*. For example, one macro defined in the standard libraries is= "abs", which returns the absolute value of its parameter. Let us define our own version of *ABS* as shown below. Note that we are defining it in upper case not only to avoid conflicting with the existing "abs".
#define ABS(my_val) ((my_val) < 0)? -(my_val): (my_val)
#define can also be given arguments which are used in its replacement. The definitions are then called macros. Macros work rather like functions, but with the following minor differences:

- Since macros are implemented as a textual substitution, by this the performance of program improves compared to functions.

- Recursive macros are generally not a good idea.

- Macros do not care about the type of their arguments. Hence macros are a good choice where we want to operate on reals, integers or a mixture of the two. Programmers sometimes call such type flexibility polymorphism.

- Macros are generally fairly small

Examples Based on Macros

Example 1

Write a program to declare constants and macro functions using #*define*

Solution

```
/* Program to illustrate the macros */
#include <stdio.h>
#include<conio.h>
#define STR1 "A macro definition!\n"
#define STR2 "must be all on one line!\n"
#define EXPR1 1+2+3
#define EXPR2 EXPR1+5
#define ABS(x) (((x) < 0)? - (x):(x))
#define MAX (p,q) ((p < q)? (q):(p))
#define BIGGEST(p,q,r) (MAX(p, q) < r)?(r):
(MAX(p, q))
void main()
{
clrscr();
printf(STR1);
printf(STR2);
printf("Largest number among %d, %d and
%d is %d\n",EXPR1, EXPR2, ABS (-3),
BIGGEST(1,2,3));
getch();
}
```

The output of the above code is as follows:
.

A macro definition
must be all on one line!
Largest number among 6, 11 and 3 is 3

The explanation of the above code segment is as follows:

The macro STR1 is replaced with "A macro definition \n" in the first *printf ()* function. The macro STR2 is replaced with "must be all on one line! \n" in the second *printf* functions. The macro EXPR1 is replaced with 1+2+3 in third *printf* statement. The macro EXPR2 is replaced with EXPR1 +5 in fourth *printf* statement. The macro ABS (–3) is replaced with (– 3<0)? – (– 3): 3. And evaluation 3 is replaced. The largest among the three numbers is displayed.

Example 2

Write a program to find out square and cube of any given number using macros.

Solution

```
# include<stdio.h>
#include<conio.h>
# define sqr(x) (x * x)
# define cub(x) (sqr(x) * x)
void main()
{
int num;
clrscr();
printf("Enter a number: ");
scanf("%d", &num);
printf(" \n Square of the number is %d",
sqr(num));
printf(" \n Cube of the number is %d\n",
cub(num));
getch();
}
```

The output of the above code segment is as follows:

Enter a number: 4
Square of the number is 16
Cube of the number is 64

Note

Multi-line macros can be defined by placing a backward slash (\ ) at end of each line except the last line. This feature permits a single macro (i.e. a single identifier) to represent a compound statement.

Example 3

Write a macro to display the string HELLO in the following fashion.

```
H
HE
HEL
HELL
HELLO
HELLO
HELL
HEL
HE
H
```

Solution

```
/* Program to display the string as given in
the problem*/
# include<stdio.h>
# define LOOP for(x=0; x<5; x++) \
{y=x+1; \
printf("%-5.*s\n", y, string); } \
for(x=4; x>=0; x—) \
{
y=x+1; \
printf("%-5.*s \n", y, string);}
void main()
{
int x, y;
clrscr();
static char string[ ] = "HELLO";
printf("\n");
LOOP;
getch();
}
```

When the above program is executed the reference to macro (loop) is replaced by the set of statements contained within the macro definition.

The output of the above code segment is as follows:

```
H
HE
HEL
HELL
HELLO
HELLO
HELL
HEL
HE
H
```

Caution in using macros

You should be very careful in using macros. In particular, the textual substitution means that arithmetic expressions are liable to be corrupted by the order of evaluation rules (precedence rules). Here is an example of a macro, which will not work.

#define DOUBLE (n) n + n

Now if we have a statement,

z = DOUBLE (p) * q;

This will be expanded to

z = p + p * q;

And since * has a higher priority than +, the compiler will treat it as.

z = p + (p * q);

The problem can be solved using a more robust definition of DOUBLE

#define DOUBLE (n) (n + n)

Here, the braces around the definition force the expression to be evaluated before any surrounding operators are applied. This should make the macro more reliable.

#UNDEF

This preprocessor is used to undefine a preprocessor previously defined preprocessor to prevent bugs and misuse. Also they are necessary for proper use of #ifdef and #ifndef preprocessor. This preprocessor in other word is used for localization of MACROs to only to the section of code which requires them. Once undefined the macro cannot be no longer referenced in code after #undef statement.

General form of #undef preprocessor is as follows:
#undef MACRO_NAME
The following example shows the working of #undef.
#include <stdio.h>
#include<conio.h>
#define PIE 3.14
#define AREA(x) (PIE)*(x)*(x)
#undefined PIE //PIE cannot be longer used but Area can be.
void main ()
{
clrscr();
printf ("%f", AREA (10)); //Prints 314.159265
... as answer.
getch();
}

#ERROR

The use of this preprocessor is in debugging. Whenever this preprocessor is encountered during compilation the compiler would stop compilation and display the error message associated with the preprocessor in compilation output/result Window. Depending upon compiler any other debugging information will also accompany your error message.

The general syntax of #error is as follows:
#error message
Also note that message does not need to be in double quotes.

The following example shows the working of #error.
#include <stdio.h>
#include<conio.h>
void main ()
{
clrscr();
#error I am error and while i am here i will not let this program compile
getch();
}

READING FROM OTHER FILES USING # INCLUDE

The preprocessor directive #include is an instruction to read in the entire contents of another file at that point. This is generally used to read in header files for library functions. Header files contain details of functions and types used within the library. They must be included before the program can make use of the library functions.

The general syntax of #include is as follows:
#include <filename.h>
Or
#include "filename.h"
The above instruction causes the contents of the file "filename.h" to be read, parsed, and compiled at that point. The difference between the suing of # and " " is that, where the preprocessor searches for the *filename.h*. For the files enclosed in < > (less than and greater than symbols) the search will be done in standard directories (include directory) where the libraries are stored. And in case of files enclosed in " " (double quotes) search will be done in "current directory" or the directory containing the source file. Therefore, " " is normally used for header files you have written, and # is normally used for headers which are provided for you (which someone else has written). Library header file names are enclosed in angle brackets, < >. These tell the preprocessor to look for the header file in the standard location for library definitions.

This is */usr/include* for most UNIX systems. And *c:/tc/include* for turbo compilers on DOS/WINDOWS based systems. Use of #include for the programmer in multi-file programs,

where certain information is required at the beginning of each program file. This can be put into a file by name "globals.h" and included in each program file by the following line:
#include "globals.h"

If we want to make use of inbuilt functions related to input and output operations, no need to write the prototype and definition of the functions. We can simply include the file by writing
#include <stdio.h>
And call the functions by the function calls. The standard header file *stdio.h* is a collection of function prototype (declarations) and definition related to input and output operations.

The extension ".h"', simply stands for "header" and reflects the fact that #include directives usually sit at the top (head) of your source files. ".h" extension is not compulsory–you can name your own header files anything you wish to, but .h is traditional, and is recommended. Placing common declarations and definitions into header files means that if they always change, they only have to be changed in one place, which is a much more feasible system.

You can put the following in the header files
- External declarations of global variables and functions.
- Structure definitions.
- Typedef declarations.

However, there are a few things *not* to put in header files:
- Defining instances of global variables. If you put these in a header file, and include the header file in more than one source file, the variable will end up multiply defined.
- Function bodies (which are also defining instances), may not be put in header files. Since these headers may end you up with multiple copies of the function and hence "multiply defined" errors. People sometimes put commonly-used functions in header files and then use #include to bring them (once) into each program where they use that function, or use #include to bring together the several source files making

up a program, but both of these are not good practice. It is much better to learn how to use your compiler or linker to combine together separately compiled object files.

#IF,#ELSE,#ELIF AND #ENDIF

#if, #else, #elif and #endif are discussed here together because they are almost always used together. They are called conditional compilation preprocessor because they allow you to selectively compile your code on basis of a bool condition. Condition here must be an constant expression because the condition is evaluate at compile and not runtime so all values in expression should be properly defined before compilation as a result variables cannot be used in the expression.

Now we evaluate the function these preprocessors one by one in detail.

The general form of using #if statement
#if condition
statement1
statement2
.
.

Statement N.
#endif
If condition specified in first statement evaluates to be true then the statements between #if and #endif are compiled. Here #endif is used to mark the end of #if block.

C source code below shows the use of #if preprocessor compiler.
#include <stdio.h>
#include<conio.h>
#define MACRO 0
void main ()
{
clrscr();
#if MACRO == 0
printf ("MACRO is false");
#endif
getch();
}

#else preprocessor directive is counterpart of #if preprocessor similar to else keyword is counter part of if keyword in conditional statements. It provides the code which compiles when the associative #if condition fails. In other words, it provides an alternative compilation path.

General form of using #else preprocessor is #if condition.

```
statement1
statement2
.
.
.
statementN.
#else
Statement N+1
Statement N+2
.
.
.
Statement N+M.
#endif
```

Following C source code shows the use of #else preprocessor.

```
#include <stdio.h>
#include<conio.h>
#define MACRO 1
void main ()
{
clrscr();
#if MACRO == 0
printf ("MACRO is false");
#else
printf ("MACRO is true");
#endif
getch();
}
```

#elif preprocessor can be expanded to #else if and is used to construct a structure similar to if-else ladder. The structure it forms allows multiple compilation option and the control flow keeps on falling on next step in ladder till the conditions are evaluated to be false but if any condition evaluates to be true rest of the steps in ladder are bypassed on the same level. Also different levels can be produced using nesting of ladders. 8 levels in C89 and 63 levels in C99 are minimum supported but like always compilers support many more levels. Also for efficiency criteria and fast

compilation you may want to limit the levels of nesting.

If at lowest step in the ladder a #else preprocessor is present then the code associated with it compiles if all condition fails. Also like always #endif preprocessor is used to mark the end of the #elif block.

General form of using #elif preprocessor is as follows

```
#if condition
statement1
statement2
.
.
.
StatementN.
#elseif
statementN+1
statementN+2
.
.
.
statementN+M.
.
.
.
. ... A levels
#else
Statement N+M+A+1
Statement N+M+A+1
.
.
Statement N+M+A+Z
#endif
```

The C source code given below shows the use of #elif preprocessor.

```
#include <stdio.h>
#include<conio.h>
#define MACRO 4
void main ()
{
clrscr();
#if MACRO == 0
printf ("MACRO has value 0");
#elseif MACRO == 1
printf ("MACRO has value 1");
#elseif MACRO == 2
```

```
printf ("MACRO has value 2");
#elseif MACRO == 3
printf ("MACRO has value 3");
#else
printf ("Macro value is not 1, 2 or 3 "); // This
statement will compile
#endif
getch();
}
```

#endif as you can see from above is use to mark of end of #if, #else and #else of blocks. It is also used to mark end of #ifdef and #ifndef given in the next section. One important property of #endif you must know that it always associate if the nearest conditional compilation preprocessor. This property is especially important when you create multiple levels using nesting feature provided in C preprocessor.

#IFDEF AND #IFNDEF

#ifdef and #ifndef are also a type of conditional compilation preprocessor but with a little twist. They work on #ifdef and #ifndef preprocessors and can be used in combination with #else and #elif preprocessor.

#ifdef stands for if defined and #ifndef for if not defined. The condition they work on is whether a MACRO is defined or not defined. Code specified by ifdef compiles if the following MACRO is defined while code specified by the #ifndef compiles if the following MACRO is not defined. Both the preprocessors use the preprocessor #endif to mark the end of the conditional compilation code.

Both #ifdef and #ifndef can be nested similar to if-else statments. It is specified by standard C that any compiler conforming to standard C must provide a minimum of at least 8 levels of minimum nesting while C99 specifies that a minimum of 63 levels of nesting must be supported.

#ifdef can also be written as #if defined and both forms are equivalent.

General form of #ifdef preprocessor is as follows:

```
#ifdef MACRO
statement1
```

```
statement2
.
.
.
StatementN.
#endif
```

General form of #ifndef preprocessor is as follows:

```
#ifndef MACRO
statement1
statement2
.
.
.
StatementN.
#endif
```

The source code belows shows the purpose of #ifdef and #ifndef preprocessor.

```
#include <stdio.h>
#include<conio.h>
#define MYMACRO 100
void main ()
{
clrscr();
#ifdef MYMACRO
printf ("MYMACRO is defined"); // This line
will compile
#else
printf ("MYMACRO is not defined");
#endif
#ifndef OURMACRO
printf ("OURMACRO is not defined"); // This
line will compile
#else
printf ("OURMACRO is defined");
#endif
#if define EVERYBODYSMACRO
printf ("EVERYBODYSMACRO is defined");
#else
printf ("EVERYBODYSMACRO is not defined");
// This line will compile
getch();
}
```

#LINE

#line preprocessor is used to change the values of 2 MACROS, _ _LINE _ _ and _ _ FILE _ _.
_ _ LINE _ _ - This a dynamic macro which stores the line number of the line presently

being compiled. Value is constantly updated as compiler moves forward.

_ _ FILE _ _ - This macro stores the file name of the source file being compiled.

General form of #line preprocessor is

#line number "filename"

Here the file name is optional. Filename string replaces the string value of _ _FILE_ _ while the number changes the value of _ _LINE_ _.

The major use of #line is in debugging and rare programming situation.

Following C source code shows the working of #line preprocessor.

```
#include <stdio.h>
#include<conio.h>
void main ()
{
printf ("\n%d", __LINE__); //Prints 6
#line 100;
printf ("\n%d",__LINE__); // Prints 101
printf ("\n%d", __FILE__);// Prints original
source file name
#line 103 "Super C"
printf ("\n%d", __FILE__); //Prints Super C
getch();
}
```

#PRAGMA

#pragma is completely compiler defined and varies almost completely from compiler to compiler. Standard C only state this pre-processor but neither defines nor support its use. Compiler is free to define its how and what.

Also C99 supports a new operator _Pragma which provides an alternative to #pragma.

AND ## PREPROCESSOR OPERATOR

and ## are unique in sense that they are not preprocessor but operators acting on pre-processors. They are a bit different in use and application and are used very rarely allowing preprocessor to handle very rare situations.

is called stringsize operator and turns the argument it precede into a quoted string. Use of # is shown in the C source code given below and should be properly studied.

```
#include <stdio.h>
#include<conio.h>
```

```
#define createstring (s) #s
void main ()
{
clrscr();
printf (createstring (C is Easy);// Interpreted
as printf ("C is Easy");
getch();
}
```

is called the pasting operator which is used to concates two tokens. Use of ## is shown in the source code shown below

```
#include <stdio.h>
#include<conio.h>
#define combine (Part1, Part2) Part1##Part2
void main ()
{
int variable = 400;
clrscr();
printf ("%d", combine (vari, able));// Interpreted
as printf ("%d",variable);
getch();
}
```

PREDEFINED MACROS IN C

As you know from the above, macros are pre-defined words which serves special purposes and are created using #define preprocessor. Along with the preprocessor you define yourself every compiler has tons of inbuilt macro you can use. Standard C defines and supports 10 very usefull inbuilt Marcos, 5 of them was added by C99.

These macros are:

(a)_ _LINE_ _

This a dynamic macro which stores the line number of the line presently being compiled. Value is constantly updated as compiler moves forward.

(b) _ _FILE_ _

This macro stores the file name of the source file being compiled.

(c) _ _DATE_ _

This macro stores the date of compilation in month/day/year format.

(d) _ _ TIME _ _

This macro stores the time of compilation in hour:minute:second format.

(e) _ _STDC_ _

This is a bool macro which have value 1 if the compiler is in accordance with standard C, 0 otherwise.

(f) _ _STDC_HOSTED_ _

This is also a bool macro which has value 1 if operating system is present, 0 if not. This was added by C99.

(g) _ _STDC_VERSION_ _

This is stores the value indicating the version of compiler. It was added by C99 and has minimum value 199901 indicating C99 version 1.

(h) _ _STDC_IEC599_ _

Also added by C99 it is a bool macro is 1 if IEC 60059 floating point arithmetic is supported by compiler, 0 if not.

(i) _ _ STDC_IEC599_Complex_ _

Introduced by C99 it is a bool macro is 1 if IEC 60059 complex arithmetic is supported by compiler, 0 if not.

(j) _ _ STDC_ISO_10646_ _

Stores Value representing the year and month of the the ISO/IEC 10646 specification supported by compiler.

Also note that some values may be larger than that supported by integer so a long integer should be used in the respective case.

C source code below shows possible use of above macros and should possibly compiled by C99 compiler-

```
#include <stdio.h>
#include<conio.h>
void main ()
{
clrscr();
printf ("\nYou are at line : %d" __LINE__);
printf ("\nFile being compiled is : %s" __FILE__);
printf ("\nDate of Compilation is : %ld" __DATE__);
printf ("\nTIME of Compilation is : %ld" __TIME__);
if (__STDC__)
{
printf ("\nCompiler used conforms to standard C");
}
else
{
printf ("\nCompiler used does not conforms to standard C");
}
printf ("\nCompiler version is %ld", __STDC_VERSION__);
if (__STDC_IEC599__)
{
printf ("\nIEC 60059 floating point arithemetic is supported by compiler");
}
else
{
printf ("\nIEC 60059 floating point arithemetic is not supported by compiler");
}
if (__STDC_IEC599_Complex__)
{
printf ("\nIEC 60059 complex arithemetic is supported by compiler");
}
else
{
printf ("\nIEC 60059 complex arithemetic is not supported by compiler");
}
printf ("\nDate of the ISO/IEC 10646 specification supported by compiler is %ld",__ __STDC_ISO_10646__);
getch();
}
```

MACROS vs FUNCTIONS

Till now we have discussed about macros. Any computations that can be done on macros can also be done on functions. But there is a difference in implementations and in some cases it will be appropriate to use macros than function and vice versa. We will see the difference between a macro and a function now.

- Macro calls are replaced with macro-expansions (meaning). In function call, the control is passed to a function definition along with arguments, and definition is processed and value may be returned to call
- Macros run programs faster but increase the program size. Functions make program size smaller and compact.
- If macro is called 100 numbers of times, the size of the program will Increase. If function is called 100 numbers of times, the program size will not increase.
- It is better to use macros, when the definition is very small in size. It is better to use functions, when the definition is bigger in size.

SUMMARY

- Preprocessor directives are lines included in the code of our programs that are not program statements but directives for the preprocessor.
- We can define the constant with the help of #define preprocessor directive.
- We can include the header files with the help of #include preprocessor directive.
- The preprocessor directives enable the programmer to write programs that are easy to develop, read, modify and transport to a different computer system.
- We can make use of various preprocessor directives such as **#define**, **#include**, **#ifdef** - **#else** - **#endif**, **#if** and **#elif** in our program.
- The directives like **#undef** and **#pragma** are also useful although they are seldom used.

Exercises

Section A

Multiple Choice Questions

1. Macros can
   (a) Be defined using #define
   (b) Lasts until undefined #undef
   (c) Accepts # and ##
   (d) All of the above

2. Conditional compilations can be performed using
   (a) #ifdef   (b) #ifndef
   (c) #elif   (d) All of the above

3. #elif directive can
   (a) Allow a section of program to be compiled only if the macro has been previously defined
   (b) Serve to specify some condition to be met in order for the portion of code they surround
   (c) Both a and b
   (d) None of the above

4. #line allows us to
   (a) Line number within code files
   (b) File name that appears when error takes place
   (c) Both a and b
   (d) None of the above

5. Preprocessor directive means
   (a) A message from compiler to the programmer
   (b) A message from compiler to the linker
   (c) A message from programmer to the preprocessor
   (d) A message from programmer to the microprocessor

6. Which of the following are correctly formed **#define** statements:
   (a) #define INCH PER FEET 12
   (b) #define SQR (X) (X * X)
   (c) #define SQR(X) X * X
   (d) #define SQR(X) (X * X)

7. A header file is:
   (a) A file that contains standard library functions
   (b) A file that contains definitions and macros
   (c) A file that contains user-defined functions
   (d) A file that is present in current working directory

8. Which of the following is not a pre-processor directive
   (a) #if   (b) #elseif
   (c) #undef   (d) #pragma

9. All macro substitutions in a program are done
   (a) Before compilation of the program
   (b) After compilation
   (c) During execution
   (d) None of the above

10. In a program the statement:
    #include "filename" is replaced by the contents of the file "filename"
    (a) Before compilation
    (b) After compilation
    (c) During execution
    (d) None of the above

State True/False

1. Preprocessor directives begins with #.
2. Preprocessor directives is terminated with the help of;
3. #include is a preprocessor directive which is used to include any header file in program.
4. #line is a preprocessor directive.
5. _ _LINE_ _ is a predefine macro in C language.
6. _ _FILE_ _ is a predefine macro in C language.
7. The preprocessor is executed before the actual compilation begins.
8. A macro must always be written in capital letters.
9. A macro should always be accommodated in a single line.
10. Macros with arguments are not allowed.

Section B

Short Answer Type Questions

1. How many **#include** directives can be there in a given program file?
2. What is the difference between the following two **#include** directives:
   #include "conio.h"
   #include <conio.h>

3. What would be the output of the following program segment?
```c
void main( )
{
int i = 2 ;
#ifdef DEF
i *= i ;
#else
printf ( "\n%d", i ) ;
#endif
}
```

4. What would be the output of the following program segment?
```c
#define PRODUCT(x) ( x * x )
void main( )
{
int i = 3, j ;
j = PRODUCT( i + 1 ) ;
printf ( "\n%d", j ) ;
}
```

5. What would be the output of the following program segment?
```c
#define PRODUCT(x) ( x * x )
void main( )
{
int i = 3, j, k ;
j = PRODUCT( i++ ) ;
k = PRODUCT ( ++i ) ;
printf ( "\n%d %d", j, k ) ;
}
```

6. What would be the output of the following program segment?
```c
# define SEMI ;
void main()
{
int p = 3 SEMI ;
printf ( "%d", p ) SEMI
}
```

7. Define the concept of preprocessor directive in C programming language.

8. Define the various predefined pre-processor directive used in C language.

9. Define the general syntax of #line preprocessor directive.

10. Define the working of #line preprocessor directive.

Section C

Long Answer Type Questions

1. Write down a macro to test whether a character entered is a small case letter or not.

2. Write down a macro to test whether a character entered is a upper case letter or not.

3. Write down a macro to test whether a character is an alphabet or not.

4. Write down a macro to input two numbers and print the maximum number between them.

5. Write down a macro to input three numbers and print the maximum number between them.

6. Write down a macro to input two numbers and print the minimum number between them.

7. Write down a macro to input three numbers and print the minimum number between them.

8. Write macro definitions with arguments for calculation of area and perimeter of a triangle, a square and a circle. Store these macro definition in a file called "areaperi.h". Include this file in your program, and call the macro definitions for computing area and perimeter for different squares, triangles and circles.

9. Write down macro definition to find arithmetic mean of two numbers.

10. Write down a macro definition to find absolute value of a number.

17   File Handling in C

The examples we have seen so far in the previous chapters deal with standard input and output. When data is stored using variables, the data is lost when the program exits unless something is done to save it. This chapter discusses methods of working with files, and a data structure to store data. C views file simply as a sequential stream of bytes. Each file ends either with an *end-of-file* marker or at a specified byte number recorded in a system maintained, administrative data structure. C supports two types of files called *binary files* and *text files*.

The difference between these two files is in terms of storage. In *text files*, everything is stored in terms of text, i.e. even if we store an integer 54; it will be stored as a 3-byte string- "54\0". In a text file certain character translations may occur. For example a *new line(\n)* character may be converted to a carriage return, linefeed pair. This is what Turbo C does. Therefore, there may not be one to one relationship between the characters that are read or written and those in the external device. A *binary file* contains data that was written in the same format used to store internally in the main memory.

For example, the integer value 1245 will be stored in 2 bytes depending on the machine while it will require 5 bytes in a text file. The fact that a numeric value is in a standard length makes binary files easier to handle. No special string to numeric conversions is necessary.

The disk I/O in C is accomplished through the use of library functions. The ANSI standard, which is followed by Turbo C, defines one complete set of I/O functions. But since originally C was written for the UNIX operating system, UNIX standard defines a second system of routines that handles I/O operations. The first method, defined by both standards, is called a buffered file system. The second is the unbuffered file system.

In this chapter, we will first discuss buffered file functions and then the unbuffered file functions in the following sections.

FILE

Abstractly, a file is a collection of bytes stored on a secondary storage device, which is generally a disk of some kind. The collection of bytes may be translated, for example, as characters, words, lines, paragraphs and pages from a textual document; fields and records belonging to a database; or pixels from a graphical image. The meaning accompanied a particular file is determined entirely by the data structures and operations used by a program to process the file. It is conceivable (and it sometimes happens) that a graphics file will be read and displayed by a program designed to process textual data. The result is that no significant output occurs (probably) and this is to be expected. A file is simply a machine readable storage media where programs and data are stored for machine usage.

CLASSIFICATION OF FILE

Essentially there are two kinds of files that programmers deal with text files and binary files.

Text Files

A text file can be a stream of characters that a computer can process sequentially. It is not only treated sequentially but only in forward direction. For this reason a text file is normally opened for only one kind of operation (reading, writing, or appending) at any given time.

Likewise, since text files only process characters, they can only read or write data one character at a time. (In C Programming Language, functions are allowed for that deal with lines of text, but these still essentially process data one character at a time.) A text stream in C is a particular kind of file. Depending on the demands of the operating system, newline characters may be converted to or from carriage-return/linefeed combinations depending on whether data is being written to, or read from, the file. Other character transitions may also occur to satisfy the storage requirements of the operating system. These translations occur transparently and they occur because the programmer has indicated the intention to process a text file.

Binary Files

A binary file is no different to a text file. It is a *accumulation* of bytes. In C Programming Language a byte and a character are same. Hence a binary file is also mentioned to as a character stream, but there are two crucial differences.

1. No special processing of the data occurs and each byte of data is changed to or from the disk unprocessed.

2. C Programming Language places no constructs on the file, and it may be read from, or written to, in any manner chosen by the computer programmer.

Binary files can be either treated sequentially or, depending on the needs of the application, they can be processed using random access techniques. In C Programming Language, processing a file using random access techniques needs moving the current file position to an appropriate place in the file before reading or writing data. This indicates a second characteristic of binary files—they a generally processed using read and write operations simultaneously.

For example, a database file will be produced and treated as a binary file. A record update operation will involve locating the appropriate record, reading the record into memory, changing it in some way, and finally writing the record back to disk at its appropriate location in the file. These kinds of operations are usual to many binary files, but are rarely found in applications that process text files.

FILE HANDLING IN C USING FILE POINTERS

We know that a sequential stream of bytes ending with an *end-of-file* marker is what is called a *file*. When the file is opened the stream is associated with the file. By default, three files and their streams are automatically opened when program execution begins—the *standard input*, *standard output*, and the *standard error*. Streams provide communication channels between files and programs.

For example, the standard input stream enables a program to read data from the keyboard, and the standard output stream enables to write data on the screen. Opening a file returns a pointer to a FILE structure (defined in <stdio.h>) that contains information, such as size, current file pointer position, type of file, etc. to execute operations on the file. This structure also contains an integer called a *file descriptor* which is an index into the table maintained by the operating system namely, the *open file table*. Each element of this table contains a block called *file control block (FCB)* used by the operating system to administer a particular file.

The standard input, standard output and the standard error are manipulated using file pointers *stdin, stdout* and *stderr*. The set of functions which we are now going to discuss come under the category of buffered file system. This file system is referred to as buffered because, the routines maintain all the disk buffers required for reading/writing automatically.

To access any file, we need to declare a pointer to FILE structure and then associate it with the particular file. This pointer is referred as a *file pointer* and the general syntax of declaring a file pointer is as follows:
FILE *fp;

Open a File Using the Function *fopen()*

Once a file pointer variables has been declared, the next step is to open a file. The *fopen()* function opens a stream for use and links a file with that stream. This function returns a file pointer.

The general syntax of fopen function is as follows:
FILE *fopen(char *filename,*mode);
where *mode* is a string, containing the desired open status. The filename must be a string of characters that provide a valid file name for the operating system and may include a path specification.

The following table represents the various modes used in File Handling:

Mode	Meaning
"r" / "rt"	opens a text file for read only access
"w" / "wt"	creates a text file for write only access
"a" / "at"	text file for appending to a file
"r+t"	open a text file for read and write access
"w+t"	creates a text file for read and write access
"a+t"	opens or creates a text file and read access
"rb"	opens a binary file for read only access
"wb"	create a binary file for write only access
"ab"	binary file for appending to a file
"r+b"	opens a binary or read and write access
"w+b"	creates a binary or read and write access,
"a+b"	open or binary file and read access

If the file has been successfully opened the function will return a pointer to a FILE object that is used to identify the stream on all further operations involving it. Otherwise, a null pointer is returned.

Consider the following statements.
FILE *p1, *p2;
p1=fopen("data","r");
p2=fopen("results","w");
In these statements the p1 and p2 are created and assigned to open the files data and results respectively the file data is opened for reading and result is opened for writing. In case the results file already exists, its stuffing are deleted and the files are opened as a new file. If data file does not exist error will occur. The following code fragment explains how to open a file for reading.

```
#include <stdio.h>
#include<conio.h>
#include<stdlib.h>
void main ()
{
FILE *fp;
clrscr();
if ((fp=fopen("file1.dat", "r"))==NULL)
{
printf("FILE DOES NOT EXIST\n");
exit(0);
}
getch();
}
```

The value returned by the *fopen( )* function is a file pointer. If any error occurs while opening the file, the value of this pointer is *NULL*, a constant declared in *<stdio.h>*. Always check for this possibility as shown in the above example.

Close a File Using the Function fclose( )

When the processing of the file is finished, the file should be closed using the fclose() function.

The general syntax of fclose() function is as follows:
int fclose(FILE *fptr);

This function flushes any unwritten data for stream, discards any unread buffered input, frees any automatically allocated buffer, and then closes the stream. The return value is 0 if the file is closed successfully or a constant *EOF*, an end-of file marker, if an error occurred. This constant is also defined in *<stdio.h>*. If the function *fclose()* is not called explicitly, the operating system normally will close the file when the program execution terminates. Observe the following code segment.

....

```
FILE *p1 *p2;
p1=fopen ("Input","w");
p2=fopen ("Output","r");
....

...
fclose(p1);
fclose(p2)
```

The above program opens two files and closes them after all operations on them are completed, once a file is closed its file pointer can be reversed on other file.

The following code fragment explains how to close a file.

```
# include <stdio.h>
#include<conio.h>
void main ( )
{
FILE *fp;
if ((fp=fopen("file1.dat", "r"))==NULL)
{
printf("FILE DOES NOT EXIST\n");
exit(0);
}
................
................
................
................
/* close the file */
fclose(fp);
}
```

Once the file is closed, it cannot be used further. If required it can be opened in the same or another mode.

INPUT AND OUTPUT USING FILE POINTERS

After opening the file, the next thing needed is the way to read or write the file. There are several functions and macros defined in *<stdio.h>* header file for reading and writing the file. These functions can be categorized according to the form and type of data read or written on to a file.

These functions are classified as follows

- Character input/output functions
- Integer input/output functions

- String input/output functions
- Formatted input/output functions
- Block input/output functions.

Character Input and Output in Files

ANSI C provides a set of functions for reading and writing character by character or one byte at a time. These functions are defined in the standard library. They are listed and described as follows:

- getc()
- putc()

getc( ) is used to read a character from a file and *putc( )* is used to write a character to a file.

The general syntax of putc() and getc() are as follows:

```
int putc(int ch, FILE *stream);
int getc(FILE *stream);
```

The file pointer indicates the file to read from or write to. The character *ch* is formally called an integer in *putc( )* function but only the low order byte is used. On success *putc( )* returns a character(in integer form) written or EOF on failure. Similarly getc( ) returns an integer but only the low order byte is used. It returns *EOF* when end-of-file is reached. *getc( )* and *putc( )* are defined in <stdio.h> as macros not functions.

fgetc() and fputc()

Apart from the above two macros, C also defines equivalent functions to read/write characters from/to a file.

These are as follows:

```
int fgetc(FILE *stream);
int fputc(int c, FILE *stream);
```

To check the end of file, C includes the function feof( ) whose prototype is as follows

```
int feof(FILE *fp);
```

It returns **1** if end of file has been reached or **0** if not.

The following code fragment explains the use of these functions.

Problem 1

Write a program to copy one file to another.

Solution

```
/*Program to copy one file to another */
#include <stdio.h>
```

```
main( )
{
FILE *fp1;
FILE *fp2;
int ch;
if((fp1=fopen("f1.dat","r")) == NULL)
{
printf("Error opening input file\n");
exit(0);
}
if((fp2=fopen("f2.dat","w")) == NULL)
{
printf("Error opening output file\n");
exit(0);
}
while (!feof(fp1))
{
ch=getc(fp1);
putc(ch,fp2);
}
fclose(fp1);
fclose(fp2);
}
```

The output of the above code segment is as follows:

If the file "f1.dat" is not present, then the output would be:

Error opening input file

If the disk is full, then the output would be:

Error opening output file

If there is no error, then "f2.dat" would contain whatever is present in "f1.dat" after the execution of the program, if "f2.dat" was not empty earlier, then its contents would be overwritten.

Problem 2

Write a program which reads an existing file called myfile.txt character by character and uses the n variable to count how many dollar characters ($) does the file contain with the help of getc().

Solution

```
#include <stdio.h>
#include<conio.h>
void main ()
{
FILE * pFile;
```

```
int c;
int n = 0;
clrscr();
pFile=fopen ("myfile.txt","r");
if (pFile==NULL)
printf("Error opening file");
else
{
do
{
c = getc (pFile);
if (c == '$')
n++;
} while (c != EOF);
fclose (pFile);
printf ("File contains %d$.\n",n);
}
getch();
}
```

Problem 3

Write a program to create a file called alphabet.txt and writes ABCDEFGHIJKLM NOPQRSTUVWXYZ to it with the help of putc().

Solution

```
#include <stdio.h>
#include<conio.h>
void main ()
{
FILE * pFile;
char c;
clrscr();
pFile=fopen("alphabet.txt","wt");
for (c = 'A' ; c <= 'Z' ; c++)
{
putc (c, pFile);
}
fclose (pFile);
getch();
}
```

Problem 4

Write a program which reads an existing file called myfile.txt character by character and uses the n variable to count how many dollar characters ($) does the file contain with the help of fgetc().

Solution

```
#include <stdio.h>
#include<conio.h>
void main ()
{
FILE * pFile;
int c;
int n = 0;
clrscr();
pFile=fopen ("myfile.txt","r");
if (pFile==NULL) perror ("Error opening
file");
else
{
do
{
c = fgetc (pFile);
if (c == '$')
n++;
} while (c != EOF);
fclose (pFile);
printf ("File contains %d$.\n",n);
}
getch();
}
```

Problem 5

Write a program to create a file called
alphabet.txt and writes ABCDEFGHIJKLM
NOPQRSTUVWXYZ to it with the help of
fputc().

Solution

```
#include <stdio.h>
#include<conio.h>
void main ()
{
FILE * pFile;
char c;
pFile = fopen ("alphabet.txt","w");
if (pFile!=NULL)
{
for (c = 'A' ; c <= 'Z' ; c++)
{
fputc ( (int) c, pFile );
}
fclose (pFile);
}
getch();
}
```

Integer Input/Output Functions

**In this category we will discuss getw() and
putw() functions.**
The getw() and putw() function
These are integer-oriented functions. They are
similar to get c and putc functions and are
used to read and write integer values. These
functions would be useful when we deal with
only integer data.

putw() function is used to print the integer
numbers to the file streams and the general
syntax of putw() function is as follows
    putw(integer,FILEPOINTER);
    getw() function is used to read the integer
value from the FILE POINTER. And the
general syntax of getw() function is as follows:
    getw(FILEPOINTER);

Problem

Write a program to create a data file which
consists the numbers from 1 to 30 and then
read this data file and then write all the even
numbers in the EVEN file and all the add
numbers in the ODD file and then print the
contents of EVEN and ODD file.

Solution

```
/*Example program for using getw and putw
functions*/
#include< stdio.h >
#include<conio.h>
void main()
{
FILE *f1,*f2,*f3;
int number I;
printf("Contents of the data file\n\n");
f1=fopen("DATA","W");
for(I=1;I< 30;I++)
{
scanf("%d",&number);
if(number==-1)
break;
putw(number,f1);
}
fclose(f1);
f1=fopen("DATA","r");
f2=fopen("ODD","w");
f3=fopen("EVEN","w");
while((number=getw(f1))!=EOF)/* Read from
data file*/
```

```
{
if(number%2==0)
putw(number,f3);/*Write to even file*/
else
putw(number,f2);/*write to odd file*/
}
fclose(f1);
fclose(f2);
fclose(f3);
f2=fopen("ODD","r");
f3=fopen("EVEN","r");
printf("\n\nContents of the odd file\n\n");
while(number=getw(f2))!=EOF)
printf("%d",number);
printf("\n\nContents of the even file");
while(number=getw(f3))!=EOF)
printf("%d",number);
fclose(f2);
fclose(f3);
getch();
}
```

String Input/Output Functions

If we want to read a whole line in the file then each time we will need to call character input function, instead C provides some string input/output functions with the help of which we can read/write a set of characters at one time. These are defined in the standard library and are discussed below:

- fgets( )
- fputs( )

These functions are used to read and write strings. Their syntax is as follows:

int fputs(char *str, FILE *stream);
char *fgets(char *str, int num, FILE *stream);

The integer parameter in *fgets( )* is used to indicate that at most num-1 characters are to be read, terminating at end-of-file or end-of-line. The end-of-line character will be placed in the string *str* before the string terminator, if it is read. If end-of-file is encountered as the first character, EOF is returned, otherwise str is returned. The f*puts( )* function returns a non-negative number or EOF if unsuccessful.

Problem

Write a program read a file and count the number of lines in the file, assuming that a line can contain at most 80 characters.

Solution

```
/*Program to read a file and count the number
of lines in the file */
#include<stdio.h>
#include<conio.h>
#include<process.h>
void main()
{
FILE *fp;
int cnt=0;
char str[80];
clrscr();
/* open a file in read mode */
if ((fp=fopen("lines.dat","r"))== NULL)
{ printf("File does not exist\n");
exit(0);
}
/* read the file till end of file is encountered */
while(!(feof(fp)))
{ fgets(str,80,fp); /*reads at most 80 characters
in str */
cnt++; /* increment the counter after reading
a line */
}
}/* print the number of lines */
printf("The number of lines in the file is
:%d\n",cnt);
fclose(fp);
getch();
}
```

The output of the above code segment is as follows:

Let us assume that the contents of the file *"lines.dat"* are as follows:

SRMSCET is one of the best engineering college of India
SRMSCET is one of the best engineering college of India
SRMSCET is one of the best engineering college of India
SRMSCET is one of the best engineering college of India

After the execution the output would be
The number of lines in the file is: 4

Formatted Input/Output Functions

If the file contains data in the form of digits, real numbers, characters and strings, then character input/output functions are not enough as the values would be read in the form of characters. Also if we want to write data in some specific format to a file, then it is not possible with the above described functions. Hence C provides a set of formatted input/output functions. These are defined in standard library and are discussed as follows fscanf() and fprintf()

These functions are used for formatted input and output. These are identical to *scanf()* and *printf()* except that the first argument is a file pointer that specifies the file to be read or written, the second argument is the format string.

The general syntax of these functions are as follows:

int fscanf(FILE *fp, char *format,. . .);
int fprintf(FILE *fp, char *format,. . .);

Both these functions return an integer indicating the number of bytes actually read or written.

Problem

Write a program to read formatted data (account number, name and balance) from a file and print the information of clients with zero balance, in formatted manner on the screen.

Solution

```
#include<stdio.h>
#include<conio.h>
void main()
{
int account;
char name[30];
double bal;
FILE *fp;
if((fp=fopen("bank.dat","r"))== NULL)
printf("FILE not present \n");
else
do{
fscanf(fp,"%d%s%lf",&account,name,&bal);
if(!feof(fp))
{
if(bal==0)
printf("%d %s %lf\n",account,name,bal);
}
}while(!feof(fp));
```

```
fclose(fp);
getch();
}
```

The output of the above code is as follows:

This program opens a file *"bank.dat"* in the read mode if it exists, reads the records and prints the information (account number, name and balance) of the zero balance records.

Let the file be as follows:

101	Manish	1200
102	Vineet	1500
103	Swathi	0
104	Ajay	1600
105	Pragati	0

The output would be as follows:

103	Swathi	0
105	Pragati	0

Block Input/Output Functions

Block Input/Output functions read/write a block (specific number of bytes from/to a file. A block can be a record, a set of records or an array. These functions are also defined in standard library and are described as follows:
- fread( )
- fwrite( )

These two functions allow reading and writing of blocks of data. Their syntax is as follows:

int fread(void *buf, int num_bytes, int count, FILE *fp);

int fwrite(void *buf, int num_bytes, int count, FILE *fp);

In case of *fread()*, buf is the pointer to a memory area that receives the data from the file and in *fwrite()*, it is the pointer to the information to be written to the file. n*um_bytes* specifies the number of bytes to be read or written. These functions are quite helpful in case of binary files. Generally these functions are used to read or write array of records from or to a file. The use of the above functions is shown in the following program.

Problem

Write a program using *fread( )* and *fwrite()* to create a file of records and then read and print the same file.

Solution

```c
/* Program to illustrate the fread() and fwrite()
functions*/
#include<stdio.h>
#include<conio.h>
#include<process.h>
#include<string.h>
void main()
{
struct stud
{
char name[30];
int age;
int roll_no;
}s[30],st;
int i;
FILE *fp;
clrscr();
/*opening the file in write mode*/
if((fp=fopen("sud.dat","w"))== NULL)
{
printf("Error while creating a file\n");
exit(0);
}
/* reading an array of students */
for(i=0;i<30;i++)
scanf("%s %d %d",s[i].name,s[i].age,s[i].roll_
no);
/* writing to a file*/
fwrite(s,sizeof(struct stud),30,fp);
fclose(fp);
/* opening a file in read mode */
fp=fopen("stud.dat","r");
/* reading from a file and writing on the screen
*/
while(!feof(fp))
{
fread(&st,sizeof(struct stud),1,fp);
fprintf("%s %d %d",st.name,st.age,st.roll_no);
}
fclose(fp);
getch();
}
```

The output of the above code segment is as follows:

This program reads 30 records (name, age and roll_number) from the user, writes one record at a time to a file. The file is closed and then reopened in read mode; the records are again read from the file and written on to the screen.

SEQUENTIAL vs RANDOM ACCESS FILES

We already know that C supports two type of files—text and binary files, also two types of file systems—buffered and unbuffered file system. We can also differentiate in terms of the type of file access as sequential access files and random access files. Sequential access files allow reading the data from the file in sequential manner which means that data can only be read in sequence. All the above examples that we have considered till now in this unit are performing sequential access.

Random access files allow reading data from any location in the file. To achieve this purpose, C defines a set of functions to manipulate the position of the file pointer. Now we will discuss the concept of random access to files.

RANDOM ACCESS TO FILES

Sometimes it is required to access only a particular part of the file and not the complete file.

If we want to access the file randomly then we have to use the concept of random access to files in C language. In this concept we will discuss the following functions

- fseek function
- ftell function
- rewind function
- fgetpos() function
- fsetpos() function

(a) fseek Function

fseek function is used to reposition stream position in the files that means we can say that with the help of fseek function we can change the position of the file pointer.

The general syntax of fseek() function is as follows:

int fseek (FILE * stream, long int offset, int origin);

The above function sets the position indicator associated with the *stream* to a new position defined by adding *offset* to a reference position specified by *origin*.

The End-of-File internal indicator of the stream is clear after a call to this function, and all effects from previous calls to ungetc are dropped.

When using fseek on text files with *offset* values other than zero or values retrieved with ftell, bear in mind that on some platforms some format transformations occur with text files which can lead to unexpected re-positioning.

On streams open for update (read+write), a call to fseek allows to switch between reading and writing.

Parameters

The following are the parameters that are passed in the fseek() function.

(i) *stream*

Pointer to a FILE object that identifies the stream.

(ii) *offset*

Number of bytes to offset from *origin*.

(iii) *origin*

Position from where *offset* is added. It is specified by one of the following constants defined in <stdio.h>:

SEEK_SET	Beginning of file
SEEK_CUR	Current position of the file pointer
SEEK_END	End of file

Return value

If successful, the function returns a zero value. Otherwise, it returns nonzero value.

Example

The following example demonstrates the concept of fseek() function.

```
#include <stdio.h>
#include <conio.h>
void main ()
{
FILE * pFile;
clrscr();
pFile = fopen ( "myfile.txt", "w" );
fputs ( "This is an apple.", pFile );
fseek ( pFile, 9, SEEK_SET );
fputs ( " sam", pFile );
fclose ( pFile );
getch();
}
```

After this code is successfully executed, the file example.txt contains:
This is a sample.

(b) *ftell Function*

ftell function is used to get current position in file stream.

The general syntax of ftell() function is as follows:

```
long int ftell ( FILE * stream );
```

The above function returns the current value of the position indicator of the *stream*.

For binary streams, the value returned corresponds to the number of bytes from the beginning of the file. For text streams, the value is not guaranteed to be the exact number of bytes from the beginning of the file, but the value returned can still be used to restore the position indicator to this position using fseek.

Parameters

The following is the parameter that can be passed in ftell() function.

Stream

Pointer to a FILE object that identifies the stream.

Return value

On success, the current value of the position indicator is returned.

If an error occurs, -1L is returned, and the global variable errno is set to a positive value. This value can be interpreted by perror.

Example

```
#include <stdio.h>
#include<conio.h>
void main ()
{
FILE * pFile;
long size;
pFile = fopen ("myfile.txt","rb");
if (pFile==NULL) perror ("Error opening file");
else
{
fseek (pFile, 0, SEEK_END);
size=ftell (pFile);
fclose (pFile);
printf ("Size of myfile.txt: %ld bytes.\n",size);
```

}
getch();
}
This program prints out the size of myfile.txt in bytes.

(c) *rewind() Function*

Rewind() function is used to set position indicator to the beginning.

The general syntax of rewind function is as follows:

void rewind (FILE * rewind);

The above function sets the position indicator associated with *stream* to the beginning of the file.

A call to rewind is equivalent to:

fseek (stream, 0L, SEEK_SET);

except that, unlike fseek, rewind clears the error indicator.

On streams open for update (read+write), a call to rewind allows to switch between reading and writing.

Parameters

The following is the parameter which can be passed in the rewind() function.

stream

Pointer to a FILE object that identifies the stream.

Return value

This function does not return any value.

Example

```
#include <stdio.h>
#include<conio.h>
void main ()
{
int n;
FILE * pFile;
char buffer [27];
pFile = fopen ("myfile.txt","w+");
for ( n='A' ; n<='Z' ; n++)
fputc ( n, pFile);
rewind (pFile);
fread (buffer,1,26,pFile);
fclose (pFile);
buffer[26]='\0';
puts (buffer);
getch();
}
```

The explanation of the above program segment is as follows:

A file called myfile.txt is created for reading and writing and filled with the alphabet. The file is then rewinded, read and its content is stored in a buffer, that then is written to the standard output:

ABCDEFGHIJKLMNOPQRSTUVWXYZ

(d) *fsetpos*

fsetpos is used to set position indicator of file stream.

The general syntax of fsetpos is as follows:

int fsetpos (FILE * stream, const fpos_t * pos);

The above function changes the internal file position indicator associated with *stream* to a new position. The *position* parameter is a pointer to an fpos_t object whose value shall have been previously obtained with a call to fgetpos.

The End-of-File internal indicator of the stream is cleared after a call to this function, and all effects from previous calls to ungetc are dropped.

On streams open for update (read+write), a call to fsetpos allows to switch between reading and writing.

Parameters

The following are the parameters which can be passed in the fsetpos() function

(i) *stream*

Pointer to a FILE object that identifies the stream.

(ii) *position*

Pointer to a fpos_t object containing a position previously obtained with fgetpos.

Return value

If successful, the function returns a zero value.

Otherwise, it returns a nonzero value and sets the global variable errno to a positive value, which can be interpreted with perror.

Example

```
#include <stdio.h>
#include<conio.h>
void main ()
{
```

```
FILE * pFile;
fpos_t position;
pFile = fopen ("myfile.txt","w");
fgetpos (pFile, &position);
fputs ("That is a sample",pFile);
fsetpos (pFile, &position);
fputs ("This",pFile);
fclose (pFile);
getch();
}
```

After this code is successfully executed, a file called myfile.txt will contain:
This is a sample

fgetpos() Function

fgetpos() function is used to get current position in file stream.

The general syntax of fgetpos() is as follows:
int fgetpos (FILE * stream, fpos_t * position);

The above function gets the information needed to uniquely identify the current value of the *stream*'s position indicator and stores it in the location pointed by *position*.

The parameter *position* should point to an already allocated object of the type fpos_t, which is only intended to be used as a parameter in future calls to fsetpos.

To retrieve the value of the internal file position indicator as an integer value, use ftell function instead.

Parameters

The following are the parameters which can be passed in the fgetpos() function.

(i) *stream*
Pointer to a FILE object that identifies the stream.

(ii) *position*
Pointer to a fpos_t object.

Return value

The function return a zero value on success, and a non-zero value in case of error.

Example

```
#include <stdio.h>
#include<conio.h>
void main ()
{
```

```
FILE * pFile;
int c;
int n;
fpos_t pos;
clrscr();
pFile = fopen ("myfile.txt","r");
if (pFile==NULL) perror ("Error opening file");
else
{
c = fgetc (pFile);
printf ("1st character is %c\n",c);
fgetpos (pFile,&pos);
for (n=0;n<3;n++)
{
fsetpos (pFile,&pos);
c = fgetc (pFile);
printf ("2nd character is %c\n",c);
}
fclose (pFile);
}
getch();
}
```

The explanation of the above program is as follows:

This example opens myfile.txt, reads the first character once and then it reads 3 times the same second character.

Output

```
1st character is A
2nd character is B
2nd character is B
2nd character is B
```

Programming Problems of File Handling

Problem 1

Write a program to display the contents of a file on the screen?

Solution

```
#include <stdio.h>
#include<conio.h>
void main()
{
    FILE *fopen(), *fp;
    int c ;
    fp = fopen( "prog.c", "r" );
    c = getc( fp ) ;
```

```
while ( c != EOF )
{
putchar( c );
c = getc ( fp );
}
fclose( fp );
getch();
}
```

Problem 2

Write a program to input the file name and then display the contents of that file.

Solution

```
#include <stdio.h>
#include<conio.h>
void main()
{
    FILE *fopen(), *fp;
    int c ;
    char filename[40] ;
    clrscr();
    printf("Enter file to be displayed: ");
    gets( filename ) ;
    fp = fopen( filename, "r"):
    c = getc( fp ) ;
    while ( c != EOF )
    {
    putchar(c);
    c = getc ( fp );
    }
    fclose( fp );
    getch();
}
```

Problem 3

Write a program to count the number of lines and characters in a file.

Solution

```
#include <stdio.h>
#include<conio.h>
void main()
{
    FILE *fopen(), *fp;
    int c, nc, nlines;
    char filename[40] ;
    clrscr();
    nlines = 0 ;
    nc = 0;
```

```
printf("Enter file name: ");
gets( filename );
fp = fopen( filename, "r" );
if ( fp == NULL )
{
printf("Cannot open %s for reading \n",
filename );
exit(1); /* terminate program */
}
c = getc( fp ) ;
while ( c != EOF )
{
if ( c == '\n' )
nlines++ ;
nc++ ;
c = getc ( fp );
}
fclose( fp );
if ( nc != 0 )
{
printf("There are %d characters in %s \n",
nc, filename );
printf("There are %d lines \n", nlines );
}
else
printf("File: %s is empty \n", filename );
getch();
}
```

Problem 4

Write a program to display file contents 20 lines at a time. The program pauses after displaying 20 lines until the user presses either Q to quit or return to display the next 20 lines.

Solution

```
#include <stdio.h>
#include<conio.h>
void main()
{
    FILE *fopen(), *fp;
    int c, linecount;
    char filename[40], reply[40];
    printf("Enter file name: ");
    gets( filename );
    fp = fopen( filename, "r" ); /* open for
reading */
    if ( fp == NULL ) /* check does file exist etc */
    {
```

```
    printf("Cannot open %s for reading \n",
filename );
    exit(); /* terminate program */
    }
    linecount = 1 ;
    reply[0] = '\0' ;
    c = getc( fp ) ; /* Read 1st character if any */
    while ( c != EOF && reply[0] != 'Q' &&
reply[0] != 'q')
    {
    putchar( c ) ;/* Display character */
    if ( c == '\n' )
    linecount = linecount+ 1 ;
    if ( linecount == 20 )
    {
    linecount = 1 ;
    printf("[Press Return to continue, Q to
quit]");
    gets( reply ) ;
    }
    c = getc ( fp );
    }
    fclose( fp );
    getch();
}
```

Problem 5

Write a program to compare two files specified by the user, displaying a message indicating whether the files are identical or different. This is the basis of a compare command provided by most operating systems.

Solution

```
#include <stdio.h>
void main()
{
    FILE *fp1, *fp2, *fopen();
    int ca, cb;
    char fname1[40], fname2[40] ;
    printf("Enter first filename:") ;
    gets(fname1);
    printf("Enter second filename:");
    gets(fname2);
    fp1 = fopen( fname1, "r" ); /* open for
reading */
    fp2 = fopen( fname2, "r" ) ; /* open for
writing */
    if ( fp1 == NULL ) /* check does file exist
etc */
    {
    printf("Cannot open %s for reading \n",
fname1 );
    exit(1); /* terminate program */
    }
    else if ( fp2 == NULL )
    {
    printf("Cannot open %s for reading \n",
fname2 );
    exit(1); /* terminate program */
    }
    else /* both files opened successfully */
    {
    ca = getc( fp1 ) ;
    cb = getc( fp2 ) ;
    while ( ca != EOF && cb != EOF && ca == cb)
    {
    ca = getc( fp1 ) ;
    cb = getc( fp2 ) ;
    }
    if ( ca == cb )
    printf("Files are identical \n");
    else if ( ca != cb )
    printf("Files differ \n" );
    fclose ( fp1 );
    fclose ( fp2 );
    }
    getch();
}
```

Problem 6

Write a file copy program which copies the file "prog.c" to "prog.old"

Solution

```
#include <stdio.h>
#include<conio.h>
void main()
{
    FILE *fp1, *fp2, *fopen();
    int c ;
    fp1 = fopen( "prog.c", "r" ); /* open for
reading */
    fp2 = fopen( "prog.old", "w" ) ; ../* open
for writing */
    if ( fp1 == NULL ) /* check does file exist etc
*/
    {
    printf("Cannot open prog.c for reading \n" );
```

```
exit(1); /* terminate program */
}
else if ( fp2 == NULL )
{
printf("Cannot open prog.old for writing
\n");
exit(1); /* terminate program */
}
else  /* both files O.K. */
{
c = getc(fp1) ;
while ( c != EOF)
{
putc( c, fp2); /* copy to prog.old */
c = getc( fp1 ) ;
}
fclose ( fp1 );/* Now close files */
fclose ( fp2 );
printf("Files successfully copied \n");
}
getch();
}
```

Problem 7

The above program only copies the specific file prog.c to the file prog.old. We can make it a general purpose program by prompting the user for the files to be copied and opening them appropriately.

Solution

```
#include <stdio.h>
#include<conio.h>
void main()
{
    FILE *fp1, *fp2, *fopen();
    int c ;
    char fname1[40], fname2[40] ;
    clrscr();
    printf("Enter source file:") ;
    gets(fname1);
    printf("Enter destination file:");
    gets(fname2);
    fp1 = fopen( fname1, "r" ); /* open for
reading */
    fp2 = fopen( fname2, "w" ) ; ../* open for
writing */
    if ( fp1 == NULL )/* check does file exist etc */
    {
```

```
printf("Cannot open %s for reading \n",
fname1 );
exit(1); /* terminate program */
}
else if ( fp2 == NULL )
{
printf("Cannot open %s for writing \n",
fname2 );
exit(1); /* terminate program */
}
else  /* both files O.K. */
{
c = getc(fp1) ;          /* read from source */
while ( c != EOF)
{
putc( c, fp2); /* copy to destination */
c = getc( fp1 ) ;
}
fclose ( fp1 );/* Now close files */
fclose ( fp2 );
printf("Files successfully copied \n");
}
getch();
}
```

NUMERIC LIBRARY

cmath declares a set of functions to compute common mathematical operations and transformations.

TRIGONOMETRIC FUNCTIONS

In the trigonometric functions we will discuss the following functions:

cos()	Compute cosine
sin()	Compute sine
tan()	Compute tangent
acos()	Compute arc cosine
asin()	Compute arc sine
atan()	Compute arc tangent
atan2()	Compute arc tangent with two parameters

(a) cos()

cos() function is used to compute cosine.

The general syntax of cos() function is as follows:

double cos (double x);

float cos (float x);

long double cos (long double x);

The above function returns the cosine of an angle of x radians.

Parameters

Floating point value representing an angle expressed in radians.

Return value

The above function returns cosine of x.

Portability

In C, only the double version of this function exists with this name.

Example

```
#include <stdio.h>
#include <math.h>
#define PI 3.14159265
void main ()
{
double param, result;
param = 60.0;
result = cos (param*PI/180);
printf ("The cosine of %lf degrees is %lf.\n",
param, result );
getch();
}
```

The output of the above program is as follows:
The cosine of 60.000000 degrees is 0.500000.

(b) sin()

sin() function is used to compute sine.
The general syntax of sin() function is as follows:

```
double sin (double x);
float sin (float x);
long double sin (long double x);
```

The above function returns the sine of an angle of x radians.

Parameters

Floating point value representing an angle expressed in radians.

Return value

The above function returns sine of x.

Portability

In C, only the double version of this function exists with this name.

Example

```
#include <stdio.h>
#include <math.h>
#define PI 3.14159265
void main ()
{
double param, result;
clrscr();
param = 30.0;
result = sin (param*PI/180);
printf ("The sine of %lf degrees is %lf.\n",
param, result );
getch();
}
```

The output of the above program is as follows:
The sine of 30.000000 degrees is 0.500000.

(c) tan()

tan() function is used to compute tangent.
The general syntax of tan() function is as follows

```
double tan (double x );
float tan (float x );
long double tan ( long double x );
```

The above function returns the tangent of an angle of x radians.

Parameters

Floating point value representing an angle expressed in radians.

Return value

The above function returns tangent of x.

Portability

In C, only the double version of this function exists with this name.

Example

```
#include <stdio.h>
#include <math.h>
#define PI 3.14159265
void main ()
{
double param, result;
param = 45.0;
result = tan (param*PI/180);
printf ("The tangent of %lf degrees is %lf.\n",
param, result );
getch();
}
```

The output of the above program is as follows:
The tangent of 45.000000 degrees is 1.000000.

(d) acos()

acos() function is used to compute arc cosine. The general syntax of acos() function is as follows:
double acos (double x);
float acos (float x);
long double acos (long double x);
The above function returns the principal value of the arc cosine of x, expressed in radians.

In trigonometrics, arc cosine is the inverse operation of cosine.

Parameters

Floating point value in the interval [-1,+1]. If the argument is out of this interval, a domain error occurs, setting the global variable errno to the value EDOM.

Return value

Principal arc cosine of x, in the interval [0,pi] radians.

Portability

In C, only the double version of this function exists with this name.

Example

```
#include <stdio.h>
#include <math.h>
#define PI 3.14159265
int main ()
{
double param, result;
param = 0.5;
result = acos (param) * 180.0 / PI;
printf ("The arc cosine of %lf is %lf degrees.\n", param, result );
getch();
}
```

The output of the above program is as follows:

The arc cosine of 0.500000 is 60.000000 degrees.

(e) asin()

asin() function is used to compute arc sine. The general syntax of asin() function is as follows:

double asin (double x);
float asin (float x)
long double asin (long double x);
The above function returns the principal value of the arc sine of x, expressed in radians.

In trigonometrics, arc sine is the inverse operation of sine.

Parameters

Floating point value in the interval [-1,+1]. If the argument is out of this interval, a domain error occurs, setting the global variable errno to the value EDOM.

Return value

Arc sine of x, in the interval [-pi/2,+pi/2] radians.

Portability

In C, only the double version of this function exists with this name.

Example

```
#include <stdio.h>
#include <math.h>
#define PI 3.14159265
void main ()
{
double param, result;
clrscr();
param = 0.5;
result = asin (param) * 180.0 / PI;
printf ("The arc sine of %lf is %lf degrees\n", param, result );
getch();
}
```

The output of the above program is as follows:
The arc sine of 0.500000 is 30.000000 degrees.

(f) atan()

atan() function is used to compute arc tangent. The general syntax of atan() function is as follows
double atan (double x);
float atan (float x);
long double atan (long double x);
The above function returns the principal value of the arc tangent of x, expressed in radians.

In trigonometrics, arc tangent is the inverse operation of tangent.

Notice that because of the sign ambiguity, a function cannot determine with certainty in which quadrant the angle falls only by its tangent value. You can use atan2 if you need to determine the quadrant.

Parameters

Floating point value.

Return value

Principal arc tangent of *x*, in the interval [-pi/2,+pi/2] radians.

Portability

In C, only the double version of this function exists with this name.

Example

```
#include <stdio.h>
#include <math.h>
#define PI 3.14159265
void main ()
{
double param, result;
clrscr();
param = 1.0;
result = atan (param) * 180 / PI;
printf ("The arc tangent of %lf is %lf degrees\n", param, result );
getch();
}
```

The output of the above program is as follows: The arc tangent of 1.000000 is 45.000000 degrees.

(g) *atan2()*

atan2() function is used to compute arc tangent with two parameters.

The general syntax of atan2() function is as follows:

```
double atan2 ( double y, double x );
long double atan2 ( long double y, long double x );
float atan2 ( float y, float x );
```

The above function returns the principal value of the arc tangent of *y/x*, expressed in radians.

To compute the value, the function uses the sign of both arguments to determine the quadrant.

Parameters

y floating point value representing an y-coordinate.

x floating point value representing an x-coordinate.

If both arguments passed are zero, a domain error occurs, which sets the global variable ERRNO to the EDOM value.

Return value

Principal arc tangent of *y/x*, in the interval [–pi, +pi] radians.

Portability

In C, only the double version of this function exists with this name.

Example

```
#include <stdio.h>
#include <math.h>
#include<conio.h>
#define PI 3.14159265
void main ()
{
double x, y, result;
clrscr();
x = -10.0;
y = 10.0;
result = atan2 (y,x) * 180 / PI;
printf ("The arc tangent for (x=%lf, y=%lf) is %lf degrees\n", x, y, result );
getch();
}
```

The output of the above program is as follows: The arc tangent for (x=-10.000000, y=10.000000) is 135.000000 degrees.

HYPERBOLIC FUNCTIONS

Cosh	Compute hyperbolic cosine
Sinh	Compute hyperbolic sine
Tanh	Compute hyperbolic tangent

(a) *cosh()*

cosh() function is used to compute hyperbolic cosine.

The general syntax of cosh() function is as follows:

```
double cosh (double x );
float cosh (float x );
long double cosh ( long double x );
```

The above function returns the hyperbolic cosine of *x*.

Parameters

x-floating point value.

Return value

The above function will return hyperbolic cosine of *x*.

If the magnitude of the result is so large that it cannot be represented in an object of the return type, the function returns HUGE_VAL, and the value of the global variable errno is set to the ERANGE value.

Portability

In C, only the double version of this function exists with this name.

Example

```
#include <stdio.h>
#include<conio.h>
#include <math.h>
void main ()
{
double param, result;
clrscr();
param = log(2.0);
result = cosh (param);
printf ("The hyperbolic cosine of %lf is
%lf.\n", param, result );
getch();
}
```

The output of the above program is as follows:
The hyperbolic cosine of 0.693147 is 1.250000.

(b) sinh()

sinh() function is used to compute hyperbolic sine.

The general syntax of sinh() function is as follows:

double sinh (double x);
float sinh (float x);
long double sinh (long double x);

The above function returns the hyperbolic sine of *x*.

Parameters

x-floating point value.

Return value

The above function returns hyperbolic sine of *x*.

If the magnitude of the result is so large that it cannot be represented in an object of the return type, the function returns HUGE_VAL with the same sign as the correct value of the operation, and the value of the global variable errno is set to the ERANGE value.

Portability

In C, only the double version of this function exists with this name.

Example

```
#include <stdio.h>
#include<conio.h>
#include <math.h>
void main ()
{
double param, result;
clrscr();
param = log(2.0);
result = sinh (param);
printf ("The hyperbolic sine of %lf is %lf.\n",
param, result );
getch();
}
```

The output of the above program is as follows:
The hyperbolic sine of 0.693147 is 0.750000.

(c) tanh()

tanh() function is used to compute hyperbolic tangent.

The general syntax of tanh() function is as follows:

double tanh (double x);
float tanh (float x);
long double tanh (long double x);

The above function returns the hyperbolic tangent of *x*.

Parameters

x floating point value.

Return value

Hyperbolic tangent of *x*.

Portability

In C, only the double version of this function exists with this name. Example

```
#include <stdio.h>
#include<conio.h>
```

```
#include <math.h>
void main ()
{
double param, result;
clrscr();
param = log(2.0);
result = tanh (param);
printf ("The hyperbolic tangent of %lf is %lf.
\n", param, result );
getch();
}
```

The output of the above function is as follows:
The hyperbolic tangent of 0.693147 is 0.600000.

EXPONENTIAL AND LOGARITHMIC FUNCTIONS

Exp	Compute exponential function
log	Compute natural logarithm
log10	Compute base-10 logarithm
Modf	Break into fractional and integral parts

(a) exp()

exp() function is used to compute exponential function.

The general syntax of exp() function is as follows:

double exp (double x);
float exp (float x);
long double exp (long double x);

The above function returns the exponential function of x, which is the e number raised to the power x.

Parameters

x-floating point value.

Return value

The above function returns exponential value of x.

If the magnitude of the result is so large that it cannot be represented in an object of the return type, the function returns HUGE_VAL, and the value of the global variable errno is set to the ERANGE value.

Portability

In C, only the double version of this function exists with this name.

Example

#include <stdio.h>

```
#include <math.h>
#include<conio.h>
void main ()
{
double param, result;
param = 5.0;
clrscr();
result = exp (param);
printf ("The exponential value of %lf is
%lf.\n", param, result );
getch();
}
```

The output of the above program is as follows:
The exponential value of 5.000000 is 148.413159.

(b) log()

log() function is used to compute natural logarithm.

The general syntax of log() function is as follows

double log (double x);
float log (float x);
long double log (long double x);

The above function returns the natural logarithm of x.

The natural logarithm is the base-e logarithm, the opposite of the natural exponential function (exp). For base-10 logarithms, a specific function log10 exists.

Parameters

x-floating point value.

If the argument is negative, a domain error comes, setting the global variable errno to the value EDOM.

If it is zero, the function returns a negative HUGE_VAL and adjusts the value of the global variable errno to the ERANGE value.

Return value

Natural logarithm of x.

Portability

In C, only the double version of this function exists with this name.

Example

#include <stdio.h>
#include <math.h>
#include<conio.h>

```
void main ()
{
double param, result;
clrscr();
param = 5.5;
result = log (param);
printf ("ln(%lf) = %lf\n", param, result );
getch();
}
```
The output of the above program is as follows:
ln(5.500000) = 1.704748.

(c) log10

log10 function is used to compute base-10 logarithm
The general syntax of log10 is as follows:
```
double log10 (double x );
float log10 (float x );
long double log10 ( long double x );
```
The above function returns the base-10 logarithm of x.

Parameters

x-floating point value.
If the argument is negative, a domain error comes, setting the global variable errno to the value EDOM.
If it is zero, the function returns a negative HUGE_VAL and determines the value of the global variable errno to the ERANGE value.

Return value

Base-10 logarithm of x, for values of x greater than zero.

Portability

In C, only the double version of this function exists with this name.

Example
```
#include <stdio.h>
#include <math.h>
#include<conio.h>
void main ()
{
double param, result;
param = 1000.0;
clrscr();
result = log10 (param);
printf ("log10(%lf) = %lf\n", param, result );
```

```
getch();
}
```
The output of the above program is as follows:
log10(1000.000000) = 3.000000

(d) modf

modf is a function which is used to break into fractional and integral parts.
The general syntax of modf function is as follows
```
double modf ( long double x, long double
* intpart );
long double modf ( long double x, long
double * intpart );
float modf ( float x, float * intpart );
```
The above function Breaks x into two parts: the integer part (stored in the object pointed by intpart) and the fractional part (returned by the function). Each component has the same sign as x.

Parameters

The following are the parameters which we have to pass in the modf() function.

x-floating point value

intpart

Pointer to an object where the integral part is to be stored.

Return value

The above function returns the fractional part of x, with the same sign.

Portability

In C, only the double version of this function exists with this name.

Example
```
#include <stdio.h>
#include <math.h>
#include<conio.h>
void main ()
{
double param, fractpart, intpart;
clrscr();
param = 3.14159265;
fractpart = modf (param, &intpart);
printf ("%lf = %lf + %lf \n", param, intpart,
fractpart);
```

```
getch();
}
```
The output of the above program is as follows:
3.141593 = 3.000000 + 0.141593

POWER FUNCTIONS

pow	Raise to power
sqrt	Compute square root

(a) pow()

pow() function is used to raise to power.

The general syntax of pow() function is as follows

```
double pow (double base, double exponent);
long double pow (long double base, long double exponent);
float pow (float base,float exponent);
```

The above function returns *base* raised to the power *exponent*: base^{exponent}

Parameters

The following are the parameters which is passed in the pow() function.

Base

Floating point value.

Exponent

Floating point value.

Return value

The result of raising *base* to the power *exponent*. If the magnitude of the outcome is so large that it cannot be represented in an object of the return type, a range error occurs, returning HUGE_VAL with the appropiate sign and setting the value of the global variable errno to the ERANGE value.

If *base* is negative and *exponent* is not an essential value, or if *base* is zero and *exponent* is negative, a domain error occurs, setting the global variable errno to the value EDOM.

Portability

In C, only the double version of this function exists with this name.

Example

```
#include <stdio.h>
#include <math.h>
#include<conio.h>
```

```
void main ()
{
printf ("7 ^ 3 = %lf\n", pow (7,3));
printf ("4.73 ^ 12 = %lf\n", pow (4.73,12));
printf ("32.01 ^ 1.54 = %lf\n", pow (32.01, 1.54));
getch();
}
```

The output of the above program is as follows

7 ^ 3 = 343.000000
4.73 ^ 12 = 125410439.217423
32.01 ^ 1.54 = 208.036691

(b) sqrt()

sqrt() function is used to compute square root. The general syntax of sqrt() function is as follows:

```
double sqrt ( double x );
float sqrt (float x );
long double sqrt ( long double x );
```

The above function returns the square root of x.

Parameters

x floating point value.
If the argument is negative, a domain error comes, setting the global variable errno to the value EDOM.

Return value

The above function returns the square root of x.

Portability

In C, only the double version of this function exists with this name.

Example

```
#include <stdio.h>
#include <math.h>
#include<conio.h>
void main ()
{
double param, result;
param = 1024.0;
clrscr();
result = sqrt (param);
printf ("sqrt(%lf) = %lf\n", param, result );
getch();
}
```

The output of the above program is as follows:
sqrt(1024.000000) = 32.000000

ROUNDING, ABSOLUTE VALUE AND REMAINDER FUNCTIONS

ceil	Round up value
fabs	Compute absolute value
floor	Round down value
fmod	Compute remainder of division

(a) ceil

ceil function is used to find the smallest integral value that is not less than x.
The general syntax of ceil function is as follows
double ceil (double x);
float ceil (float x);
long double ceil (long double x);
The above function returns the smallest integral value that is not less than x.

Parameters

x-floating point value.

Return value

The smallest integral value not less than x.

Portability

In C, only the double version of this function exists with this name.

Example

```
#include <stdio.h>
#include <math.h>
#include<conio.h>
void main ()
{
printf ("ceil of 2.3 is %.1lf\n", ceil (2.3) );
printf ("ceil of 3.8 is %.1lf\n", ceil (3.8) );
printf ("ceil of -2.3 is %.1lf\n", ceil (-2.3) );
printf ("ceil of -3.8 is %.1lf\n", ceil (-3.8) );
getch();
}
```

The output of the above program is as follows:
ceil of 2.3 is 3.0
ceil of 3.8 is 4.0
ceil of -2.3 is -2.0
ceil of -3.8 is -3.0

(b) fabs()

fabs() function is used to compute absolute value.

The general syntax of fabs() function is as follows:
double fabs (double x);
float fabs (float x);
long double fabs (long double x);
The above function returns the absolute value of x ($|x|$).

Parameters

x-floating point value.

Return value

The absolute value of x.

Portability

In C, only the double version of this function exists with this name.

Example

```
#include <stdio.h>
#include <math.h>
#include <conio.h>
void main ()
{
    clrscr();
    printf ("The absolute value of 3.1416 is %lf\n", fabs (3.1416) );
    printf ("The absolute value of -10.6 is %lf\n", fabs (-10.6) );
    getch();
}
```

The output of the above program is as follows:
The absolute value of 3.1416 is 3.141600
The absolute value of -10.6 is 10.600000

(c) floor

floor function is used to round down value.
The general syntax of floor function is as follows
double floor (double x);
float floor (float x);
long double floor (long double x);
The above function returns the largest integral value that is not greater than x.

Parameters

x-floating point value.

Return value

The largest integral value not greater than x.

Portability

In C, only the double version of this function exists with this name.

Example

```
#include <stdio.h>
#include <math.h>
#include<conio.h>
void main ()
{
clrscr();
printf ("floor of 2.3 is %.1lf\n", floor (2.3) );
printf ("floor of 3.8 is %.1lf\n", floor (3.8) );
printf ("floor of -2.3 is %.1lf\n", floor (-2.3) );
printf ("floor of -3.8 is %.1lf\n", floor (-3.8) );
getch();
}
```

The output of the above program is as follows:
floor of 2.3 is 2.0
floor of 3.8 is 3.0
floor of -2.3 is -3.0
floor of -3.8 is -4.0

(d) fmod()

fmod() function is used to compute remainder of division.

The general syntax of fmod() function is as follows

```
double fmod (double numerator,double denominator );
float fmod (float numerator, float denominator);
long double fmod (long double numerator, long double denominator);
```

The above function returns the floating-point remainder of *numerator/denominator*.

The remainder of a division operation is the result of subtracting the integral quotient multiplied by the denominator from the numerator:

remainder = numerator - quotient * denominator

Parameters

The following are the parameters that can be passed in fmod() function.

numerator

Floating point value with the division numerator.

denominator

Floating point value with the division denominator.

Return value

The above function returns the remainder of dividing the arguments.

Portability

In C, only the double version of this function exists with this name.

Example

```
/* fmod example */
#include <stdio.h>
#include <math.h>
#include<conio.h>
void main ()
{
    clrscr();
    printf ("fmod of 5.3 / 2 is %lf\n", fmod (5.3,2) );
    printf ("fmod of 18.5 / 4.2 is %lf\n", fmod (18.5,4.2) );
    getch();
}
```

The output of the above program is as follows:
fmod of 5.3 / 2 is 1.300000
fmod of 18.5 / 4.2 is 1.700000

OTHER STANDARD C FUNCTION

The following are the other standard C function

(a) abort

The function abort() terminates the current program. Depending on the execution, the return value can indicate failure.

The general syntax of abort function is as follows:

void abort(void);

(b) assert

The assert() macro is used to test for errors. If *exp* measures to zero, assert() writes information to **stderr** and exits the program. If the macro NDEBUG is determined, the assert() macros will be ignored.

The general syntax of assert is as follows:

assert(exp);

(c) exit

The exit() function stops the program. *exit_code* is passed on to be the return value of the program, where normally zero indicates success and non-zero indicates an error.

The general syntax of exit() function is as follows:

void exit(int exit_code);

(d) *system*

The system() function runs the given *command* by passing it to the default command interpreter.

The return value is normally zero if the command executed without errors. If *command* is **NULL**, system() will test to see if there is a command interpreter accessible. Non-zero will be gave back if there is a command interpreter available, zero if not.

The general syntax of system() function is as follows:

int system(const char *command);

SUMMARY

- If we want to store the data permanently then we have to use the concept of file handling in C language.
- FILE *p is a file pointer which stores the name of the file.
- fgetc() and fputc() functions are used to read and write into the file character by character.
- fscanf() and fprintf() are the predefined functions which are used to input and output statements to the file.
- File is divided into two categories (a) text files, (b) binary files
- For binary files we have to use the fread() and frite() functions for reading and writing to the files.
- For random access we have to use fseek(), ftell(), rewind() functions.

Exercises

Section A

Multiple Choice Questions

1. This could not be a mode in which the file can be opened
   (a) w+
   (b) a+b
   (c) t
   (d) None of the above

2. If we have to open a file in binary mode to write in it then we have to use
   (a) a+b   (b) wb+
   (c) w+a   (d) All of the above

3. format of closing a file with fclose function is
   (a) int fclose(FILE * stream);
   (b) fclose(FILE * stream);
   (c) Both a and b
   (d) None of these

4. Specifier representing unsigned hexadecimal integer
   (a) u   (b) x
   (c) h   (d) i

5. fgets() takes the following parameters
   (a) Pointer to array
   (b) Max no. of char
   (c) Pointer to file object
   (d) All of the above

6. fwrite() function takes as parameters
   (a) Pointer to the array
   (b) Size in bytes
   (c) Number of elements
   (d) All of the above

7. In order to randomly access we can use
   (a) fseek function
   (b) ftell function
   (c) fgetpost function
   (d) All of the above

8. asin() function does not have a return type as
   (a) float   (b) long double
   (c) int   (d) double

9. atan2() function has
   (a) 1 parameter and 2 return arguments
   (b) 2 parameters and 1 return arguments
   (c) 1 parameter and 1 return arguments
   (d) No parameter and no return type

10. Offset can be added by
    (a) SEEK_SET   (b) SEEK_CUR
    (c) Both a and b   (d) BEGIN

State True/False

1. fprintf() is used for writing a set of data values to a file.

2. ftell() tells the type of file specified as parameter.

3. "r+" creates a file for reading and writing.
4. fgetc() and getc() functions are used to get characters from file stream.
5. putc has two parameters character to be written and stream where character is to be written.
6. getw() and putw() are used for integer values only.
7. ceil function is used to round up values.
8. assert() function runs the given command by passing it to the default command Interpreter.
9. ftell() function takes file stream as parameter.
10. In C only the double version of cosh() function exists.

Section B

Short Answer Type Questions

1. Define the concept of file handling in C language.
2. Define the classification of files used in C language.
3. Differentiate between sequential access and random access files with suitable example.
4. Write a program in C to read and store the bio-data into file.
5. Write a program in C to read the bio from file and print to screen.
6. Write a program in C to read the biodata until $ is given for name.
7. Write a program in C to read the bio from file and print to screen.
8. Write a program in C to read a character one by one until $ is given.
9. Write a program in C to display the text form file.
10. Write a program in C to merge two files.
11. Write a program in C to convert upper case letters into lower case.
12. Write a program in C to encraph the given file.
13. Write a program in C to decraph the given file.

14. Write a program in C to copy the contents of array into file.
15. Write a program in C to copy the file contents into array.

Section C

Long Answer Type Questions

1. Write a program in C to display line by line of the file contents.
2. Write a program in C to display line by line of the file contents including line number.
3. Write a program in C to display page by page of the file contents including line number.
4. Write a program in C to display line by line and count the no of upper and lower case letters and numbers.
5. Write a program in C to count the number of upper and lowercase letters and numbers in file.
6. Write a program in C to count the number of lines and words and characters in file.
7. Write a program in C to count the given pattern.
8. Write a program in C to count the given pattern using command line arguments.
9. Write a program in C to print the particular line when pattern is occurred using command line arguments.
10. Write a program in C to print the particular line with line number when pattern is occured using command line arguments.
11. Write a program in C to print the particular line with line number when pattern is occured using command line arguments in all files.
12. Write a program in C to print the particular line with line number when pattern dost not occur using command line arguments in all files?
13. Write a program in C to read and store the bio-data into file.

18 Android Operating System

Mobile users demand more choices, more opportunities and more functionalities to customize their phones. Mobile operators want to provide value-added content to their subscribers in a manageable and lucrative way. Mobile developers want the freedom to develop the powerful mobile applications users demand with minimal roadblocks to success. Finally, handset manufacturers want a stable, secure, and affordable platform to power their devices. Up until now single mobile platform has adequately addressed the needs of all the parties. Enter Android, which is a potential game-changer for the mobile development community. An innovative and open platform, Android is well positioned to address the growing needs of the mobile marketplace.

THE OPEN HANDSET ALLIANCE

The company's initial forays into mobile were beset with all the problems you would expect. The freedoms Internet users enjoyed were not shared by mobile phone subscribers. Internet users can choose from the wide variety of computer brands, operating systems, Internet service providers, and Web browser applications. Nearly all Google services are free and ad driven. Many applications in the Google Labs suite would directly compete with the applications available on mobile phones. The applications range from simple calendars and calculators to navigation with Google Maps and the latest tailored news from News Alerts-not to mention corporate acquisitions like Blogger and YouTube.

When this approach did not yield the intended results, Google decided to a different approach to revamp the entire system upon which the wireless application development was based, hoping to provide a more open environment for users and developers: the Internet model. The Internet model allows users to choose between freeware, shareware, and paid software. This enables free market competition among services.

Forming of the Open Handset Alliance

With its user-centric, democratic design philosophies, Google has led a movement to turn the existing closely guarded wireless market into one where phone users can move between carriers easily and have unfettered access to applications and services. With its vast resources, Google has taken a broad approach, examining the wireless infrastructure from the FCC wireless spectrum policies to the handset manufacturers' requirements, application developer needs, and mobile operator desires.

Next, Google joined with other like-minded members in the wireless community and posed the following question: What would it take to build a better mobile phone? The open handset alliance (OHA) was formed in November 2007 to answer that very question. The OHA is a business alliance comprised of many of the largest and most successful mobile companies on the planet. Its members include chip makers, handset manufacturers, software developers, and service providers.

Working together, OHA members began developing a nonproprietary open standard platform that would aim to alleviate the aforementioned problems hindering the mobile community. They called it the Android project.

Google's involvement in the Android project has been extensive. The company hosts the open source project and provides online documentation, tools, forums, and the software development kit (SDK).

Manufacturers: Designing the Android Handsets

More than half the members of the OHA are handset manufacturers, such as Samsung, Motorola, HTC, and LG, and semiconductor companies, such as Intel, Texas Instruments, NVIDIA, and Qualcomm. These companies are helping design the first generation of Android handsets.

The first shipping Android handset—the T-Mobile G1-was developed by handset manufacturer HTC with service provided by T-Mobile. It was released in October 2008. Many other Android handsets are slated for 2009 and early 2010.

Content Providers: Developing Android Applications

Google has led the pack, developing Android applications, many of which, like the email client and Web browser, are the core features of the platform. OHA members, such as eBay, are also working on Android application integra-tion with their online auctions.

The first Android developer challenge received 1,788 submissions—all newly developed Android games, productivity helpers, and a slew of location-based services (LBS). We also saw humanitarian, social networking, and mash-up applications. Many of these applications have debuted with users through the Android Market-Google's software distribution mechanism for Android.

Mobile Operators: Delivering the Android Experience

After you have the phones, you have to get them out to the users. Mobile operators from Asia, North America, Europe, and Latin America have joined the OHA, ensuring a market for the Android movement. With almost half a billion subscribers, telephony giant China Mobile is a founding member of the alliance. Other operators have signed on as well.

Taking Advantage of All Android has to Offer

Android's open platform has been embraced by much of the mobile development community—extending far beyond the members of the OHA. As Android phones and applications become more readily available, many in the techcommunity anticipate other mobile operators and handset manufacturers will jump on the chance to sell Android phones to their subscribers, especially given the cost benefits compared to proprietary platforms. Already, North American operators, such as Verizon Wireless and AT&T, have shown an interest in Android, and T-Mobile already provides handsets.

If the open standard of the Android platform results in reduced operator costs in licensing and royalties, we could see a migration to open handsets from proprietary platforms such as BREW, Windows Mobile, and even the Apple iPhone. Android is well suited to fill this demand.

ANDROID PLATFORM DIFFERENCES

Android is hailed as "the first complete, open, and free mobile platform."

- **Complete:** The designers took a comprehensive approach when they developed the Android platform. They began with a secure operating system and built a robust software framework on the top that allows for rich application development opportunities.
- **Open:** The Android platform is provided through open source licensing. Developers have unprecedented access to the handset features when developing applications.
- **Free:** Android applications are free to develop. There are no licensing or

royalty fees to develop on the platform also no membership fees, testing fees and signing or certification fees are required. Android applications can be distributed and commercialized in a variety of ways.

ANDROID: A NEXT GENERATION PLATFORM

Although Android has many innovative features not available in existing mobile platforms, its designers also leveraged many tried-and-true approaches proven to work in the wireless world. It is true that many of these features appear in existing proprietary platforms, but Android combines them in a free and open fashion, while simultaneously addressing many of the flaws on these competing platforms.

The Android mascot is a little green robot, shown in Figure. You will see this little guy (girl?) often used to depict Android-related materials. Android is the first in a new generation of mobile platforms, giving its platform developers a distinct edge on the competition. Android's designers examined the benefits and drawbacks of existing platforms and then incorporate their most successful features.

At the same time, Android's designers avoided the mistakes others suffered in the past.

Fig. 18.1: Android mascot

The various features of Android are as follows:

a. Free and Open Source

Android is an open source platform. Neither developers nor handset manufacturers pay royalties or license fees to develop the platform. The underlying operating system of Android is licensed under GNU General Public License Version 2 (GPLv2), a strong "copyleft" license where any third-party improvements must continue to fall under the open source licensing agreement terms. The Android framework is distributed under the Apache Software License (ASL/Apache2), which allows for the distribution of both open and closed source derivations of the source code. Commercial developers (handset manufacturers especially) can choose to enhance the platform without having to provide their improvements to the open source community. Instead, developers can profit from enhancements such as handset-specific improvements and redistribute their work under whatever licensing they want.

Android application developers have the ability to distribute their applications under whatever licensing scheme they prefer. Developers can write open source freeware or traditional licensed applications for profit and everything in between.

b. Familiar and Inexpensive Development Tools

Unlike some proprietary platforms that require developer registration fees, vetting, and expensive compilers, there are no upfront costs to developing Android applications.

c. Freely Available Software Development Kit

The Android SDK and tools are freely available. Developers can download the Android SDK from the Android Web site after agreeing to the terms of the Android Software Development Kit License Agreement.

d. Familiar Language, Familiar Development Environments

Developers have several choices when it comes to integrated development environments (IDEs). Many developers choose the popular and freely available eclipse IDE to design and develop Android applications. Eclipse is the most popular IDE for Android development and there is an Android plug-in available for facilitating Android development.

Android applications can be developed on the following operating systems:

- Windows XP or Vista
- Mac OS × 10.4.8 or later (×86 only)

- Linux (tested on Linux Ubuntu 6.06 LTS, Dapper Drake)

e. *Reasonable Learning Curve for Developers*
Android applications are written in a well-respected programming language: Java. The Android application framework includes traditional programming constructs, such as threads and processes and specially designed data structures to encapsulate objects commonly used in mobile applications. Developers can rely on familiar class libraries, such as java.net and java.text. Specialty libraries for tasks like graphics and database management are implemented using well-defined open standards like OpenGL Embedded Systems (OpenGL ES) or SQLite.

f. *Enabling Development of Powerful Applications*
In the past, handset manufacturers often established special relationships with trusted third-party software developers (OEM/ODM relationships). This elite group of software developers wrote native applications, such as messaging and Web browsing, which shipped on the handset as part of the phone's core feature set. To design these applications, the manufacturer would grant the developer privileged inside access and knowledge of a handset's internal software framework and firmware.

On the Android platform, there is no distinction between native and third-party applications, enabling healthy competition among application developers. All Android applications use the same libraries. Android applications have unprecedented access to the underlying hardware, allowing developers to write much more powerful applications.

Applications can be extended or replaced altogether. For example, Android developers are now free to design email clients tailored to specific email servers such as Microsoft Exchange or Lotus Notes.

g. *Rich, Secure Application Integration*
If you recall the bat story I previously shared, you will note that I accessed a wide variety of phone applications in the course of a few moments: text messaging, phone dialer, camera, email, picture messaging, and the browser. Each was a separate application running on the phone—some built-in and some purchased. Each had its own unique user interface. None were truly integrated. Not so with Android. One of the Android platform's most compelling and innovative features is well-designed application integration. Android provides all the tools necessary to build a better "bat trap," if you will, by allowing developers to write applications that leverage core functionality such as Web browsing, mapping, contact management, and messaging seamlessly. Applications can also become content providers and share their data among each other in a secure fashion.

Platforms like Symbian have suffered from setbacks due to malware. Android's vigorous application security model helps protect the user and the system from malicious software.

h. *No Costly Obstacles to Publication*
Android applications have none of the costly and time-intensive testing and certification programs required by other platforms such as BREW and Symbian.

i. *A "Free Market" for Applications*
Android developers are free to choose any kind of revenue model they want. They can develop freeware, shareware, or trial-ware applications, ad-driven, and paid applications. Android was designed to fundamentally change the rules about what kind of wireless applications could be developed. In the past, developers faced many restrictions that had little to do with the application functionality or features:

- Store limitations on the number of competing applications of a given type.
- Store limitations on pricing, revenue models, and royalties.
- Operator unwillingness to provide applications for smaller demographics.

With Android, developers can write and successfully publish any kind of application they want. Developers can tailor applications to small demographics, instead of just large-scale money-making ones often insisted upon by mobile operators. Vertical market applications can be deployed to specific, targeted users.

Because developers have a variety of application distribution mechanisms to choose from, they can pick the methods that work for them instead of being forced to play by others' rules. Android developers can distribute their applications to users in a variety of ways.

- Google developed the Android market, a generic Android application store with a revenue-sharing model.
- Handango.com added Android applications to its existing catalogue using their billing models and revenue sharing model.
- Developers can come up with their own delivery and payment mechanisms.

Mobile operators are still free to develop their own application stores and enforce their own rules, but it will no longer be the only opportunity developers have to distribute their applications.

Android might be the next generation in mobile platforms, but the technology is still in its early stages. Early Android developers have had to deal with the typical roadblocks associated with a new platform: frequently revised SDKs, lack of good documentation, and market uncertainties. There are only a handful of Android handsets available to consumers at this time.

On the other hand, developers diving into Android development now benefit from the first-to-market competitive advantages we have seen on other platforms such as BREW and Symbian. Early developers who give feedback are more likely to have an impact on the long-term design of the Android platform and what features will come in the next version of the SDK. Finally, the Android forum community is lively and friendly.

THE ANDROID PLATFORM

Android is an operating system and a software platform upon which applications are developed. A core set of applications for everyday tasks, such as Web browsing and email, are included on Android handsets.

As a product of the Open Handset Alliance's vision for a robust and open source development environment for wireless, Android is an emerging mobile development platform. The platform was designed for the sole purpose of encouraging a free and open market that all mobile applications phone users might want to have and software developers might want to develop.

EVOLUTION OF ANDROID OPERATING SYSTEM

Android's Underlying Architecture

The Android platform is designed to be more fault-tolerant than many of its predecessors. The handset runs a Linux operating system,

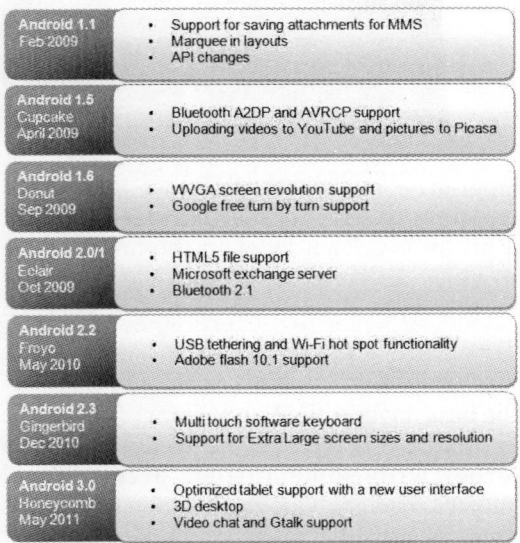

upon which Android applications are executed in a secure fashion. Each Android application runs in its own virtual machine. Android applications are managed code; therefore, they are much less likely to cause the phone to crash, leading to fewer instances of device corruption (also called "bricking" the phone, or rendering it useless).

The above figure shows the diagram of Android architecture. The Android OS can be referred to as a software stack of different layers, where each layer is a group of several

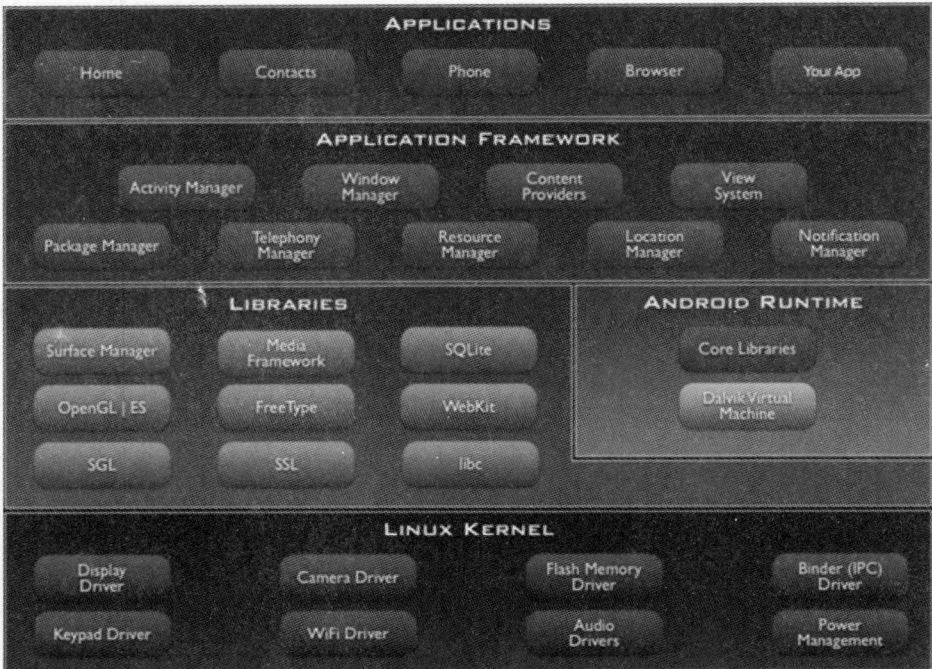

program components. Together it includes operating system, middleware and important applications. Each layer in the architecture provides different services to the layer just above it. We will examine the features of each layer in detail.

a. *Linux Kernel*

The basic layer is the Linux kernel. The whole Android OS is built on top of the Linux 2.6 kernel with some further architectural changes made by Google. It is this Linux that interacts with the hardware and contains all the essential hardware drivers. Drivers are programs that control and communicate with the hardware. For example, consider the Bluetooth function. All devices have Bluetooth hardware in it. Therefore, the kernel must include a Bluetooth driver to communicate with the Bluetooth hardware. The Linux kernel also acts as an abstraction layer between the hardware and other software layers. Android uses the Linux for all its core functionality such as memory management, Process management, networking, security settings, etc. As the Android is built on a most

popular and proven foundation, it made the porting of Android to variety of hardware, a relatively painless task.

b. *Libraries*

The next layer is the Android's native libraries. It is the layer that enables the device to handle different types of data. These libraries are written in C or C++ language and are specific for a particular hardware. Some of the important native libraries include the following:

a. **Surface manager:** It is used for compositing window manager with off-screen buffering. Off-screen buffering means you cannot directly draw into the screen, but your drawings go to the off-screen buffer. There it is combined with other drawings and form the final screen the user will see. This off screen buffer is the reason behind the transparency of windows.

b. **Media framework:** Media framework provides different media codecs allowing the recording and playback of different media formats.

c. **SQLite:** SQLite is the database engine used in android for data storage purposes.

d. **WebKit:** It is the browser engine used to display HTML content.

e. **OpenGL:** Used to render 2D or 3D graphics content to the screen.

c. Android Runtime

Android Runtime consists of Dalvik Virtual machine and Core Java libraries.

a. **Dalvik virtual machine:** It is a type of JVM used in android devices to run applications and is optimized for low processing power and low memory environments. Unlike the JVM, the Dalvik Virtual Machine does not run .class files, instead it runs .dex files. .dex files are built from .class file at the time of compilation and provides higher efficiency in low resource environments. The Dalvik VM allows multiple instance of virtual machine to be created simultaneously providing security, isolation, memory management and threading support. It is developed by Dan Bornstein of Google.

b. **Core java libraries:** These are different from Java SE and Java ME libraries. However, these libraries provide most of the functionalities defined in the Java SE libraries.

d. Application Framework

These are the blocks that our applications directly interact with. These programs manage the basic functions of phone like resource management, voice call management, etc. As a developer, you just consider these are some basic tools with which we are building our applications.

Important blocks of application framework are:

- **Activity Manager:** Manages the activity life cycle of applications.

- **Content Providers:** Manage the data sharing between applications.

- **Telephony Manager:** Manages all voice calls. We use telephony manager if we want to access voice calls in our application.

- **Location Manager:** Location management, using GPS or cell tower.

- **Resource Manager:** Manage the various types of resources we use in our application.

e. Applications

Applications are the top layer in the Android architecture and this is where our applications are goanna fit. Several standard applications come pre-installed with every device, such as:

- SMS client app
- Dialer
- Web browser
- Contact manager

As a developer we are able to write an application which replace any existing system application. That is, you are not limited in accessing any particular feature. You are practically limitless and can do whatever you want to do with the android (as long as the users of your application permits it). Thus Android is opening endless opportunities to the developer.

Applications

These are the basics of Android applications:

- Android applications are composed of one or more application components (activities, services, content providers, and broadcast receivers).

- Each component performs a different role in the overall application behavior, and each one can be activated individually (even by other applications).

- The manifest file must declare all components in the application and should also declare all application requirements, such as the minimum version of Android required and any hardware configurations required.

- Non-code application resources (images, strings, layout files, etc.) should include alternatives for different device configurations (such as different strings for different languages).

APPENDIX
MODEL QUESTION PAPER

Model Question Paper 1
[MODEL QUESTION PAPER] ECS-101
B.Tech.
FIRST/SECOND SEMESTER EXAMINATION
COMPUTER CONCEPTS AND PROGRAMMING IN 'C'

M.M. 100   Time: 3 Hrs

SECTION A

This section consists of multiple choice questions/fill in the blanks/true/false.

i. The different types of memory units are
   (a) RAM   (b) ROM
   (c) PROM   (d) all of the above

ii. Which of the following is a SDLC phase
   (a) Problem definition
   (b) Problem design
   (c) Coding
   (d) All of the above

iii. The name of the first digital computer is
   _____.

iv. The 2's complement of binary number 101100 is 010100. (T/F)

v. Who developed the 'C' language
   (a) Ken Thomson
   (b) Bjarne Stroutstrup
   (c) Dennis Ritchie
   (d) Kernighan

vi. A computer splits a program into a number of _____.

vii. Compiling a C Program means translating it into _____ language.

viii. C is an example of
   (a) Object oriented language
   (b) Structured Programming language
   (c) Object based language
   (d) Component based language.

ix. The bitwise and is used for
   (a) Masking   (b) Comparison
   (c) Division   (d) Shifting byte

x. The associativity of bitwise AND, OR, XOR is
   (a) Right to left
   (b) Left to right
   (c) Option **a)** For arithmetic expression and option **b)** For pointer expression
   (d) Option **a)** For pointer expression and **b)** for arithmetic expression

xi. Header file in C contain
   (a) Compiler commands
   (b) Library functions
   (c) Header information of C programs
   (d) Operators for files

xii. Comma is used as
   (a) A separator in C
   (b) An operator in C
   (c) A delimiter in C
   (d) Terminator in C

xiii. The default return data type in function definition is
   (a) void   (b) int
   (c) float   (d) char

xiv. Identity the correct statement.
   (a) Float array can be read as a whole
   (b) Integer array can be read as a whole
   (c) Char array can be read as a whole
   (d) Double array can be read as a whole

xv. Pointers are supported in
   (a) FORTRAN   (b) PASCAL
   (c) C   (d) Both b and c

316

xvi. The pointers can be used to achieve
   (a) Call by function
   (b) Call by reference
   (c) Call by name
   (d) Call by procedure

xvii. The function used for dynamic deallocation of memory is
   (a) destroy ( )   (b) delete ( )
   (c) free ( )   (d) remove ( )

xviii. The no of arguments used in calloc () function is
   (a) 0
   (b) 1

   (c) 2
   (d) 3

xix. Preprocessor directives begin with
   (a) #
   (b) $
   (c) C: 1
   (d) None of the above

xx. The general format of file open command
   (a) fptr = fopen (filename/mode),
   (b) fptr = fopen (filename,mode)
   (c) fptr; fopen = mode,
   (d) None of the above

SECTION B

Note: *Attempt any three parts of the following*

1. Write an algorithm and flow chart to generate all numbers which are divisible by 3 but not by 7.

2. Write a C Program to find the sum of each of the following series

   (a) $1 - \dfrac{1}{2} + \dfrac{1}{3} - \dfrac{1}{4} + \dfrac{1}{5} - \dfrac{1}{6} + \ldots\ldots$ up to n terms

   (b) $1! + 2! + 3! + \ldots\ldots n!$

3. Write a C Program to compute the No. of vowels, consonants, words and lines in a given text with the help of function.

4. Write a C Program to accept a matrix and determine whether it is a
   (a) Square matrix
   (b) Unit matrix
   (c) Symmetric matrix

5. Define the various functions used by (string.h) Header file with suitable example.

SECTION C

Note: *Attempt any one part of the following*

1. What do mean by operating system? Define the various function of operating system. Compare DOS, windows and Unix operating system with suitable example.

or

Define the number system and also classify the number system with suitable example and convert the following number system according to their bases
   (a) $(243)_8 = (?)_2$
   (b) $(4AB)_{16} = (?)_2$
   (c) $(110010)_2 = (?)_8$
   (d) $(105)_8 = (?)_{16}$

2. Differentiate between
   (a) Unformatted and formatted I/o statements.
   (b) Scant ( ) and gets ( ) with reference to string input.
   (c) %s and % [1n n] specification with scan f ( ).

or

Write a C Program to find the value of y in the following equation $y = x^2 + 3x + 1$

3. Write a C Program to compute the commission on sales as per the following policy

   (a) If sales is less than Rs.1000/—no commission.

   (b) If sales is more than Rs. 1000 and less than Rs. 25,000/—then the commission is 10% of sales.

   (c) If sales is more than or equal to Rs. 25,000/—then the commission is Rs. 200/- plus 8% of sales exceeding Rs. 1000/-

or

What do you mean by modular programming? Give advantage and disadvantage with the help of diagram ad examples and also define the various states of functions.

4. Define the concept of 2-D Array. Write a program to input two 3×3 matrices and print the multiplication of both the matrices

or

Define the concept of pointer. Also define the dynamic memory allocation and various functions for dynamic memory allocation.

5. Define the concept of sorting and write a program to input 10 named and then sort that name in ascending order.

or

Differentiate between function and macros and write a program to input two numbers and print the max. Number with the help of macro.

Model Question Paper 2
[MODEL QUESTION PAPER] ECS-101
B.Tech.
FIRST/SECOND SEMESTER EXAMINATION
COMPUTER CONCEPTS AND PROGRAMMING IN 'C'

M.M. 100   Time: 3 Hrs

SECTION A

1. Attempt all 20 questions.   (20×1=20)

(a) Which of the following is evaluated first?
   i. &&   ii. ||
   iii. !

(b) Which is not valid in C?
   i. class a Class{public:int x;};
   ii. /* A comment */
   iii. char x=12;

(c) Which of the following is not a valid declaration for main()?
   i. int main()
   ii. int main(int argc, char *argv[])
   iii. They both work

(d) Evaluate as true or false: !(1 &&0 || !1)
   i. True
   ii. False
   iii. Invalid statement

(e) What operator is used to access a struct through a pointer?
   i. ->   ii. >>
   iii. *

(f) Which uses less memory?
   i. struct astruct
      {
      int x;
      float y;
      int v;
      };
   ii. union aunion
      {
      int x;
      float v;
      };
   iii. char array[10];

(g) What do non global variables default to:
   i. static   ii. auto
   iii. register

(h) Evaluate !(1&&1||1&&0)
   i. Error   ii. True
   iii. False

(i) Code:

   int z,x=5,y=-10,a=4,b=2;
   z = x++ - —y * b / a;

   What number will z in the sample code above contain?
   i. 5   ii. 6
   iii. 10   iv. 11
   v. 12

(j) What function will read a specified number of elements from a file?
   i. fileread()   ii. getline()
   iii. readfile()   iv. fread()
   v. gets()

(k) int testarray[3][2][2] = {1, 2, 3, 4, 5, 6, 7, 8, 9, 10, 11, 12};

   What value does testarray[2][1][0] in the sample code above contain?
   i. 3   ii. 5
   iii. 7   iv. 9
   v. 11

(l) Code:

   int a=10,b;
   b=a++ + ++a;
   printf("%d,%d,%d,%d",b,a++,a,++a);

what will be the output when following code is executed

i.   12,10,11,13   ii.   22,10,11,13
iii. 22,11,11,11   iv.   12,11,11,11
v.   22,13,13,13

(m) Code:

```
void myFunc (int x)
{
if (x > 0)
myFunc(−x);
printf("%d, ", x);
}
```

```
int main()
{
myFunc(5);
return 0;
}
```

What will the above sample code produce when executed?

i.   1, 2, 3, 4, 5, 5,   ii.   4, 3, 2, 1, 0, 0,
iii. 5, 4, 3, 2, 1, 0,   iv.   0, 0, 1, 2, 3, 4,
v.   0, 1, 2, 3, 4, 5,

(n) What is a proper method of opening a file for writing as binary file?

i.   FILE *f = fwrite("test.bin", "b");
ii.   FILE *f = fopenb("test.bin", "w");
iii. FILE *f = fopen("test.bin", "wb");
iv.   FILE *f = fwriteb("test.bin");
v.   FILE *f = fopen("test.bin", "bw");

(o) int x = 2 * 3 + 4 * 5;

What value will x contain in the sample code above?

i.   22   ii.   26
iii. 46   iv.   50
v.   70

(p) Code:

```
int x = 3;
if( x == 2 );
x = 0;
if( x == 3 )
x++;
else x += 2;
```

What value will x contain when the sample code above is executed?

i.   1   ii.   2
iii. 3   iv.   4
v.   5

(q) Code:

```
#include <stdio.h>
int i;
```

```
void increment( int i )
{
i++;
}
int main()
{
for( i = 0; i < 10; increment( i ) )
{

}
printf("i=%d\n", i);
return 0;
}
```

What will happen when the program above is compiled and executed?

i.   It will not compile.
ii.   It will print out: i=9.
iii. It will print out: i=10.
iv.   It will print out: i=11.
v.   It will loop indefinitely.

(r) Code:

```
int i = 4;
switch (i)
{
default:
;
case 3:
i += 5;
if ( i == 8)
{
i++;
if (i == 9) break;
i *= 2;
}
i -= 4;
break;
case 8:
i += 5;
break;
}
printf("i = %d\n", i);
```

What will the output of the sample code above be?

i.   i = 5
ii.   i = 8
iii. i = 9
iv.   i = 10
v.   i = 18

(s) Code:

```
int x = 0;
for ( ; ; )
{
```

```
if (x++ == 4)
break;
continue;
}
printf("x=%d\n", x);
```

What will be printed when the sample code above is executed?

   i. x=0   ii. x=1

  iii. x=4   iv. x=5

   v. x=6

(t) Code:

```
#include <stdio.h>
void func()
{
int x = 0;
static int y = 0;
```

```
x++; y++;
printf( "%d — %d\n", x, y );
}
int main()
{
func();
func();
return 0;
}
```

What will the code above print when it is executed?

   i. 1 — 1   ii. 1 — 1

      1 — 1   2 — 1

  iii. 1 — 1   iv. 1 — 0

      2 — 2   1 — 0

   v. 1 — 1

      1 — 2

SECTION B

2. *Attempt any three parts of the following*
               (10 × 3 = 30)

(a) Write a program to swap two numbers using pointers in C

(b) Write a program to print this triangle:

```
*
**
***
****
*****
******
*******
********
*********
**********
```

Do not use ten printf statements; use two nested loops instead.

(c) Write a program to print the first 10 Fibonacci numbers. Each Fibonacci number is the sum of the two preceding ones. The sequence starts out 0, 1, 1, 2, 3, 5, 8...

(d) Write a program to print the numbers between 1 and 10, along with an indication of whether each is even or odd, like this:

          1 is odd

         2 is even

         3 is odd

           ...

(e) Write a program to read its input, one character at a time, and print each character *and* its decimal value.

SECTION C

3. *Attempt any one part of the following*
               (10 × 1 = 10)

(a) Write a program_that will accept two strings (1 <= length <= 50) and combine them to create a third string, such that the resultant string is the shortest string containing both the input strings as substrings. In case more than one such string is possible, print the string that contains the first string first.

(b) Write a program in c to find the factorial of an integer number using recursion.

4. *Attempt any one part of the following*
               (10 × 1 = 10)

(a) Write a program in c to elaborate the concept of matrix multiplication using two dimensional arrays.

(b) Write a program in c using structure to initialize 3 students record and print them.

5. *Attempt any one part of the following*
               (10 × 1 = 10)

(a) Write a function power (a,b), to calculate the value of a raised to b.

(b) Write a program in c using selection sort algorithm to sort 10 integer numbers.

6. *Attempt any one part of the following*
(10 × 1 = 10)

(a) Write a program in C to compute the sum of the digits of a given integer number.

(b) Write a program in C to read the age of 100 persons and count the number of

persons in the age group of 50 to 60.Use for and continue statement.

7. *Attempt any one part of the following*
(10 × 1 = 10)

(a) Write a function prime that returns 1 if its argument is a prime number and returns zero otherwise.

(b) Write a program in C to count the number of words in a given text file.

Model Question Paper 3
[MODEL QUESTION PAPER] ECS-101
B.Tech.
FIRST/SECOND SEMESTER EXAMINATION
COMPUTER CONCEPTS AND PROGRAMMING IN 'C'

M.M. 100   *Time: 3 Hrs*

SECTION A

1. *This question contains 20 objective type/fill in the blanks/ matching type questions. Choose/fill/match correct answer.*

(a) A set of prerecorded instructions executed by a computer is called:
i. Action   ii. Hardware
iii. I/o Device   iv. Program

(b) Which of the following is machine independent?
i. Machine language
ii. Assembly language
iii. High level language
iv. Natural language

(c) Which are the following is personal computer application software?
i. COBOL   ii. Power point
iii. Basic   iv. None of the above

(d) There are two types of bugs or errors _____ _____ and _____.

(e) Which is not a keyword in 'C'
i. cbnst   ii. main
iii. size   iv. void

(f) 'C' is a _____ programming language.

(g) Statement terminator is represented by
i. :   ii. ;
iii. n   iv. blank

(h) The operator % can be applied only to
i. Float values   ii. Integer values
iii. Double values iv. All of the above

(i) What is the output of the following code?
Main ( )
Print f ("% c", A);
i. 65   ii. a
iii. A   iv. Error

(j) Execution of print f ("4%+d",25); displays output as _____

(k) Output of the following code is _____
Main ( )
int i=0;
for (;i<ii; i++)
print f ("%d",i);
}}

(l) Output of the following program is _____.
Main ( )
Int i;
for (i=0; i<25;i++);
print f ("%d", i);
}}

(m) The _____ statement in a switch case statement skips all the remaining statements and exists the switch block.

(n) Recursive call results when
i. A function calls itself
ii. A function calls another function
iii. Option calls another function
iv. None of the above

(o) The function main ( ) is
i. A built-in function
ii. Optional
iii. User defined function
iv. All of the above

(p) Function declaration ends with a _____.

(q) Array subscripts in C always start at
i. −1   ii. 10
iii. 1   iv. Any value

(r) Related data items of different type are organized using _____.

(s) A pointer value refers to
  i. An integer constant
  ii. A float value
  iii. Any valid address in memory
  iv. Any ordinary variable

(t) fopen ( ) function used for
  i. Closing a file   ii. Opening a file
  iii. Writing a file   iv. None of these

SECTION B

2. *Attempt any three parts of the following*
   (10 × 3 = 30)

(a) i. What is computer? Draw a neat labeled block diagram of a computer and explain the function of each unit.
   ii. Write an algorithm and draw a flow chart to find the sum of digits of a given number.

(b) i. What is a data type? Explain four basic data type of C.
   ii. Write a program to read two integer values m and n and to decide and print whether m is a multiple of n.

(c) i. What is the difference between while and do-while loop? Explain with example.
   ii. Write a program to print the following outputs:
   1
   22
   333
   4444
   55555

(d) How we declare two dimensional arrays? Write a program to find the multiplication of two 3x3 matrix.

(e) i. Describe the use and limitations of the function get (1) and put (1).
   ii. Write a program to copy the contents of one file into another.

SECTION C

3. *Attempt any one part of the following*
   (10 × 5 = 50)
   (a) What is the difference between flowchart & algorithm? Write an algorithm & draw a flowchart to generate first 50 numbers in Fibonacci series.
   (b) i. write the short note on "Structured Programming"
      ii. Convert the following
      $(521.4)_{10} = ( )_{16}$
      $(13)_{10} = ( )$ BCD
      $(5A.8)_{16} = ( )_8$
      $(1011)_2 = ( )$ gray
      $(101101.11)_2 = ( )_{10}$

4. *Attempt any one part of the following*
   (a) Write a C program to read the value of X and print the result of the following equation
   $y = x^3 + 2x^2 + 3x + 5$
   (b) Differentiate between with the help of example
      i. get char (), scan f (), gets ()
      ii. put char (), print f (), puts ()

5. *Attempt any one part of the following*
   (a) distinguish between the following:
      i. Actual and formal arguments
      ii. Global and local variables
      iii. Automatic and static variables
      iv. Scope and visibility of variables
      v. & operator and * operator
   (b) Develop your own functions for performing following operations on strings:
      i. Copying one string to another
      ii. Comparing two strings
      iii. Adding a string to the end of another string

6. *Attempt any one part of the following*
   (a) What happens when an array with a specified size is assigned?
      i. With values fewer than the specified size &
      ii. With values more than the specified size.
   (b) Write a program that will compute the length of a given character string without using strlen function.

7. *Attempt any one part of the following*
   (a) i. What is sequential searching? Explain with one example.
      ii. Write a program which search a number between 1 to 100.
   (b) i. Distinguish between the following:
      • gets & gechar
      • print f & f print f ()
      ii. Write a program that appends one file at the end of another.

Index